Birkhäuser

Progress in Nonlinear Differential Equations and Their Applications

PNLDE Subseries in Control

Volume 105

The *PNLDE Subseries in Controls* is dedicated to the publication of cutting edge developments from the intersection of control theory with the analysis of differential equations. The subseries aims to include research monographs, graduate level textbooks, polished notes arising from lectures and seminars, as well as refereed proceedings of focused conferences. Topics from across the spectrum, ranging from the theoretical to the applied, will be included in the subseries.

Yacine Chitour • Paolo Mason • Mario Sigalotti

Dynamics and Stability of Continuous-Time Switched Linear Systems

Birkhäuser

Yacine Chitour
L2S, Université Paris-Saclay,
CentraleSupélec
CNRS
Gif-sur-Yvette, France

Paolo Mason
L2S, Université Paris-Saclay,
CentraleSupélec
CNRS
Gif-sur-Yvette, France

Mario Sigalotti
LJLL, Sorbonne Université
Inria
Paris, France

ISSN 1421-1750 ISSN 2374-0280 (electronic)
Progress in Nonlinear Differential Equations and Their Applications
ISSN 2731-7366 ISSN 2731-7374 (electronic)
PNLDE Subseries in Control
ISBN 978-3-031-82981-9 ISBN 978-3-031-82982-6 (eBook)
https://doi.org/10.1007/978-3-031-82982-6

Mathematics Subject Classification: 34D08, 93B05, 93C30, 93C05

This book is published under the imprint Birkhäuser, www.birkhauser-science.com by the registered company Springer Nature Switzerland AG
The registered company address is: Gewerbestrasse 11, 6330 Cham, Switzerland

To our families

Preface

Linear switched systems are a fascinating field of research, bringing forth theoretical questions arising from applications which require sophisticated mathematical tools for their resolution. The subject is by now mature but a significant proportion of these results (especially those dealing with continuous-time dynamics) are still only available in research articles, despite a unity in the statements and sometimes in the methodological approaches displayed in these papers. This book therefore stems from the authors' desire to bring together these results in order to highlight the aforementioned unity. The authors hope that the present work will be useful to researchers who will need a solid reference for certain results now considered standard. On the other hand, students wishing to learn the mathematical basics of linear systems with continuous-time commutations will find the main concepts covered in this book.

Besides fundamental tools such as common Lyapunov functions, converse Lyapunov theorems, and maximal Lyapunov exponents, the authors also put forward the concept of the Barabanov norm, which is now well understood from a theoretical point of view, but which does not seem to be much used by a community closer to applications. One of the reasons may be that this fundamental object was developed in the context of arbitrary switches but has no immediate equivalent for classes of switching signals subject to various constraints (dwell-time, persistent excitation, etc.), the latter being the typical setup of control engineers. One of the aims of this book is to bridge this gap as far as possible by explaining how the main features of Barabanov norms can be generalized for classes of constrained switchings.

The authors have tried to keep a general point of view, rather than treating classes of switching signals separately, by developing an axiomatic approach and identifying structural properties of these classes that allow crucial aspects of the Barabanov norm to be extended.

Acknowledgments The authors would like to thank Matteo Della Rossa, Raphaël Jungers, Fritz Colonius, Ihab Haidar, and the anonymous reviewers of the first draft of this book for their numerous and precious feedbacks as well as Ugo Boscain, Antoine Chaillet, and Jamal Daafouz for having introduced us to many interesting questions in this domain.

Gif-sur-Yvette, France — Yacine Chitour
Gif-sur-Yvette, France — Paolo Mason
Paris, France — Mario Sigalotti
November 5, 2024

Introduction

Consider a non-autonomous linear system

$$\dot{x}(t) = M(t)x(t) \tag{1}$$

where $x(t)$ is a vector in $\mathbb{R}^n$ and $M(t)$ is an $n \times n$ matrix with real coefficients. One of the most natural and studied issues for such a system is to understand its asymptotic behavior as t tends to infinity and, in case it is asymptotically stable, to characterize at which rate its trajectories tend to the origin.

Assume now that the value of the function $t \mapsto M(t)$ is not known at every time t, but only partial information on $M(\cdot)$ is available such as its range or the rules affecting its behavior. Contrarily to what a hasty intuition could suggest, even if the range of $M(\cdot)$ is contained in the set of Hurwitz matrices (i.e., matrices N such that $\lim_{t\to+\infty} e^{tN} = 0$), the convergence of $x(t)$ to the origin as t goes to infinity is not guaranteed. This is illustrated in Fig. 1, which depicts an exponentially diverging trajectory corresponding to a periodic piecewise constant function $M(\cdot)$ whose values M_1, M_2 are 2×2 Hurwitz matrices. Even if the trajectories of each of the autonomous systems $\dot{x}(t) = M_1x(t)$ and $\dot{x}(t) = M_2x(t)$ converge to the origin following elliptic spirals (left and center subfigures), one can build a concatenation of such spirals which diverges to infinity (right subfigure).

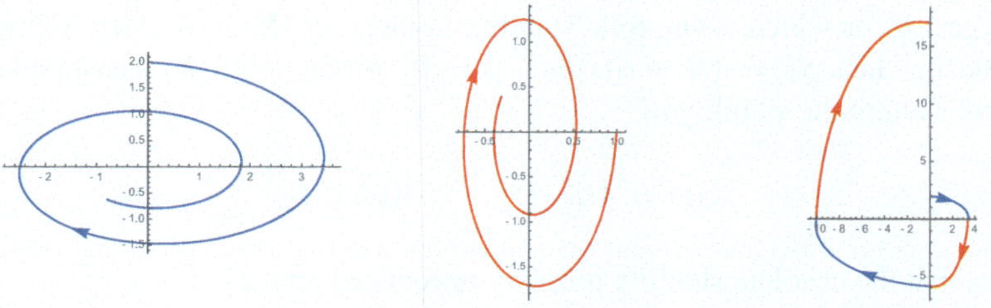

Fig. 1 Left: trajectory of $\dot{x}(t) = M_1x(t)$, converging to the origin; Center: trajectory of $\dot{x}(t) = M_2x(t)$, converging to the origin; Right: trajectory of $\dot{x}(t) = M(t)x(t)$, diverging at infinity

Notice that the range of $M(\cdot)$ does not provide enough information for characterizing the stability. For instance, in the previous example, if one further imposes that each interval on which $M(\cdot)$ is constant is large enough, then the origin is asymptotically stable. (Indeed, for $t > 0$ large enough and $i \in \{1, 2\}$, one has $\|e^{tM_i}\| < 1$.)

This simple example already highlights how the asymptotic behavior depends on a nontrivial interplay between the range of the function $M(\cdot)$ and the rules that its dependence on time is assumed to follow.

The switched system approach consists in taking into account the uncertainty about the function $M(\cdot)$ by embedding System (1) into a parametric family of non-autonomous systems

$$\dot{x}(t) = A(t)x(t), \qquad A(\cdot) \in \mathcal{S}, \tag{2}$$

where the parameter is the function $A(\cdot)$ and the class $\mathcal{S}$ contains all functions satisfying the constraints (on the range and on the dependence on time) that $M(\cdot)$ must fulfill. The elements of $\mathcal{S}$ are called *switching laws* or *switching signals* and the corresponding switched linear system (2) is denoted $(\Sigma_{\mathcal{S}})$. In this approach, the issue of determining the asymptotic stability of the origin for System (1) is replaced by that of establishing the asymptotic stability of the origin for (2) uniformly over the class $\mathcal{S}$. Similarly, the question of characterizing the convergence rate for (1) becomes that of characterizing the so-called *maximal Lyapunov exponent over* $\mathcal{S}$ that describes the worst possible asymptotic behavior over the family of non-autonomous systems (2). The notion of maximal Lyapunov exponent over a class $\mathcal{S}$ considered in this book is borrowed from the theory of dynamical systems (see, e.g., [CK00]). The study of its qualitative properties is the main purpose of this monograph.

The study of the asymptotic stability of continuous-time switched systems can be traced back to the notion of absolute stability introduced in the 1940s by Lur'e and Postnikov [LP44], as nicely evoked in the survey [Lib01]. In a nutshell, the absolute stability problem (also called Lur'e problem) consists in determining whether the origin is asymptotically stable for the single-input linear control system $\dot{x} = Nx + bu$ fed by $u = \varphi(c^\top x)$, where φ is any continuous function satisfying $0 \leq s\varphi(s) \leq ks^2$ for every $s \in \mathbb{R}$. Here, the Hurwitz matrix N, the vectors b and c, and the positive scalar k are given. One is actually often interested in determining the largest k for which asymptotic stability holds, see [Mar06]. After setting, for $k > 0$, the class $\mathcal{S}_k = \{N + \alpha(t)bc^\top \mid \alpha\colon [0, +\infty) \to [0, k] \text{ measurable}\}$, the uniform asymptotic stability of

$$\dot{x}(t) = A(t)x(t), \qquad A(\cdot) \in \mathcal{S}_k,$$

implies that the absolute stability property introduced above holds true.

More generally, switched systems of the form (2) can be used as *relaxations* of dynamical systems whose analysis is more difficult to handle. For instance, if we consider $\dot{x} = A(x)x$ with $A\colon \mathbb{R}^n \to \{M_1, \ldots, M_m\}$ piecewise constant on positive convex cones, the existence of trajectories defined for all times is not guaranteed and Zeno phenomena may occur (infinitely many discontinuities of $t \mapsto A(x(t))$ in a finite interval of time). Considering the switched system $(\Sigma_{\mathcal{S}})$ associated with the class $\mathcal{S}$ of all measurable functions $t \mapsto A(t)$ taking values in $\{M_1, \dots, M_m\}$, one can see that Carathéodory solutions of $(\Sigma_{\mathcal{S}})$ are defined for all times, and trajectories of $\dot{x} = A(x)x$, when they are well defined, can be seen as trajectories of $(\Sigma_{\mathcal{S}})$. Hence asymptotic stability of the origin for $(\Sigma_{\mathcal{S}})$ implies asymptotic stability of the origin for $\dot{x} = A(x)x$.

Another source of motivation stems from the fact that switched systems can be seen as particular instances of hybrid systems, which are dynamical systems exhibiting an interaction between continuous and discrete dynamics [GST12, vdSS00]. The continuous dynamics is that of the state x, while the discrete dynamics drives the variation of the mode. An early model for such a system is given in [Wit66]. In this framework the set of modes is usually finite and the discrete dynamics is determined, for instance, by a timed automaton.

With respect to the first important monograph on switched systems with a mathematical perspective, namely [Lib03], our work focuses more specifically on stability and it provides more mathematical developments. In fact, the present monograph can be seen as the continuous-time counterpart of the book [Jun09], which only deals with discrete-time switched linear systems. It also addresses issues already tackled in [SG05, SG11, LA22], with the notable differences that our study, on the one hand, concentrates mainly on stability rather than control design and, on the other hand, aims at structural characterization by emphasizing the role of extremal norms.

We now turn to a description of the material covered by this book. We refer to the last section of each chapter for a selection of the main historical references relative to the subjects discussed in the chapter.

In the first chapter we introduce the mathematical setting for switched linear systems. We provide, in particular, the most commonly used classes $\mathcal{S}$ of signals studied in the literature, which are subsets of $\mathcal{S}_{\mathrm{arb}}(\mathcal{M})$, defined as the set of measurable functions taking values in a bounded set $\mathcal{M}$ of $n \times n$ real-valued matrices. Such classes are invariant by positive time-shift. We recall the basic existence results for Carathéodory solutions and we consider the main notions of stability associated with $(\Sigma_{\mathcal{S}})$. We then compare these properties and relate them to those of $(\Sigma_{\mathcal{S}^\star})$, where the class $\mathcal{S}^\star$ stands for the weak-$\star$ closure of $\mathcal{S}$. For that purpose, we rely on an approximation property for trajectories of $(\Sigma_{\mathcal{S}^\star})$ by those of $(\Sigma_{\mathcal{S}})$, referred to as *long time approximation*. Such a property holds true under a mild technical condition satisfied by rather general classes of signals. We then consider two asymptotic growth rates associated with a class $\mathcal{S}$. The first one is called the *maximal Lyapunov exponent* of $\mathcal{S}$ and is given by

$$\hat{\lambda}(\mathcal{S}) = \sup_{A \in \mathcal{S}} \limsup_{t \to \infty} \frac{\log \|\Phi_A(t, 0)\|}{t},$$

where $\Phi_A(t, 0)$ denotes the fundamental matrix from time 0 to time t associated with the signal A. The second Lyapunov exponent $\lambda(\mathcal{S})$ is called the *uniform exponential rate* of $\mathcal{S}$ and is given by

$$\lambda(\mathcal{S}) = \limsup_{t\to\infty} \sup_{A\in\mathcal{S}} \frac{\log \|\Phi_A(t, 0)\|}{t}.$$

Even though $\hat{\lambda}(\mathcal{S})$ is a natural extension to switched linear system of the concept of maximal Lyapunov exponent of the non-autonomous system (1), it turns out that $\lambda(\mathcal{S})$ is more suitable to capture uniform stability properties of $(\Sigma_{\mathcal{S}})$. In particular, $(\Sigma_{\mathcal{S}})$ is uniformly exponentially stable if and only if $\lambda(\mathcal{S}) < 0$. It is now straightforward to see from the definitions that $\hat{\lambda}(\mathcal{S}) \leq \lambda(\mathcal{S})$ and one of the results of the chapter establishes the equality if the class $\mathcal{S}$ satisfies the long-time approximation property mentioned above. The argument relies on the classical Fenichel lemma, which states that attractivity of the origin is equivalent to uniform exponential stability provided that $\mathcal{S}$ is weak-$\star$ compact, that is, $\mathcal{S} = \mathcal{S}^\star$.

Recall that, given any $n \times n$ matrix, seeking eigenvectors yields to the block reduction in Jordan form. Similarly, given a set $\mathcal{M}$ of matrices, the existence of nontrivial subspaces invariant by every element of $\mathcal{M}$ leads to specific space decompositions, which yields a block reduction result and a family of reduced switched linear systems. It turns out that $\lambda(\mathcal{S})$ is equal to the maximum among the maximal Lyapunov exponents of the reduced switched systems.

The second chapter is devoted to a complete characterization of uniform stability properties for switched linear systems of the form $(\Sigma_{\mathcal{S}_{\mathrm{arb}}(\mathcal{M})})$ with $\mathcal{M}$ made of two 2×2 matrices M_1, M_2. More precisely, necessary and sufficient conditions for several stability concepts such as uniform exponential stability and quadratic stability are expressed in terms of explicit conditions on the coefficients of M_1 and M_2. The main technical tool is the concept of *worst-case trajectory*, which roughly speaking are trajectories of the switched system identifying regions of the plane in which all other trajectories of the systems are trapped. Note that the concept of worst-case trajectory is not anymore valid in dimension larger than two. Actually, it seems even hopeless in that case to explicitly characterize the uniform stability properties of switched linear systems in terms of their parameters.

The third chapter describes techniques to address stability of switched linear systems, which are based on the use of Lyapunov functions. In the first part of the chapter we discuss the concept of *common Lyapunov function*, namely a continuous, positive-definite, real-valued function defined on $\mathbb{R}^n$ which is decreasing along non-trivial trajectories of the switched linear system. We also provide a characterization of the monotone decreasing property in a convex setting. Then, in the second part of the chapter, we specifically consider the question of existence of *common quadratic Lyapunov functions*, i.e., common Lyapunov functions of the form $x^\top Px$ with P real symmetric positive-definite. For planar switched linear systems $(\Sigma_{\mathcal{S}_{\mathrm{arb}}(\mathcal{M})})$ with $\mathcal{M}$ made of two matrices M_1, M_2, we characterize the existence of a common quadratic Lyapunov function. Finally, in the third part of the chapter, we establish converse Lyapunov theorems, that is, we provide several constructions for common Lyapunov functions associated with uniform exponentially stable switched linear systems. Moreover, we study families of candidate common Lyapunov functions called *universal classes*, that is, families among which any uniformly exponentially stable system admits a common Lyapunov function. We provide necessary conditions and sufficient conditions for a family to be universal. As an application, we recover the fact that the family of polynomial functions is universal but this is no longer true if one imposes a uniform bound on the degree of the polynomial.

In the fourth chapter, given a switched linear system $(\Sigma_{\mathcal{S}_{\mathrm{arb}}(\mathcal{M})})$ with no nontrivial invariant subspace, we identify an adapted norm on $\mathbb{R}^n$, which turns out to be a common Lyapunov function for the system if the latter is uniformly asymptotically stable. Such a norm is called a *Barabanov norm* and is one of the central notions of this book. We next embark into the systematic study of *extremal trajectories*, which are trajectories of $(\Sigma_{\mathcal{S}_{\mathrm{arb}}(\mathcal{M})})$ (up to the renormalization $\mathcal{M} \mapsto \mathcal{M} - \lambda(\mathcal{S}_{\mathrm{arb}}(\mathcal{M}))\mathrm{Id}_n$) with constant Barabanov norm. This is done relying on tools from optimal control theory. We then use extremal trajectories to investigate properties of Barabanov norms such as strict convexity, regularity, and uniqueness. Finally the last part of the chapter gathers additional properties on switched linear systems whose arguments rely crucially on Barabanov norms. In particular, we prove a Berger–Wang-type formula, expressing that the maximal Lyapunov exponent can be equivalently computed if one replaces in its definition the norm of the fundamental matrix $\Phi_A(t, 0)$ by its spectral radius. Another such property is the continuity of $\lambda(\mathcal{S}_{\mathrm{arb}}(\mathcal{M}))$ with respect to $\mathcal{M}$, for the Hausdorff distance between compact sets of $n \times n$ matrices. Finally, we provide bounds on the degree of the polynomial divergence of $(\Sigma_{\mathcal{M}-\lambda(\mathcal{S}_{\mathrm{arb}}(\mathcal{M}))\mathrm{Id}_n})$ in terms of the block reduction in invariant subspaces and of a suitable notion of resonance between diagonal blocks.

Chapter 5 focuses on the behavior of extremal trajectories in the case of switched linear systems $(\Sigma_{\mathcal{S}_{\mathrm{arb}}(\mathcal{M})})$ in $\mathbb{R}^3$ with no nontrivial invariant subspace. The results of the previous chapter guarantee that a Barabanov norm exists for such a system, and its spheres are two-dimensional (Lipschitz) manifolds. Let us fix any such Barabanov sphere S. Then, up to the renormalization $\mathcal{M} \mapsto \mathcal{M} - \lambda(\mathcal{S}_{\mathrm{arb}}(\mathcal{M}))\mathrm{Id}_n$, $(\Sigma_{\mathcal{S}_{\mathrm{arb}}(\mathcal{M})})$ induces a differential inclusion on S whose trajectories are the extremal trajectories living in S. The goal of the chapter is to develop a Poincaré–Bendixson analysis of the differential inclusion defined on S. This can be done, as shown in the first part of the chapter, under a technical assumption satisfied for general sets $\mathcal{M}$ of modes. In the second part of the chapter, we study the special case where $\mathcal{M}$ is the convex hull of two 3×3 matrices whose difference is of rank one. In that case, we show that extremal trajectories which are periodic (whose existence can be deduced from the Poincaré–Bendixson analysis) correspond to piecewise periodic switching signals with at most four pieces.

In the sixth chapter we extend our analysis to classes $\mathcal{S}$ of switching signals other than $\mathcal{S}_{\mathrm{arb}}(\mathcal{M})$. One of the main difficulties when dealing with these classes is that they are not closed under concatenation, which prevents one, in particular, from extending the construction of a Barabanov norm. To circumvent this obstacle, we introduce the notion of *skeleton* which is a family of signals defined on finite intervals and whose concatenations belong to the original class $\mathcal{S}$. When the skeleton is large enough, we show that it retains the main asymptotic properties of $(\Sigma_{\mathcal{S}})$ and in particular one can extend the concept of extremal trajectories. We also show that the main classes of signals $\mathcal{S}$ considered in the literature admit such a skeleton. As a consequence, we extend to such classes the Berger–Wang-type formula proved in Chap. 4 in the case of arbitrary switching. Moreover, when a skeleton of a class $\mathcal{S}$ is large enough, one can characterize the maximal Lyapunov exponent using only trajectories of $(\Sigma_{\mathcal{S}})$ with periodic angular component. The last part of the chapter is devoted to the study of switched linear systems whose set of modes $\mathcal{M}$ admits a quadratic weak common Lyapunov function. In particular, we identify a class of switching signals such that the corresponding switched system is attractive.

The seventh and closing chapter of this monograph gathers material on switched linear control systems of the type $\dot{x} = A(t)x + B(t)u$, $y = C(t)x$, where $(A(\cdot), B(\cdot), C(\cdot))$ plays the role of switching signal, while u and y denote, respectively, the input and the observation. The first part of the chapter studies the controllability of such systems for different classes of switching signals, exploiting as in the previous chapter the notion of skeleton to handle classes that are not closed under concatenation. By duality we also obtain results on the observability of switched linear systems. Based on the controllability and observability analysis just mentioned, we then formalize the notion of minimal realization for switched linear control systems. Minimal realization is used to characterize switched linear control systems for which the L^2-gain (i.e., the norm of the input-output response in the L^2 functional setting) is bounded. A special occurrence of switched linear control systems is that of system of the type $\dot{x} = Ax + \alpha(t)Bu$, with $\alpha(\cdot)$ having values in $[0, 1]$ and satisfying a *persistent excitation* condition. We study the stabilizability of such systems by signal-independent linear feedbacks. In the last part of the chapter we analyze the problem of designing switching laws capable of stabilizing a switched linear system $(\Sigma_{\mathcal{S}_{\text{arb}}(\mathcal{M})})$. In particular, stabilizability by a switching law depending on the initial condition may be characterized via a "spectral" index analogous to the Lyapunov exponents introduced in Chap. 1, referred to as the *pointwise stabilization radius*. We further investigate the equivalence between stabilization in open loop and feedback stabilization, and we provide a converse Lyapunov theorem.

The following scheme aims at helping the reader navigate between the chapters of the book. A solid arrow connecting Chapter X to Chapter Y indicates that the results from the former play an important role in the latter, while a dotted line is used when the results from Chapter X are just used in examples within Chapter Y.

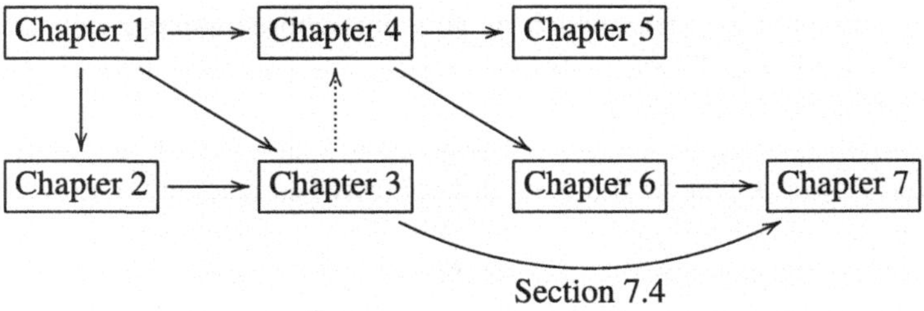

We conclude this introduction by pointing out that the selection of results presented in the book is subjective and is far from covering the immense literature on switched systems of the last 30 years. Let us mention some of the main topics that are *not* treated in this monograph.

First of all, we do not consider in what follows any applicative aspect relative to switched linear systems, nor any computational issue, e.g., associated with the numerical estimation of the maximal Lyapunov exponent. Let us just mention that any numerical procedure to test the stability of switched system can be turned into an algorithm to provide upper bounds for the maximal Lyapunov exponent, simply by probing the stability of the switched system with signals $t \mapsto A(t) + \nu \mathrm{Id}_n$, $A \in \mathcal{S}$, using $\nu \in \mathbb{R}$ as a parameter. Many numerical procedures to test the stability of switched system have been developed, based on linear matrix inequalities or sum of squares programs [GC10, PP03, CCG+12], but also on homogeneous rational Lyapunov functions [CC17]. A technique giving both upper and lower bounds on the maximal Lyapunov exponent is the invariant polytope algorithm [GWZ05, GP13, Mej20, CGPS21].

From a more fundamental point of view, several important dynamical systems are not addressed: above all the discrete-time and hybrid-time settings are not discussed here, as well switched systems with infinite-dimensional state space or those with nonlinear, differential-algebraic, impulsive, or event-triggered dynamics. Similarly, we restrict ourselves to a deterministic setting, excluding the probabilistic point of view, where, for instance, the switching signal can be considered as a random variable.

Contents

List of Symbols

$\mathbb{N}, \mathbb{R}, \mathbb{R}_+, \mathbb{C}$	positive integers, real, nonnegative real, and complex numbers
$\mathrm{sign}(x)$	sign of a real number x, that is, $\mathrm{sign}(x) = \begin{cases} +1 & \text{if } x > 0 \\ 0 & \text{if } x = 0 \\ -1 & \text{if } x < 0 \end{cases}$
$\lfloor x \rfloor$	largest integer smaller than or equal to a real number x
$\lceil x \rceil$	smallest integer larger than or equal to a real number x
$\Re(z)$	real part of a complex number z
$\Im(z)$	imaginary part of a complex number z
$\bar{z}$	conjugate of a complex number z
$F(Z)$	image $\{F(z) \mid z \in Z\}$ of a subset $Z \subset X$ for a function $F: X \to Y$
$F^{-1}(Z)$	pre-image $\{x \mid F(x) \in Z\}$ of a subset $Z \subset Y$ for a function $F: X \to Y$
$F^{-1}(y)$	pre-image $F^{-1}(\{y\})$ of a point $y \in Y$ for a function $F: X \to Y$
$\mathrm{Im}(F)$	image $F(X)$ of a function $F: X \to Y$
$\mathrm{Ker}(F)$	kernel $\{x \in X \mid F(x) = 0\}$ of a linear function $F: X \to Y$
$M_{n,m}(\mathbb{K})$	space of $n \times m$ matrices with entries in the field $\mathbb{K}$ ($\mathbb{K} = \mathbb{R}$ or $\mathbb{K} = \mathbb{C}$). The space of column vectors $M_{n,1}(\mathbb{K})$ is identified with $\mathbb{K}^n$
$M_n(\mathbb{K})$	space of $n \times n$ matrices with entries in $\mathbb{K}$, that is, $M_n(\mathbb{K}) = M_{n,n}(\mathbb{K})$
$\\|\cdot\\|$	Euclidean norm in $\mathbb{R}^n$ and induced operator norm in $M_n(\mathbb{R})$
$E^{\perp}$	orthogonal complement to a subspace E of $\mathbb{R}^n$
$\sum_{k \in \mathcal{K}} E_k$	sum over $k \in \mathcal{K}$ of subspaces E_k of a common vector space, i.e., the smallest vector space containing all E_k
$\mathrm{span}(S)$	smallest linear subspace of a vector space V containing $S \subset V$
Id_n	$n \times n$ identity matrix
$M^{\top}$	transpose of a matrix M
M^*	conjugate transpose of a matrix $M \in M_{n,m}(\mathbb{C})$, i.e., $M^* = (\bar{m}_{ji})_{j=1,\dots,m,\ i=1,\dots,n}$, where $M = (m_{ij})_{i=1,\dots,n,\ j=1,\dots,m}$

$\rho(M)$	spectral radius of a square matrix M, i.e., maximum of the moduli of its eigenvalues
$\det(M)$	determinant of a square matrix M
$\mathrm{tr}(M)$	trace of a square matrix M
$\|M\|_F$	Frobenius norm of a real matrix M equal to $\mathrm{tr}(M^\top M)^{1/2}$
e^M	matrix exponential of a square matrix M
$\Pi_{j=1}^N M_j$	matrix product $M_N \cdots M_1$ among square matrices
$\mathrm{GL}(n)$	group (for the matrix product) made by the invertible matrices of $M_n(\mathbb{R})$
$\mathrm{O}(n)$	subgroup of $\mathrm{GL}(n)$ made of orthogonal matrices, i.e., matrices M in $M_n(\mathbb{R})$ such that $M^\top M = MM^\top = \mathrm{Id}_n$
$\mathrm{SO}(n)$	subgroup of $\mathrm{GL}(n)$ made of orthogonal matrices whose determinant is equal to 1
$\mathbb{RP}^{n-1}$	projective space $\{\mathrm{span}\{x\} \mid x \in \mathbb{R}^n \setminus \{0\}\}$ endowed with the finest topology that makes the canonical projection $\Pi\colon \mathbb{R}^n \setminus \{0\} \ni x \mapsto \mathrm{span}\{x\} \in \mathbb{RP}^{n-1}$ continuous
S^{n-1}	unit sphere of $\mathbb{R}^n$
$B(x_0, r)$	closed ball in $\mathbb{R}^n$ of center $x_0 \in \mathbb{R}^n$ and radius $r > 0$
$P \succ Q$ $(P \succeq Q)$	relation between symmetric square matrices, meaning that $P - Q$ is positive-definite (positive semi-definite)
$P \prec Q$ $(P \preceq Q)$	relation between symmetric square matrices, meaning that $P - Q$ is negative-definite (negative semi-definite)
$d(M, \mathcal{M})$	distance between a matrix $M \in M_n(\mathbb{R})$ and a subset $\mathcal{M}$ of $M_n(\mathbb{R})$, defined as $d(M, \mathcal{M}) = \inf_{N \in \mathcal{M}} \|N - M\|$
$d_H(\mathcal{M}_1, \mathcal{M}_2)$	Hausdorff distance between two closed subsets $\mathcal{M}_1$ and $\mathcal{M}_2$ of $M_n(\mathbb{R})$, that is $$d_H(\mathcal{M}_1, \mathcal{M}_2) = \max\left(\max_{M_1 \in \mathcal{M}_1} d(M_1, \mathcal{M}_2), \max_{M_2 \in \mathcal{M}_2} d(M_2, \mathcal{M}_1)\right)$$
$[\![M_1, M_2]\!]$	segment $\{\xi M_1 + (1 - \xi) M_2 \mid \xi \in [0, 1]\}$ between two matrices M_1 and M_2
$[M_1, M_2]$	Lie bracket (or, commutator) between two matrices M_1 and M_2, that is, $[M_1, M_2] = M_1 M_2 - M_2 M_1$
$\mathrm{Lie}(\mathcal{M})$	Lie algebra generated by a subset $\mathcal{M}$ of $M_n(\mathbb{R})$, that is, the minimal subspace of $M_n(\mathbb{R})$ containing $\mathcal{M}$ and such that $[M_1, M_2] \in \mathrm{Lie}(\mathcal{M})$ for every $M_1, M_2 \in \mathrm{Lie}(\mathcal{M})$
$G_{\mathcal{M}}$	matrix group generated by a subset $\mathcal{M}$ of $M_n(\mathbb{R})$, that is, the minimal subgroup of $\mathrm{GL}(n)$ containing e^{tM} for every $t \in \mathbb{R}$ and $M \in \mathcal{M}$
$\Omega + v$	translation by an element v of a vector space E of a subset Ω of E, that is, $\Omega + v = \{\omega + v \mid \omega \in \Omega\}$
$\mathcal{K}v$	multiplication of a vector v by a set of scalars or matrices of suitable dimensions, i.e., $\mathcal{K}v = \{Kv \mid K \in \mathcal{K}\}$
$\mathrm{co}(\Omega)$	convex hull of a subset Ω of a vector space E

$\overline{\text{co}}(\Omega)$	closed convex hull of a subset Ω of a normed vector space E
$\|f\|_p$	L^p-norm, for $1 \le p \le +\infty$, of a measurable function $f: I \to \mathbb{R}^n$ defined on a (possibly unbounded) interval I of $\mathbb{R}$, that is, $$\|f\|_p = \begin{cases} (\int_I \|f(t)\|^p dt)^{1/p} & \text{if } 1 \le p < +\infty \\ \text{ess sup}_{t\in I} \|f(t)\| & \text{if } p = \infty \end{cases}$$
$L^p(I, \mathbb{R}^n)$	space of measurable functions $f: I \to \mathbb{R}^n$ such that $\|f\|_p < +\infty$
$L^p_{\text{loc}}(I, \mathbb{R}^n)$	space of measurable functions $f: I \to \mathbb{R}^n$ such that for every compact interval $K \subset I$, $\|f\|_K\|_p < +\infty$
$C(\Omega, \mathbb{R}^n)$	space of continuous functions $f: \Omega \to \mathbb{R}^n$
$C^\ell(\Omega, \mathbb{R}^n)$	space of ℓ times differentiable functions $f: \Omega \to \mathbb{R}^n$ with continuous derivatives
$A_1 * A_2$	concatenation of two functions $A_1: [0, t_1] \to Y$, $A_2: [0, t_2] \to Y$, that is, $A_1 * A_2: [0, t_1+t_2] \to Y$ coincides with $A_1(\cdot)$ on $[0, t_1]$ and with $A_2(\cdot - t_1)$ on $(t_1, t_1+t_2]$. The notation readily extends to the case where A_2 is defined on $[0, +\infty)$ and to the concatenation of a sequence of functions $(A_j: [0, t_j] \to Y)_{j\in\mathbb{N}}$
$\dot{w}(t)$	time-derivative of a function $w: \mathbb{R} \to \mathbb{R}^n$ at time t
$\nabla w(x)$	gradient of a function $w: \mathbb{R}^n \to \mathbb{R}$ at $x \in \mathbb{R}^n$
$\frac{d}{dx} w(x)$	Jacobian matrix of a function $w: \mathbb{R}^n \to \mathbb{R}^m$ at $x \in \mathbb{R}^n$
$\partial w(x)$	subdifferential of a function $w: \mathbb{R}^n \to \mathbb{R}$ at $x \in \mathbb{R}^n$, that is, $\partial w(x) = \{l \in \mathbb{R}^n \mid l^\top(y - x) \le w(y) - w(x) \text{ for every } y \in \mathbb{R}^n\}$
$\text{Hess}_x w$	Hessian matrix of a function $w: \mathbb{R}^n \to \mathbb{R}$ at $x \in \mathbb{R}^n$
$T_p\mathcal{M}$	tangent space to a manifold $\mathcal{M}$ at a point $p \in \mathcal{M}$
$\Phi_A(t, s)$	fundamental matrix from time s to time t associated with a signal A (see Definition 1.2)
$\omega(x(\cdot))$	ω-limit set of a curve $x(\cdot)$ (see Definition 4.23)

Standing Notation (cf. Chap. 1 and the Appendix for Detailed Definitions)

$\mathscr{M}$	bounded subset of $M_n(\mathbb{R})$ (set of modes)
$\mathcal{S}$	positive time-shift-invariant subset of $L^\infty(\mathbb{R}_+, \mathscr{M})$ (class of signals). The main examples of classes of signals are introduced in Sect. 1.1.1
$\mathcal{S}^\star$	weak-$\star$ closure of $\mathcal{S}$
$\mathcal{S}_{\text{arb}}(\mathscr{M})$	class of all measurable signals with values in $\mathscr{M}$
$(\Sigma_\mathcal{S})$	switched system associated with the class of signals $\mathcal{S}$
$(\Sigma_\mathscr{M})$	switched system associated with the class $\mathcal{S}_{\text{arb}}(\mathscr{M})$

$\hat{\lambda}(\mathcal{S})$	maximal Lyapunov exponent of a class of signals $\mathcal{S}$ (see Definition 1.17)
$\lambda(\mathcal{S})$	uniform exponential rate of a class of signals $\mathcal{S}$ (see Definition 1.17)
$\lambda(\mathscr{M})$ $(\hat{\lambda}(\mathscr{M}))$	uniform exponential rate (maximal Lyapunov exponent) of the class of arbitrary signals with values in $\mathscr{M}$ (they actually coincide by Remark 1.27)
$\mu(\mathcal{S})$	joint spectral exponent of a class of signals $\mathcal{S}$ (see Sect. 1.3.3)

Chapter 1
Switched Linear Systems: Definitions and Basic Facts

In this chapter, we provide the definition of *a switched linear system* as a time-varying linear system on $\mathbb{R}^n$ of the form $\dot{x}(t) = A(t)x(t)$, where the switching signal $A(\cdot)$ belongs to a time-shift invariant class $\mathcal{S}$ of $\mathscr{M}$-valued measurable functions, where $\mathscr{M}$ is a bounded set of $n \times n$ matrices. The class $\mathcal{S}$ needs not be closed under concatenation and it describes the possible constraints imposed on the switching signal such as periodicity, piecewise constancy, dwell-time etc. We review the main stability properties that can be satisfied by a switched linear system, from the weakest one referred to as *weak attractivity* to the strongest one, called *uniform exponential stability*. To characterize these properties, we consider two real numbers which quantitatively measure the stability of the switched linear system, namely the *maximal Lyapunov exponent* $\hat{\lambda}(\mathcal{S})$ and the *uniform exponential rate* $\lambda(\mathcal{S})$. In general it holds that $\hat{\lambda}(\mathcal{S}) \leq \lambda(\mathcal{S})$ and, even if $\hat{\lambda}(\mathcal{S})$ is more intuitive from the dynamical point of view, it turns out that $\lambda(\mathcal{S})$ is more adapted for robustness and computational issues. As a matter of fact, a major issue throughout this book will be to derive conditions on $\mathcal{S}$ ensuring equality between $\hat{\lambda}(\mathcal{S})$ and $\lambda(\mathcal{S})$. In that direction, we start with Fenichel's lemma, which gives a necessary condition on $\mathcal{S}$ for equivalence between weak attractivity and uniform exponential stability for the switched linear system associated with $\mathcal{S}$. We also provide block reduction results when all matrices of $\mathscr{M}$ can be put under upper block diagonal structure by the same linear change of basis. We close this chapter by defining two auxiliary switched systems of interest, the lifted and projected switched systems associated with a switched linear system, the first one being a switched linear system in the set of $n \times n$ matrices and the second one a switched system on the projective space $\mathbb{RP}^{n-1}$.

Y. Chitour et al., *Dynamics and Stability of Continuous-Time Switched Linear Systems*, PNLDE Subseries in Control 105,
https://doi.org/10.1007/978-3-031-82982-6_1

1.1 Switched Linear Dynamics

Throughout this book, we deal with linear systems $(\Sigma_{\mathcal{S}})$ of the type

$$\dot{x}(t) = A(t)x(t), \qquad t \geq 0, \quad A(\cdot) \in \mathcal{S}, \quad x \in \mathbb{R}^n, \qquad (\Sigma_{\mathcal{S}})$$

where $\mathcal{S}$ is a set of $\mathcal{M}$-valued functions defined on $[0, +\infty)$, with $\mathcal{M}$ a given bounded subset of $M_n(\mathbb{R})$, and $(\Sigma_{\mathcal{S}})$ is viewed as a family of non-autonomous systems parameterized by $A(\cdot) \in \mathcal{S}$. The elements of $\mathcal{S}$ are called *switching signals* or *switching laws*. Each matrix $M \in \mathcal{M}$ (and, by extension, each associated linear system $\dot{x} = Mx$) is referred to as a *mode* of $(\Sigma_{\mathcal{S}})$, and $\mathcal{M}$ is called the *set of modes*.

The set $\mathcal{S}$ is always assumed to be *positive time-shift-invariant* (*shift-invariant* for short), i.e., for every $h \geq 0$ and $A(\cdot) \in \mathcal{S}$, the signal $A(\cdot + h)\colon [0, +\infty) \to \mathcal{M}$ also belongs to $\mathcal{S}$.

Let us comment on the assumptions made on $\mathcal{M}$ and $\mathcal{S}$. The boundedness of $\mathcal{M}$ ensures that all trajectories of $(\Sigma_{\mathcal{S}})$ have at most exponential growth, with a common upper bound on their exponential rates, whereas the shift-invariance of $\mathcal{S}$ encodes the fact that the switching laws obey rules which do not change over time.

A crucial property for a class of signals, which may or may not hold, is closure under concatenation.

Definition 1.1 Let $\mathcal{S}$ be a shift-invariant class. We say that $\mathcal{S}$ is

- *closed under finite concatenation* if for every $k \geq 2$, $A_1, \dots, A_k \in \mathcal{S}$ and $t_1, \dots, t_{k-1} > 0$ the concatenation $A_1|_{[0,t_1]} * A_2|_{[0,t_2]} * \cdots * A_{k-1}|_{[0,t_{k-1}]} * A_k$ is in $\mathcal{S}$;
- *closed under concatenation* if for every sequence $(A_k)_{k\in\mathbb{N}}$ in $\mathcal{S}$ and $(t_k)_{k\in\mathbb{N}}$ in $(0, +\infty)$ such that $\sum_{k\in\mathbb{N}} t_k = +\infty$, the concatenation $A_1|_{[0,t_1]} * \cdots * A_k|_{[0,t_k]} * \dots$ is in $\mathcal{S}$.

Note that closure under concatenation implies closure under finite concatenation (due to the shift-invariance of $\mathcal{S}$) but the converse is not true. For example, consider the class of piecewise constant signals switching between two distinct matrices $M_0, M_1 \in M_n(\mathbb{R})$ and which are eventually (i.e., for all times large enough) equal to M_1. Such a class is shift-invariant, closed under finite concatenation, but not closed under concatenation.

1.1.1 Classes of Switching Signals

We describe in this section classes of switched linear systems $(\Sigma_{\mathcal{S}})$ that are commonly used in the literature. These classes are characterized by the choice of the set of modes $\mathcal{M}$ (the admissible dynamics) and by the switching rules defining $\mathcal{S}$ (the class of signals).

A typical situation encountered in the literature is that where $\mathcal{M}$ is finite. Since several qualitative results require that $\mathcal{M}$ is convex, one is naturally led to consider also the case where $\mathcal{M}$ is polytopic, that is, $\mathcal{M}$ is the convex hull of finitely many matrices.

For modeling purposes one might consider subsets $\mathcal{M}$ of $M_n(\mathbb{R})$ with other specific features. For instance, $\mathcal{M}$ can be made of matrices with positive entries or Metzler matrices, which is suitable for describing the evolution of non-negative quantities (e.g., chemical concentrations, size of populations, stocks...). More generally, one might consider sets of matrices whose corresponding flows satisfy a cone invariance condition.

Another type of constraints on the set of modes $\mathcal{M}$ might be motivated by a cascade structure in the physical model, in which case $\mathcal{M}$ is chosen as a subset of block triangular matrices with prescribed block structure.

Additional restrictions on the behaviour of the switched system can be introduced by specifying the switching rules. When there is no such a restriction, we speak of *arbitrary* (or *unconstrained*) *switching* and we use $\mathcal{S}_{\mathrm{arb}}(\mathcal{M})$ to denote the class of arbitrarily switching signals, i.e., $\mathcal{S}_{\mathrm{arb}}(\mathcal{M}) = L^\infty([0,\infty),\mathcal{M})$. In that case, we simply write $(\Sigma_{\mathcal{M}})$ to denote $(\Sigma_{\mathcal{S}_{\mathrm{arb}}(\mathcal{M})})$.

In many practical situations, however, constraints are imposed on the switching rule and we speak in this case of *restrained switching*. An assumption that is commonly considered in the literature requires signals to be piecewise constant, i.e., signals whose restriction to every finite time-interval is the concatenation of finitely many constant signals. The class of all such signals with values in $\mathcal{M}$ is denoted in this book by $\mathcal{S}_{\mathrm{pc}}(\mathcal{M})$. One of the main features of $\mathcal{S}_{\mathrm{pc}}(\mathcal{M})$ consists in the non-occurrence of the Zeno phenomenon, i.e., the discontinuities of a signal in $\mathcal{S}_{\mathrm{pc}}(\mathcal{M})$ do not accumulate in finite time.

A more restrictive constraint is that each mode is activated for at least a fixed minimal amount of time τ, called the *dwell-time*. The corresponding class of switching signals, denoted by $\mathcal{S}_{\mathrm{dw},\tau}(\mathcal{M})$, is the class of piecewise constant signals such that the distance between two discontinuity times is at least τ. (Notice that $\mathcal{S}_{\mathrm{pc}}(\mathcal{M})$ can be identified with $\mathcal{S}_{\mathrm{dw},0}(\mathcal{M})$.)

We list below some other classes which appear in the literature and which are relevant both for applications and theoretical purposes:

$\mathcal{S}_{\mathrm{av-dw},\tau,N_0}(\mathcal{M})$ is the class of piecewise constant signals which satisfy the (τ, N_0) *average dwell-time* condition, with $\tau > 0$ and N_0 a positive integer: for every $s, t \geq 0$, the number of switching times in $[s, s+t]$ is bounded from above by $N_0 + \frac{t}{\tau}$;

$\mathcal{S}_{\mathrm{per}}(\mathcal{M})$ is the class of periodic signals with no restriction on the period;

$\mathcal{S}_{\mathrm{cont}}(\mathcal{M})$ is the class of continuous signals;

$\mathcal{S}_{\mathrm{lip},L}(\mathcal{M})$ is the class of Lipschitz signals with Lipschitz constant $L > 0$;

$\mathcal{S}_{\mathrm{bv},T,\nu}(\mathcal{M})$ is the class of signals of (T, ν)-*bounded variation*, with $T, \nu > 0$, i.e., signals whose restriction to every interval of length T has total variation at most ν, that is, $A \in \mathcal{S}_{\mathrm{bv},T,\nu}(\mathcal{M})$ if and only if

$$\sup_{\substack{t\geq 0, k\in\mathbb{N} \\ t=t_0<t_1<\cdots<t_k=t+T}} \sum_{i=1}^{k} \|A(t_i) - A(t_{i-1})\| \leq \nu;$$

$\mathcal{S}_{\mathrm{pe},T,\mu}(\mathcal{M})$ is the class of (T, μ)-*persistently exciting* signals, with $0 < \mu < T$ and

$$\mathcal{M} = \left\{ M_\ell = N_0 + \sum_{i=1}^{m} \ell_i N_i \mid \ell = (\ell_1, \dots, \ell_m) \in [0, 1]^m \right\}$$

with $N_0, \dots, N_m \in M_n(\mathbb{R})$, i.e.,

$$\mathcal{S}_{\mathrm{pe},T,\mu}(\mathcal{M}) = \Big\{ t \mapsto M_{\alpha(t)} \mid \alpha \in L^\infty([0, \infty), [0, 1]^m) \\ \text{such that } \int_t^{t+T} \alpha_i(s) ds \geq \mu \\ \text{for all } t \geq 0 \text{ and } i = 1, \dots, m \Big\}.$$

Notice that all such classes are shift-invariant.

We stress that we do not require the class of signals $\mathcal{S}$ to contain all constant signals corresponding to the modes $M \in \mathcal{M}$. For instance, when $m = 1$ and $N_0 \neq N_1$ the class $\mathcal{S}_{\mathrm{pe},T,\mu}(\mathcal{M})$ does not contain the signal constantly equal to $M_0 = N_0$.

Among the classes introduced above, one can easily check that $\mathcal{S}_{\mathrm{arb}}(\mathcal{M})$ and $\mathcal{S}_{\mathrm{pc}}(\mathcal{M})$ are closed under concatenation and all the others are not even closed under finite concatenation, as soon as $\mathcal{M}$ contains at least two distinct matrices.

1.2 Basic Properties of Switched Systems

In this section, we provide elementary properties of switched linear systems, showing in particular that solutions are defined for all nonnegative times and that sequences of solutions converge to a solution, up to a subsequence. We also consider two types of concatenations for a class $\mathcal{S}$ of signals besides the standard definition, which turn out to be useful first to determine the weak-$\star$ closure $\mathcal{S}^\star$ of $\mathcal{S}$ and second to put forward a strong approximation property of trajectories of $(\Sigma_{\mathcal{S}^\star})$ by those of $(\Sigma_{\mathcal{S}})$.

1.2.1 Existence and Regularity of Solutions

Existence and uniqueness of Lipschitz continuous solutions for linear non-autonomous equations of the type $\dot{x}(t) = A(t)x(t)$, $x_0 \in \mathbb{R}^n$, for $A \in L^\infty([t_0, t_1], M_n(\mathbb{R}))$, $0 \leq t_0 < t_1 < +\infty$, follows from classical results on Carathéodory solutions (Theorem A.5). Moreover, for almost every $t \in [t_0, t_1]$,

$$- \|A\|_\infty \|x(t)\| \leq \frac{d}{dt}\|x(t)\| \leq \|A\|_\infty \|x(t)\|,$$

yielding

$$e^{-(t-t_0)\|A\|_\infty} \|x_0\| \leq \|x(t)\| \leq e^{(t-t_0)\|A\|_\infty} \|x_0\|, \qquad \text{for every } t \in [t_0, t_1]. \tag{1.1}$$

As a consequence of Theorem A.5, one can define the flow of $\dot{x}(t) = A(t)x(t)$ as follows.

Definition 1.2 (Fundamental Matrix and Flow) Given $A \in L^\infty([t_0, t_1), M_n(\mathbb{R}))$ with $0 \leq t_0 < t_1 \leq +\infty$, for every $s \in [t_0, t_1)$ ($s \in [t_0, t_1]$ when $t_1 < +\infty$) let $\Phi_A(\cdot, s) \in M_n(\mathbb{R})$ be the unique locally Lipschitz solution of

$$\frac{\partial}{\partial t}\Phi_A(t, s) = A(t)\Phi_A(t, s), \qquad \text{for almost every } t \in [t_0, t_1),$$
$$\Phi_A(s, s) = \mathrm{Id}_n.$$

The matrix $\Phi_A(t, s)$ and the linear mapping $x_0 \mapsto \Phi_A(t, s)x_0$ are called the *fundamental matrix from s to t* and the *flow from s to t* associated with $\dot{x} = Ax$, respectively.

For every τ_1, τ_2, the functions $t \mapsto \Phi_A(t, \tau_1)$ and $t \mapsto \Phi_A(t, \tau_2)\Phi_A(\tau_2, \tau_1)$ are solutions of the same Cauchy problem at $t = \tau_2$, yielding the relation

$$\Phi_A(\tau_3, \tau_1) = \Phi_A(\tau_3, \tau_2)\Phi_A(\tau_2, \tau_1), \qquad \text{for every } \tau_1, \tau_2, \tau_3. \tag{1.2}$$

In particular, by choosing $\tau_1 = \tau_3 = s$ and $\tau_2 = t$ in the previous equation, and by taking into account the initial condition in Definition 1.2, one deduces that the fundamental matrix is invertible and $\Phi_A(s, t) = \Phi_A(t, s)^{-1}$ for every s, t.

We next prove a continuity result for the dependence of the fundamental matrix with respect to the signal. For the definition of weak-$\star$ convergence, see Definition A.9.

Proposition 1.3 *Let $(A_k)_{k\in\mathbb{N}}$ be a bounded sequence in $L^\infty([0, +\infty), M_n(\mathbb{R}))$ weak-$\star$ converging to $A \in L^\infty([0, +\infty), M_n(\mathbb{R}))$. Then the sequence $(\Phi_{A_k}(\cdot, 0))_{k\in\mathbb{N}}$ converges to $\Phi_A(\cdot, 0)$ uniformly on compact subintervals of $[0, +\infty)$.*

Proof Fix $T > 0$ and set $E_k(t) = \Phi_{A_k}(t, 0) - \Phi_A(t, 0)$ for $k \in \mathbb{N}$ and $t \in [0, T]$. Let us prove that $\max_{t\in[0,T]} \|E_k(t)\|$ tends to zero as k tends to infinity. One has $E_k(0) = 0$ and

$$\dot{E}_k(t) = A_k(t)E_k(t) + (A_k(t) - A(t))\Phi_A(t, 0), \quad \text{for almost every } t \in [0, T].$$

Then,

$$\|E_k(t)\| \leq C \int_0^t \|E_k(s)\| ds + \|h_k(t)\|, \quad \text{for every } t \in [0, T],$$

where $C = \sup_{k\in\mathbb{N}} \|A_k\|_\infty$ and $h_k(t) = \int_0^t (A_k(s) - A(s))\Phi_A(s, 0)ds$. Note that the family of functions $(h_k)_{k\in\mathbb{N}}$ is uniformly equicontinuous and uniformly bounded on $[0, T]$. Moreover, for every $t \in [0, T]$, the sequence $(h_k(t))_{k\in\mathbb{N}}$ tends to zero by weak-$\star$ convergence of A_k to A. Applying Ascoli–Arzelà theorem (Theorem A.12) to every subsequence of $(h_k)_{k\in\mathbb{N}}$, we deduce that the sequence $H_k := \max_{t\in[0,T]} \|h_k(t)\|$ converges to zero as k tends to infinity. The conclusion now follows by using Grönwall lemma (Theorem A.7). □

We close this section by recalling the following compactness result: any bounded subset of $L^\infty([0, +\infty), M_n(\mathbb{R}))$ is relatively sequentially compact for the weak-$\star$ topology. This is a consequence of the Banach–Alaoglu theorem (Theorem A.11). Based on such a property and the previous proposition we have the following corollary.

Corollary 1.4 *Let $(A_k)_{k\in\mathbb{N}}$ and $(x_k)_{k\in\mathbb{N}}$ be two bounded sequences in the spaces $L^\infty([0, +\infty), M_n(\mathbb{R}))$ and $\mathbb{R}^n$, respectively. Then there exists a subsequence $(A_{k_r})_{r\in\mathbb{N}}$ weak-$\star$ converging to some $A \in L^\infty([0, +\infty), M_n(\mathbb{R}))$ such that $(x_{k_r})_{r\in\mathbb{N}}$ converges to some $x \in \mathbb{R}^n$ and $(t \mapsto \Phi_{A_{k_r}}(t, 0)x_{k_r})_{r\in\mathbb{N}}$ converges to $t \mapsto \Phi_A(t, 0)x$ uniformly on compact subintervals of* $[0, +\infty)$.

Regarding a general class $\mathcal{S}$ of $\mathcal{M}$-valued signals, with $\mathcal{M}$ bounded, Theorem A.11 implies that $\mathcal{S}$ is relatively sequentially compact for the weak-$\star$ topology, i.e., the weak-$\star$ closure $\mathcal{S}^\star$ of $\mathcal{S}$ is weak-$\star$ compact.

When $\mathcal{M}$ is compact, weak-$\star$ compactness is clearly satisfied by the classes $\mathcal{S}_{\mathrm{dw},\tau}(\mathcal{M})$ (for $\tau > 0$), $\mathcal{S}_{\mathrm{av-dw},\tau,N_0}(\mathcal{M})$, $\mathcal{S}_{\mathrm{lip},L}(\mathcal{M})$, $\mathcal{S}_{\mathrm{bv},T,\nu}(\mathcal{M})$ and $\mathcal{S}_{\mathrm{pe},T,\mu}(\mathcal{M})$. As regards the other classes, we will characterize their weak-$\star$ closures in Corollary 1.8.

1.2.2 Long-Time Approximation and Convexification of the Set of Modes

As remarked at the end of the previous section, the weak-$\star$ closure $\mathcal{S}^\star$ of $\mathcal{S}$ has better topological properties, noticeably the fact that it is weak-$\star$ compact. We study in this section up to which extent we can relate the asymptotic behavior of the trajectories of $(\Sigma_{\mathcal{S}})$ to those of $(\Sigma_{\mathcal{S}^\star})$. We will be able to characterize $\mathcal{S}^\star$ for certain classes $\mathcal{S}$, and to derive asymptotic properties of $(\Sigma_{\mathcal{S}})$. We start by introducing the following definition extending the notion of closure under concatenation (cf. Definition 1.1).

Definition 1.5 We say that $\mathcal{S}$ is

- *closed under finite weak concatenation* if for every $k \geq 1$, $A_1, \dots, A_k \in \mathcal{S}$, $t_1, \dots, t_k > 0$, and $h_1, \dots, h_{k-1} > 0$, there exists $B_1, \dots, B_k \in \mathcal{S}$ such that the concatenation $A_1|_{[0,t_1]} * B_1|_{[0,h_1]} * A_2|_{[0,t_2]} * \cdots * B_{k-1}|_{[0,h_{k-1}]} * A_k|_{[0,t_k]} * B_k$ is in $\mathcal{S}$;
- *closed under weak concatenation* if for every sequence $(A_k)_{k\in\mathbb{N}}$ in $\mathcal{S}$ and $(t_k)_{k\in\mathbb{N}}$, $(h_k)_{k\in\mathbb{N}}$ in $(0, +\infty)$ with $\sum_{k=1}^{\infty} t_k = +\infty$, there exists a sequence $(B_k)_{k\in\mathbb{N}}$ in $\mathcal{S}$ such that the concatenation $A_1|_{[0,t_1]} * B_1|_{[0,h_1]} * \cdots * A_k|_{[0,t_k]} * B_k|_{[0,h_k]} * \dots$ is in $\mathcal{S}$.

Remark 1.6 Closure under concatenation clearly implies closure under weak concatenation. The converse implication does not hold true: when $\mathcal{M} \subset M_n(\mathbb{R})$ is pathwise connected and has at least two elements, $\mathcal{S}_{\text{cont}}(\mathcal{M})$ is closed under weak concatenation but not closed under concatenation.

We next show how closure under finite weak concatenation allows one to characterize $\mathcal{S}^\star$. We recall that $\overline{\text{co}}(\mathcal{M})$ denotes the closure of the convex hull of the set of modes $\mathcal{M}$.

Proposition 1.7 *Assume that (i) $\mathcal{S}$ contains all constant signals with values in $\mathcal{M}$, and (ii) $\mathcal{S}$ is closed under finite weak concatenation. Then $\mathcal{S}^\star = \mathcal{S}_{\text{arb}}(\overline{\text{co}}(\mathcal{M}))$.*

Proof Since $\mathcal{S}_{\text{arb}}(\overline{\text{co}}(\mathcal{M}))$ is weak-$\star$ closed (Theorem A.11), it contains $\mathcal{S}^\star$. Let us now prove the converse inclusion. We first show that, for every $k \in \mathbb{N}$, $t_1, \dots, t_k > 0$, and $A_1, \dots, A_k \in \mathcal{S}$, there exists a signal $B \in \mathcal{S}^\star$ such that

$$B|_{[0,t_1+\cdots+t_k]} = A_1|_{[0,t_1]} * \cdots * A_k|_{[0,t_k]}.$$

For that purpose, given $m \in \mathbb{N}$ we set $h_j = 1/m$ for $j = 1, \dots, k-1$, and we deduce from closure under finite weak concatenation of $\mathcal{S}$ the existence of a signal $C_m \in \mathcal{S}$ such that $C_m(t) = A_j(t - T_j)$ for every $t \in [T_j, T_j + t_j)$ and every $j = 1, \dots, k$, where $T_j = \sum_{i=1}^{j-1} t_i + (j-1)/m$. Then, the sequence $(C_m)_{m\in\mathbb{N}}$ weak-$\star$ converges to the required signal B (Lemma A.10).

We next show that $\mathcal{S}^\star$ is closed under concatenation. Take a sequence $(B_l)_{l\in\mathbb{N}}$ in $\mathcal{S}^\star$ and $(\tau_k)_{k\in\mathbb{N}}$ in $(0, +\infty)$. For every $l \in \mathbb{N}$, consider a sequence $(A_l^{(j)})_{j\in\mathbb{N}}$ in $\mathcal{S}$ weak-$\star$ converging to B_l. By what precedes, for every $k \in \mathbb{N}$ there exists $B^{(k)} \in \mathcal{S}^\star$ such that $B^{(k)}|_{[0,\tau_1+\cdots+\tau_k]} = A_1^{(k)}|_{[0,\tau_1]} * A_2^{(k)}|_{[0,\tau_2]} * \cdots * A_k^{(k)}|_{[0,\tau_k]}$. By construction, $B^{(k)}$ weak-$\star$ converges to $B_1|_{[0,\tau_1]} * \cdots * B_j|_{[0,\tau_j]} * \dots$ as k tends to infinity. Thus the latter concatenation belongs to $\mathcal{S}^\star$, concluding the proof that $\mathcal{S}^\star$ is closed under concatenation.

Let us now prove that $\mathcal{S}^\star$ contains all constant signals with values in $\overline{\mathrm{co}}(\mathcal{M})$. Clearly, it is enough to prove the claim for constant signals with values in $\mathrm{co}(\mathcal{M})$. Consider any convex combination $\sum_{j=1}^k \alpha_j M^j$, where $M^1, \dots, M^k \in \mathcal{M}$, $\alpha_1 \dots, \alpha_k \in [0, 1]$, and $\sum_{j=1}^k \alpha_j = 1$. Let $(A_l)_{l\in\mathbb{N}}$ be the sequence of matrix signals defined by

$$A_l(t) = M^j \text{ if } t \in \bigcup_{r\in\mathbb{N}} \left[\frac{r}{l} + \sum_{p<j} \frac{\alpha_p}{l}, \frac{r}{l} + \sum_{p\leq j} \frac{\alpha_p}{l}\right).$$

Each A_l is the concatenation of signals in $\mathcal{S}$ (by Item (i)), hence it belongs to $\mathcal{S}^\star$, according to what precedes. The sequence $(A_l)_{l\in\mathbb{N}}$ admits a unique weak-$\star$ limit point equal to the signal constantly equal to $\sum_{j=1}^k \alpha_j M^j$ (Lemma A.10).

Since $\mathcal{S}^\star$ is closed under concatenation and contains all constant signals with values in $\overline{\mathrm{co}}(\mathcal{M})$, we deduce that $\mathcal{S}^\star$ contains $\mathcal{S}_{\mathrm{pc}}(\overline{\mathrm{co}}(\mathcal{M}))$. Moreover, the restriction to any finite interval I of $\mathcal{S}_{\mathrm{pc}}(\overline{\mathrm{co}}(\mathcal{M}))$ is dense in $L^1(I, \overline{\mathrm{co}}(\mathcal{M}))$, which implies by Lemma A.10 that the restriction to I of $\mathcal{S}_{\mathrm{pc}}(\overline{\mathrm{co}}(\mathcal{M}))$ is weak-$\star$ dense in $L^\infty(I, \overline{\mathrm{co}}(\mathcal{M}))$. In conclusion, $\mathcal{S}^\star$ contains $\mathcal{S}_{\mathrm{arb}}(\overline{\mathrm{co}}(\mathcal{M}))$. □

As an immediate consequence of the above proposition, one has the following.

Corollary 1.8 *Let $\mathcal{S}$ be one of the following classes: $\mathcal{S}_{\mathrm{arb}}(\mathcal{M})$, $\mathcal{S}_{\mathrm{pc}}(\mathcal{M})$, $\mathcal{S}_{\mathrm{per}}(\mathcal{M})$, or $\mathcal{S}_{\mathrm{cont}}(\mathcal{M})$ with $\mathcal{M}$ pathwise connected. Then $\mathcal{S}^\star = \mathcal{S}_{\mathrm{arb}}(\overline{\mathrm{co}}(\mathcal{M}))$.*

Notice that Proposition 1.3 only ensures the convergence of the flows associated with $\mathcal{S}$ to those associated with $\mathcal{S}^\star$ on finite time-intervals. This, however, is not enough to characterize the asymptotic behavior of $(\Sigma_{\mathcal{S}})$, as illustrated by the following example.

Example 1.9 Let $\mathcal{S}$ be the class of all piecewise constant signals switching between Id_n and $-\mathrm{Id}_n$ that are eventually equal to $-\mathrm{Id}_n$. Then every trajectory of $(\Sigma_{\mathcal{S}})$ is bounded and converges exponentially to the origin, whereas $\mathcal{S}^\star$ contains the constant signal equal to the identity matrix, hence there exist trajectories of $(\Sigma_{\mathcal{S}^\star})$ whose norm tends to infinity exponentially.

Motivated by the previous example, let us introduce a stronger convergence notion for the trajectories of $(\Sigma_{\mathcal{S}})$ towards those of $(\Sigma_{\mathcal{S}^\star})$.

Definition 1.10 We say that $\mathcal{S}$ *long-time approximates* $\mathcal{S}^\star$ if, for every trajectory $x^\star\colon [0, \infty) \to \mathbb{R}^n$ of $(\Sigma_{\mathcal{S}^\star})$ and every continuous function $r\colon [0, \infty) \to (0, \infty)$, there exists a trajectory $x\colon [0, \infty) \to \mathbb{R}^n$ of $(\Sigma_{\mathcal{S}})$ satisfying $\|x(t) - x^\star(t)\| < r(t)$ for every $t \in [0, \infty)$.

We next show that a class of signals that is closed under weak concatenation long-time approximates its weak-$\star$ closure.

Proposition 1.11 *Let $\mathcal{S}$ be closed under weak concatenation. Then $\mathcal{S}$ long-time approximates $\mathcal{S}^\star$.*

Proof Let $x^\star(\cdot)$ be a trajectory of $(\Sigma_{\mathcal{S}^\star})$ corresponding to a signal $A^\star \in \mathcal{S}^\star$. Fix $\tau_0 = 0$ and a closed neighborhood U_0 of $x^\star(0)$. Let us construct iteratively a sequence of triples $(U_k, A_k, \eta_k)_{k\in\mathbb{N}}$ so that, for every $k \in \mathbb{N}$, $\eta_k > 0$, $U_k \subset \mathbb{R}^n$ is a closed neighborhood of $x^\star(\tau_k)$ where

$$\tau_k = k + \sum_{j=1}^{k} \eta_j,$$

$A_k \in \mathcal{S}$, and the following holds true: for every $A \in \mathcal{S}_{\text{arb}}(\mathcal{M})$,

$$\Phi_{A_{k+1}|_{[0,1]}*A}(0, 1 + \eta_{k+1})U_{k+1} \subset U_k \tag{1.3}$$

and

$$\|\Phi_{A_{k+1}|_{[0,1]}*A}(s, 1 + \eta_{k+1})x - x^\star(\tau_k + s)\| < r(\tau_k + s),$$
$$\forall x \in U_{k+1},\ \forall s \in [0, 1 + \eta_{k+1}]. \tag{1.4}$$

Take a nonnegative integer k and assume that η_j, for $1 \le j \le k$, and U_k have been fixed. According to Proposition 1.3, there exists a sequence $(B_\ell)_{\ell\in\mathbb{N}}$ in $\mathcal{S}$ such that $\Phi_{B_\ell}(t, 0)$ converges to $\Phi_{A^\star}(\tau_k + t, \tau_k)$ uniformly with respect to $t \in [0, 1]$ as $\ell \to \infty$. As a consequence, there exists $\ell_1 \in \mathbb{N}$ such that for every $\eta_{k+1}, \varepsilon_{k+1} > 0$ small enough the closed ball $V_k = B(x^\star(\tau_k), \varepsilon_{k+1})$ is contained in U_k and, for every $A \in \mathcal{S}_{\text{arb}}(\mathcal{M})$,

$$\|\Phi_{B_\ell|_{[0,1]}*A}(t, 0)x - x^\star(\tau_k + t)\| < r(\tau_k + t),$$
$$\forall x \in V_k,\ \forall t \in [0, 1 + \eta_{k+1}],\ \forall \ell \ge \ell_1. \tag{1.5}$$

Let $M_k = \Phi_{A^\star}(\tau_k + 1, \tau_k)$ and

$$U_{k+1} = M_k\Big(x^\star(\tau_k) + B\Big(0, \frac{\varepsilon_{k+1}}{2}\Big)\Big) = x^\star(\tau_k + 1) + M_k\Big(B\Big(0, \frac{\varepsilon_{k+1}}{2}\Big)\Big).$$

We claim that there exist a neighborhood $\mathcal{N}_k$ of M_k such that $N^{-1}U_{k+1} \subset V_k$ for every $N \in \mathcal{N}_k$. Indeed, it suffices to choose $\mathcal{N}_k$ of the form $\{N \mid \|N^{-1}M_k - \text{Id}_n\| < \delta\}$ for δ small enough.

For η_{k+1} small enough, U_{k+1} is a neighborhood of $x^\star(\tau_{k+1})$ and there exist $\ell_2 \in \mathbb{N}$ such that

$$U_{k+1} \subset \Phi_{B_\ell|_{[0,1]}*A}(1 + \eta_{k+1}, 0)V_k, \qquad \text{for every } \ell \ge \ell_2,\ A \in \mathcal{S}_{\text{arb}}(\mathcal{M}).$$

Fix $\hat{\ell} \ge \max(\ell_1, \ell_2)$ and set $A_{k+1} = B_{\hat{\ell}}$. Since $V_k \subset U_k$, (1.3) follows at once, while (1.4) is a consequence of (1.5). This concludes the construction of the sequence $(U_k, A_k, \eta_k)_{k\in\mathbb{N}}$.

We next select a signal A in $\mathcal{S}$ such that $A(t) = A_{k+1}(t - \tau_k)$ for every $t \in [\tau_k, \tau_k + 1)$ and every nonnegative integer k. The existence of A is guaranteed by the closure under weak concatenation of $\mathcal{S}$. By Cantor's intersection theorem and (1.3), the set $\cap_{k=0}^{\infty} \Phi_A(0, \tau_k)U_k$ is nonempty. Pick some x_0 in it and set $x(t) = \Phi_A(t, 0)x_0$ for $t \geq 0$. By construction of the sequence $(U_k, A_k, \eta_k)_{k\in\mathbb{N}}$ and, in particular, (1.4), it now turns out that $\|x(t) - x^\star(t)\| < r(t)$ for every $t \in [0, \infty)$, concluding the proof. □

In particular, one deduces that $\mathcal{S}_{\rm arb}(\mathcal{M})$, $\mathcal{S}_{\rm pc}(\mathcal{M})$, and $\mathcal{S}_{\rm cont}(\mathcal{M})$ (when $\mathcal{M}$ is pathwise connected) long-time approximate $\mathcal{S}_{\rm arb}(\overline{\rm co}(\mathcal{M}))$. Notice that this does not hold for $\mathcal{S}_{\rm per}(\mathcal{M})$.

1.3 Stability of Switched Systems

This section introduces the main stability notions associated with a switched linear system, distinguishing between properties holding for each trajectory and properties holding instead uniformly with respect to all trajectories. We provide the basic properties of the maximal Lyapunov exponent and of the uniform exponential rate and we prove Fenichel's lemma.

1.3.1 Main Stability Notions

Definition 1.12 The switched system $(\Sigma_{\mathcal{S}})$ is said to be

- *unbounded* if there exists a trajectory $x(\cdot)$ of $(\Sigma_{\mathcal{S}})$ such that $\limsup_{t\to+\infty} \|x(t)\| = +\infty$;
- *bounded* if every trajectory $x(\cdot)$ of $(\Sigma_{\mathcal{S}})$ is bounded, that is, $\sup_{t\in[0,+\infty)} \|x(t)\| < +\infty$;
- *uniformly bounded* (or, equivalently, *uniformly stable*) if there exists $C > 0$ such that, for every $A \in \mathcal{S}$ and $t \geq 0$, $\|\Phi_A(t, 0)\| \leq C$;
- *weakly attractive* if $\liminf_{t\to+\infty} \|x(t)\| = 0$ for every trajectory $x(\cdot)$ of $(\Sigma_{\mathcal{S}})$;
- *attractive* if every trajectory of $(\Sigma_{\mathcal{S}})$ converges to the origin as time goes to $+\infty$;
- *uniformly attractive* if for every $\varepsilon > 0$ there exists $T > 0$ such that, for every $A \in \mathcal{S}$ and every $t > T$, $\|\Phi_A(t, 0)\| < \varepsilon$;
- *exponentially attractive* if every trajectory converges exponentially in norm to zero as time goes to $+\infty$;
- *uniformly exponentially stable* if there exist $C, \gamma > 0$ such that, for every $A \in \mathcal{S}$ and $t \geq 0$, $\|\Phi_A(t, 0)\| \leq Ce^{-\gamma t}$.

Remark 1.13 When studying more general dynamical systems, one usually distinguishes among stability properties that are local or global with respect to the initial condition. We do not need this distinction here, due to the linearity of the dynamics that we consider.

Let us discuss the relationship between the stability properties stated in Definition 1.12.

Some implications are trivial, as, for instance, the fact that uniform exponential stability implies uniform attractivity. Since $\mathcal{S}$ is assumed to be shift-invariant and the dynamics is linear, it turns out that these two properties are actually equivalent. Indeed, uniform attractivity implies that there exists $T > 0$ so that, for every $t \geq 0$ and $A \in \mathcal{S}$, one has $\|\Phi_A(T+t,t)\| \leq 1/2$. Using the fact that, according to (1.1), $C = 2\max_{A\in\mathcal{S},\, t\in[0,T]} \|\Phi_A(t,0)\|$ is finite, one easily deduces that $\|\Phi_A(t,0)\| \leq C2^{-t/T}$ for every $t \geq 0$, i.e., uniform exponential stability. No other equivalence holds between the properties above and we next provide examples illustrating this fact.

On the one hand, the class $\mathcal{S}$ provided in Example 1.9 shows that exponential attractivity (and therefore attractivity and boundedness) does not imply uniform boundedness (and therefore uniform attractivity and uniform exponential stability).

On the other hand, weak attractivity is distinct from attractivity as shown by the following one-dimensional example. For $T > 0$, let $A_T : [0,T] \to \mathbb{R}$ be equal to 1 on $[0, T/3]$ and to -1 on $(T/3, T]$. The switched system associated with the time-shift closure of the class of signals made of infinite concatenations of functions in $\{A_T \mid T > 0\}$ is then weakly attractive but not attractive.

As a final remark, attractivity does not imply exponential attractivity, as shown by considering the switched system $(\Sigma_\mathcal{S})$ in $\mathbb{R}$ with $\mathcal{S} = \{A_h \mid h \geq 0\}$, where $A_h : t \mapsto -\frac{1}{1+h+t}$.

In order to ensure that system $(\Sigma_\mathcal{S})$ satisfies properties that are uniform with respect to the switching law, one needs to identify additional conditions on the class of signals $\mathcal{S}$ beside its shift-invariance. A preliminary result in that direction is the following.

Proposition 1.14 *Let $\mathcal{S}^\star$ be the weak-$\star$ closure of a class of signals $\mathcal{S}$. Then the following holds true.*

(i) $(\Sigma_\mathcal{S})$ is uniformly bounded (respectively, uniformly attractive, uniformly exponentially stable) if and only if $(\Sigma_{\mathcal{S}^\star})$ is.
(ii) Assume that $\mathcal{S}$ long-time approximates $\mathcal{S}^\star$. Let (P) be any of the properties from Definition 1.12. Then (P) holds true for $(\Sigma_\mathcal{S})$ if and only if it holds true for $(\Sigma_{\mathcal{S}^\star})$.

Proof Item **(i)** is a direct application of Proposition 1.3 and Item **(ii)** follows immediately from the definition of long-time approximation. □

As a corollary of the above proposition, we obtain the following result.

Corollary 1.15 *Let (P) be any of the properties from Definition 1.12 and $\mathcal{S}_{\mathrm{label}}(\mathcal{M})$ be one of the following classes: $\mathcal{S}_{\mathrm{arb}}(\mathcal{M})$, $\mathcal{S}_{\mathrm{pc}}(\mathcal{M})$, $\mathcal{S}_{\mathrm{per}}(\mathcal{M})$, or $\mathcal{S}_{\mathrm{cont}}(\mathcal{M})$ with $\mathcal{M}$ pathwise connected. Then (P) holds true for $(\Sigma_{\mathcal{S}_{\mathrm{label}}(\mathcal{M})})$ if and only if it holds true for $(\Sigma_{\mathcal{S}_{\mathrm{label}}(\overline{\mathrm{co}}(\mathcal{M}))})$.*

Proof First notice that the weak-$\star$ closure of both $\mathcal{S}_{\text{label}}(\mathcal{M})$ and $\mathcal{S}_{\text{label}}(\overline{\text{co}}(\mathcal{M}))$ is equal to $\mathcal{S}_{\text{arb}}(\overline{\text{co}}(\mathcal{M}))$ by Corollary 1.8. Hence the conclusion follows from Proposition 1.11 and Item **(ii)** of Proposition 1.14. □

Corollary 1.15 does not hold for general classes of switching signals, as illustrated by the following example.

Example 1.16 Let $\mathcal{M} = \{M_0, M_1\}$ with

$$M_0 = \begin{pmatrix} 0 & -1 \\ 1 & 0 \end{pmatrix}, \qquad M_1 = \begin{pmatrix} -1 & 3 \\ 0 & -1 \end{pmatrix}.$$

For $t \in \mathbb{R}$, e^{tM_0} is a rotation and e^{tM_1} tends to zero as t goes to $+\infty$. As a consequence there exists $\tau > 0$ such that $(\Sigma_{\mathcal{S}_{\text{dw},\tau}(\mathcal{M})})$ is bounded. On the other hand, $\overline{\text{co}}(\mathcal{M})$ contains the matrix $(M_0 + M_1)/2$, which has an eigenvalue with positive real part. Therefore, $(\Sigma_{\mathcal{S}_{\text{dw},\tau}(\overline{\text{co}}(\mathcal{M}))})$ is unbounded.

1.3.2 Asymptotic Growth Rates and Fenichel's Lemma

We next provide quantitative measures for the asymptotic behavior of switched linear systems.

Definition 1.17 Given a class $\mathcal{S}$ of signals, we define

- the *maximal Lyapunov exponent* of $\mathcal{S}$ as

$$\hat{\lambda}(\mathcal{S}) = \sup_{A \in \mathcal{S}} \limsup_{t \to \infty} \frac{\log \|\Phi_A(t,0)\|}{t};$$

- the *uniform exponential rate* of $\mathcal{S}$ as

$$\lambda(\mathcal{S}) = \limsup_{t \to \infty} \sup_{A \in \mathcal{S}} \frac{\log \|\Phi_A(t,0)\|}{t}. \tag{1.6}$$

Remark 1.18 In Definition 1.17, we can equivalently replace the operator norm induced by the Euclidean norm by any other norm in $M_n(\mathbb{R})$.

Remark 1.19 If $\mathcal{S}_1 \subset \mathcal{S}_2$, then it immediately follows from Definition 1.17 that $\hat{\lambda}(\mathcal{S}_1) \leq \hat{\lambda}(\mathcal{S}_2)$ and $\lambda(\mathcal{S}_1) \leq \lambda(\mathcal{S}_2)$. In particular, if $\mathcal{S}$ contains the signal constantly equal to a mode $M \in \mathcal{M}$, then $\lambda(\mathcal{S})$ and $\hat{\lambda}(\mathcal{S})$ are larger than or equal to the maximal real part of the eigenvalues of M.

If $\mathcal{S} = \mathcal{S}_{\text{arb}}(\mathcal{M})$ we write $\lambda(\mathcal{M}) = \lambda(\mathcal{S})$ and $\hat{\lambda}(\mathcal{M}) = \hat{\lambda}(\mathcal{S})$. We collect in the following proposition some elementary properties of $\lambda(\mathcal{S})$ and $\hat{\lambda}(\mathcal{S})$.

Proposition 1.20 *Let $\mathcal{S}$ be a class of signals with values in a bounded set $\mathcal{M}$. The following properties hold true:*

1. *For every invertible matrix $P \in M_n(\mathbb{R})$, $\mu \in \mathbb{R}$, and $\eta \in (0, +\infty)$, we have*

$$\lambda(\{PA(\cdot)P^{-1} \mid A \in \mathcal{S}\}) = \lambda(\mathcal{S}), \qquad \hat{\lambda}(\{PA(\cdot)P^{-1} \mid A \in \mathcal{S}\}) = \hat{\lambda}(\mathcal{S}), \tag{1.7}$$

$$\lambda(\{A(\cdot) + \mu \mathrm{Id}_n \mid A \in \mathcal{S}\}) = \lambda(\mathcal{S}) + \mu, \quad \hat{\lambda}(\{A(\cdot) + \mu \mathrm{Id}_n \mid A \in \mathcal{S}\}) = \hat{\lambda}(\mathcal{S}) + \mu, \tag{1.8}$$

$$\lambda(\{\eta A(\eta\,\cdot) \mid A \in \mathcal{S}\}) = \eta\lambda(\mathcal{S}), \qquad \hat{\lambda}(\{\eta A(\eta\,\cdot) \mid A \in \mathcal{S}\}) = \eta\hat{\lambda}(\mathcal{S}); \tag{1.9}$$

2. *$\hat{\lambda}(\mathcal{S}) \leq \lambda(\mathcal{S}) = \lambda(\mathcal{S}^\star)$ (and the inequality may be strict);*
3. *$\lambda(\mathcal{S}) = \inf\{\lambda \mid \exists C > 0$ such that $\|\Phi_A(t,0)\| \leq Ce^{\lambda t}\ \forall A \in \mathcal{S}\ \forall t \geq 0\}$;*
4. *$\hat{\lambda}(\mathcal{S}) = \inf\{\hat{\lambda} \mid \forall A \in \mathcal{S},\ \exists C_A > 0$ such that $\|\Phi_A(t,0)\| \leq C_A e^{\hat{\lambda} t}\ \forall t \geq 0\}$;*
5. *$\lambda(\mathcal{S}) < 0 \Longleftrightarrow (\Sigma_{\mathcal{S}})$ is uniformly exponentially stable;*
6. *$\hat{\lambda}(\mathcal{S}) < 0 \Longrightarrow (\Sigma_{\mathcal{S}})$ is exponentially attractive (and the converse is not true);*
7. *$\hat{\lambda}(\mathcal{S}) > 0 \Longrightarrow (\Sigma_{\mathcal{S}})$ is unbounded (and the converse is not true);*
8. *For $t > 0$, set $\lambda_t(\mathcal{S}) = \sup_{A \in \mathcal{S}} \frac{\log\|\Phi_A(t,0)\|}{t}$. Then $\lambda(\mathcal{S}) = \inf_{t>0} \lambda_t(\mathcal{S})$;*
9. *Let $\mathcal{S}^\top = \{A^\top(\cdot) \mid A(\cdot) \in \mathcal{S}\}$, and assume that for every $t > 0$ and $A \in \mathcal{S}$, there exists $A_\# \in \mathcal{S}$ such that $A_\#(s) = A(t-s)$ for $s \in [0,t]$. Then $\lambda(\mathcal{S}^\top) = \lambda(\mathcal{S})$.*

Proof Equation (1.7) follows from the fact that, for every invertible matrix $P \in M_n(\mathbb{R})$ and $A \in \mathcal{S}$, one has $\Phi_{PAP^{-1}}(t,s) = P\Phi_A(t,s)P^{-1}$ for every $t, s \geq 0$. Furthermore, for every $\mu \in \mathbb{R}$ and $A \in \mathcal{S}$, one has $\Phi_{A+\mu\mathrm{Id}_n}(t,s) = e^{\mu(t-s)}\Phi_A(t,s)$ for every $t, s \geq 0$, which implies Eq. (1.8). Similarly, for every $\eta > 0$, by the time-rescaling $t \mapsto \eta t$ one deduces that $\Phi_{\eta A(\eta\cdot)}(t,s) = \Phi_A(\eta t, \eta s)$ for every $A \in \mathcal{S}$ and $t, s \geq 0$, whence (1.9). This concludes the proof of Item 1.

For a proof of the inequality $\hat{\lambda}(\mathcal{S}) \leq \lambda(\mathcal{S})$ in Item 2, first notice that for every $A \in \mathcal{S}$ and $s > 0$, it holds

$$\sup_{t \geq s} \frac{\log\|\Phi_A(t,0)\|}{t} \leq \sup_{t \geq s}\sup_{B \in \mathcal{S}} \frac{\log\|\Phi_B(t,0)\|}{t}.$$

Passing to the limit as s tends to infinity, one deduces that

$$\limsup_{t\to\infty} \frac{\log\|\Phi_A(t,0)\|}{t} \leq \lambda(\mathcal{S}).$$

One concludes by taking the sup over $A \in \mathcal{S}$. Moreover, a class $\mathcal{S}$ for which the above inequality is strict is the one introduced in Example 1.9, for which $\hat{\lambda}(\mathcal{S}) = -1$ and $\lambda(\mathcal{S}) = 1$. To get the equality in Item 2, one notices on the one hand that $\lambda(\mathcal{S}) \leq \lambda(\mathcal{S}^\star)$ by Remark 1.19. On the other hand, using Proposition 1.3 one gets that, for every $t > 0$, $\varepsilon > 0$, and $A^\star \in \mathcal{S}^\star$, there exists $B \in \mathcal{S}$ such that

$$\frac{\log \|\Phi_{A^\star}(t,0)\|}{t} \le \varepsilon + \frac{\log \|\Phi_B(t,0)\|}{t} \le \varepsilon + \sup_{A \in \mathcal{S}} \frac{\log \|\Phi_A(t,0)\|}{t}.$$

By taking the sup over $A^\star \in \mathcal{S}^\star$, one deduces that

$$\sup_{A^\star \in \mathcal{S}^\star} \frac{\log \|\Phi_{A^\star}(t,0)\|}{t} \le \varepsilon + \sup_{A \in \mathcal{S}} \frac{\log \|\Phi_A(t,0)\|}{t}.$$

The same inequality as above holds by adding $\sup_{t \ge s}$ in front of each sup term and one then proceeds as in the proof of the inequality $\hat{\lambda}(\mathcal{S}) \le \lambda(\mathcal{S})$ to obtain that $\lambda(\mathcal{S}^\star) \le \varepsilon + \lambda(\mathcal{S})$. One concludes by letting ε tend to zero.

We next prove Item 3 (Item 4 follows from similar arguments). Pick $\lambda > \lambda(\mathcal{S})$. Then, by definition, there exist $T, \alpha > 0$ such that, for every $t \ge T$, the function $F(t) = \sup_{A \in \mathcal{S}} \|\Phi_A(t,0)\| e^{-\lambda t}$ satisfies $F(t) \le e^{-\alpha t}$. On the other hand, since $\mathcal{M}$ is bounded, it turns out that $F|_{[0,T]}$ is bounded. Therefore, there exists $C > 0$ such that $F(t) \le C$ for every $t \in [0, +\infty)$. It follows that

$$\lambda(\mathcal{S}) \ge \inf\{\lambda \mid \exists C > 0 \text{ such that } \|\Phi_A(t,0)\| \le C e^{\lambda t}\ \forall A \in \mathcal{S}\ \forall t \ge 0\}.$$

Conversely, let $\lambda < \lambda(\mathcal{S})$. Then, for $\varepsilon > 0$ small enough there exist a sequence of times t_n going to infinity and a sequence of signals $A_n \in \mathcal{S}$ such that

$$\|\Phi_{A_n}(t_n, 0)\| \ge e^{\varepsilon t_n} e^{\lambda t_n}.$$

Since $e^{\varepsilon t_n}$ tends to infinity as n goes to infinity one easily concludes the proof of Item 3.

Let us now focus on Item 5. If $\lambda(\mathcal{S}) < 0$, then it follows at once from Item 3 that $(\Sigma_{\mathcal{S}})$ is uniformly exponentially stable. The converse implication is a consequence of the definition of uniform exponential stability and that of $\lambda(\mathcal{S})$.

Proceeding similarly for Item 6, the latter results from Item 4. An example showing that the converse in Item 6 is not true is given by $\mathcal{S}$ made of all constant signals with values in the interval $\mathcal{M} = (-1, 0)$.

To prove Item 7 we notice that $\hat{\lambda}(\mathcal{S}) > 0$ implies, by definition, the existence of $A \in \mathcal{S}$ and of a sequence of times t_h going to infinity such that $\lim_{h\to\infty} \|\Phi_A(t_h, 0)\| = +\infty$, that is, $\lim_{h\to\infty} \|\Phi_A(t_h, 0)x_h\| = +\infty$ for a sequence of unit vectors x_h in $\mathbb{R}^n$. Considering the canonical basis $e_1, \dots, e_n$ of $\mathbb{R}^n$ and since $\|\Phi_A(t_h, 0)x_h\| \le \sum_{i=1}^n \|\Phi_A(t_h, 0)e_i\| \le n \max_{i=1,\dots,n} \|\Phi_A(t_h, 0)e_i\|$ we obtain, up to extracting a subsequence,

$$\lim_{h\to\infty} \|\Phi_A(t_h, 0)e_i\| = +\infty$$

for some i, that is, $\limsup_{t\to\infty} \|x(t)\| = +\infty$ where $x(t) = \Phi_A(t, 0)e_i$. Then $(\Sigma_{\mathcal{S}})$ is unbounded. The converse inequality in Item 7 clearly does not hold by considering a class $\mathcal{S}$ reduced to a single constant signal equal to a nontrivial nilpotent matrix.

Let us now prove Item 8. Notice first that $\lambda(\mathcal{S}) = \limsup_{t\to+\infty} \lambda_t(\mathcal{S}) \geq \inf_{t>0} \lambda_t(\mathcal{S})$. Moreover, since the induced operator norm $\|\cdot\|$ on $M_n(\mathbb{R})$ is submultiplicative, one deduces from the definition of $\lambda_t(\mathcal{S})$ that

$$\lambda_{t+s}(\mathcal{S}) \leq \frac{t}{t+s}\lambda_t(\mathcal{S}) + \frac{s}{t+s}\lambda_s(\mathcal{S}), \qquad \text{for all } t, s > 0. \tag{1.10}$$

An immediate reasoning by induction yields that $\lambda_{mt}(\mathcal{S}) \leq \lambda_t(\mathcal{S})$ for every $t > 0$ and positive integer m. On the other hand, there exists an increasing sequence of times $(t_k)_{k\in\mathbb{N}}$ tending to infinity so that the sequence $(\lambda_{t_k}(\mathcal{S}))_{k\in\mathbb{N}}$ tends to $\lambda(\mathcal{S})$ as k tends to infinity. Fix $t > 0$. For every $k \in \mathbb{N}$, one has $t_k = m_k t + s_k$ for some integer m_k and real number $s_k \in (0, t)$. One deduces from (1.10) that for every integer k

$$\lambda_{t_k}(\mathcal{S}) \leq \frac{m_k t}{m_k t + s_k}\lambda_{m_k t}(\mathcal{S}) + \frac{s_k}{m_k t + s_k}\lambda_{s_k}(\mathcal{S}) \leq \frac{m_k t}{m_k t + s_k}\lambda_t(\mathcal{S}) + \frac{s_k}{m_k t + s_k}\lambda_{s_k}(\mathcal{S}).$$

Letting k tend to infinity, one gets that m_k tends to infinity, and then $\lambda(\mathcal{S}) \leq \lambda_t(\mathcal{S})$ for every $t > 0$. It follows that $\lambda(\mathcal{S}) \leq \inf_{t>0} \lambda_t(\mathcal{S})$, which concludes the proof of Item 8.

It finally remains to prove Item 9. The argument starts by noticing that, for every $m \in \mathbb{N}$, $t_1, \ldots, t_m > 0$, and $A_1, \ldots, A_m \in M_n(\mathbb{R})$, one has that

$$e^{t_m A_m^\top} \cdots e^{t_1 A_1^\top} = \left(e^{t_1 A_1} \cdots e^{t_m A_m}\right)^\top. \tag{1.11}$$

Consider $A \in \mathcal{S}$, $t > 0$, and a sequence $(A_k)_{k\in\mathbb{N}}$ in $\mathcal{S}_{\mathrm{pc}}(\mathcal{M})$ weak-$\star$ converging to $A(\cdot)$ on $[0, t]$ (whose existence follows from Proposition 1.7) and define $A_{k,\#}(s) = A_k(t - s)$ for $s \in [0, t]$ and $k \in \mathbb{N}$. Then, applying Proposition 1.3 and (1.11),

$$\|\Phi_{A^\top}(t, 0)\|_F = \lim_{k\to\infty} \|\Phi_{A_k^\top}(t, 0)\|_F = \lim_{k\to\infty} \|\Phi_{A_{k,\#}}(t, 0)\|_F = \|\Phi_{A_\#}(t, 0)\|_F,$$

where $\|\cdot\|_F$ stands for the Frobenius norm on $M_n(\mathbb{R})$, for which one has $\|B\|_F = \|B^\top\|_F = \mathrm{Tr}(BB^\top)^{1/2}$. This yields Item 9 by arbitrariness of $A \in \mathcal{S}$ and $t > 0$. □

Remark 1.21 By the equality in Item 2 of Proposition 1.20 and Corollary 1.8, one has $\lambda(\mathcal{M}) = \lambda(\overline{\mathrm{co}}(\mathcal{M}))$. In particular if $(\Sigma_{\mathcal{M}})$ is uniformly exponentially stable, then every matrix in $\overline{\mathrm{co}}(\mathcal{M})$ is Hurwitz. Recall that a square matrix N is Hurwitz if and only if the maximum of the real part of its eigenvalues is negative or, equivalently, if there exist two positive real numbers C and α such that $\|e^{tN}\| \leq Ce^{-\alpha t}$ for every $t \geq 0$.

Remark 1.22 If $\hat{\lambda}(\mathcal{S}) = 0$ then $(\Sigma_{\mathcal{S}})$ may be bounded or unbounded (just take $\mathcal{M} = \{M\}$ with $M = 0$ or M a nontrivial nilpotent matrix). In the latter case, the growth of the unbounded trajectories of $(\Sigma_{\mathcal{S}})$ is at most polynomial and $(\Sigma_{\mathcal{S}})$. We will discuss this situation in more detail in Sect. 4.3.2.

Remark 1.23 We claim that the upper-semicontinuity of the map $\mathcal{M} \mapsto \lambda(\mathcal{S}_{\text{label}}(\mathcal{M}))$ can be obtained for the classes of signals $\mathcal{S}_{\text{label}}(\mathcal{M})$ introduced in Sect. 1.1.1, where $\mathcal{M}$ belongs to a suitable subset of the family of compact subsets of $M_n(\mathbb{R})$ endowed with the topology of the Hausdorff distance. More precisely, if for every $t > 0$ the function $\mathcal{M} \mapsto \lambda_t(\mathcal{S}_{\text{label}}(\mathcal{M}))$ is continuous, then the claim follows from Item 8 of Proposition 1.20 and the general fact that the infimum of continuous functions is upper-semicontinuous. The continuity of $\mathcal{M} \mapsto \lambda_t(\mathcal{S}_{\text{label}}(\mathcal{M}))$ on the set of all compact subsets of $M_n(\mathbb{R})$ follows easily from Proposition 1.3 for the classes $\mathcal{S}_{\text{arb}}(\mathcal{M})$, $\mathcal{S}_{\text{pc}}(\mathcal{M})$, $\mathcal{S}_{\text{dw},\tau}(\mathcal{M})$, $\mathcal{S}_{\text{av-dw},\tau,N_0}(\mathcal{M})$, $\mathcal{S}_{\text{per}}(\mathcal{M})$, and $\mathcal{S}_{\text{pe},T,\mu}(\mathcal{M})$. For the other classes one needs to consider additional restrictions on $\mathcal{M}$: e.g. for $\mathcal{S}_{\text{cont}}(\mathcal{M})$ one can consider only compact pathwise connected sets of modes. For a further discussion on the continuity of the map $\mathcal{M} \mapsto \lambda(\mathcal{S}_{\text{label}}(\mathcal{M}))$, see Remark 6.21.

Remark 1.24 The result of Item 9 of Proposition 1.20 applies, in particular, to any of the classes $\mathcal{S}_{\text{label}}(\mathcal{M})$ considered in Sect. 1.1.1, for which, given $t > 0$ and $A \in \mathcal{S}_{\text{label}}(\mathcal{M})$, there exists a signal $A_\#$ belonging to the same class $\mathcal{S}_{\text{label}}(\mathcal{M})$ so that $A_\#(t-s) = A(s)$ for $s \in [0, t]$.

In order to check that the equality $\lambda(\mathcal{S}^\top) = \lambda(\mathcal{S})$ does not hold for every class of signals, consider

$$M_1 = \begin{pmatrix} 1 & 0 & 0 \\ 1 & 0 & 0 \\ 0 & 0 & 0 \end{pmatrix}, \quad M_2 = \begin{pmatrix} 0 & 0 & 0 \\ 0 & 1 & 0 \\ 0 & 1 & 0 \end{pmatrix}, \quad M_3 = \begin{pmatrix} 0 & 0 & 1 \\ 0 & 0 & 0 \\ 0 & 0 & 1 \end{pmatrix},$$

and let $\mathcal{S}$ be the collection of all time-shifts of the 3-periodic signal $A(\cdot)$ with $A(t) = M_i$ for $t \in [i-1, i)$, $i = 1, 2, 3$. By definition of $\lambda(\mathcal{S})$ and by using Gelfand's formula, one easily deduces that

$$\lambda(\mathcal{S}) = \limsup_{k\to\infty} \frac{\log \|(e^{M_3}e^{M_2}e^{M_1})^k\|}{3k} = \frac{1}{3}\log(\rho(e^{M_3}e^{M_2}e^{M_1})).$$

Similarly,

$$\lambda(\mathcal{S}^\top) = \frac{1}{3}\log(\rho(e^{M_3^\top}e^{M_2^\top}e^{M_1^\top})),$$

which is strictly smaller than $\frac{1}{3}\log(\rho(e^{M_3}e^{M_2}e^{M_1}))$.

We have introduced in Definition 1.17 two quantitative measures for the asymptotic behavior of a switched linear system, the maximal Lyapunov exponent of a class $\mathcal{S}$ being the natural extension of the classical notion of maximal Lyapunov exponent of a linear dynamical system while the uniform exponential rate captures uniform stability properties (cf. Item 5 of Proposition 1.20). Moreover, Item 2 of Proposition 1.20 points towards the need of determining extra conditions on $\mathcal{S}$ in order to recover the equality between $\lambda(\mathcal{S})$ and $\hat{\lambda}(\mathcal{S})$. It turns out that the crucial assumption for doing so is the weak-$\star$ compactness of $\mathcal{S}$, as we deduce below from the following classical result on dynamical systems.

Lemma 1.25 (Fenichel's Lemma) *Let $\mathcal{S}$ be weak-$\star$ compact. If $(\Sigma_{\mathcal{S}})$ is weakly attractive, then $(\Sigma_{\mathcal{S}})$ is uniformly exponentially stable.*

Proof We claim that there exists $T > 0$ such that, for every x_0 with $\|x_0\| = 1$ and every $A \in \mathcal{S}$, the trajectory of $\dot{x}(t) = A(t)x(t)$ with initial condition x_0 satisfies $\|x(t)\| < 1/2$ for some $t \le T$. Indeed, arguing by contradiction, assume that there exist a sequence $x_0^{(k)}$ of unit vectors, a sequence $A^{(k)}$ of signals in $\mathcal{S}$, and a sequence $T^{(k)}$ of times tending to infinity such that $\|\Phi_{A^{(k)}}(t,0)x_0^{(k)}\| \ge 1/2$ for every $t \in [0, T^{(k)}]$. By Corollary 1.4, up to extracting a subsequence, we can assume that $x_0^{(k)}$ converges to some unit vector x_0, $A^{(k)}$ weak-$\star$ converges to some $A \in \mathcal{S}$, and $\|\Phi_A(t,0)x_0\| \ge 1/2$ for every $t \ge 0$. This contradicts the convergence to zero of the trajectory of the switched system starting from x_0 and associated with A.

Consider now a trajectory $x(\cdot)$ of $(\Sigma_{\mathcal{S}})$. By the claim above, there exists an increasing sequence $(t_k)_{k\in\mathbb{N}}$ such that $t_1 \le T$, $\|x(t_k)\| \le 2^{-k}\|x(0)\|$ and $t_{k+1} - t_k \le T$ for every $k \in \mathbb{N}$. Moreover, according to (1.1), each $t_{k+1} - t_k$ is bounded from below by a positive quantity independent of k. In particular, the sequence $(t_k)_{k\in\mathbb{N}}$ is unbounded. Then, setting $C = \max_{A\in\mathcal{S},\, s\in[0,T]} \|\Phi_A(s,0)\|$, we have, for $t \in [t_k, t_{k+1}]$,

$$\|x(t)\| \le C\|x(t_k)\| \le C2^{-\frac{t_k}{T}}\|x(0)\| \le 2C2^{-\frac{t}{T}}\|x(0)\|.$$

Hence $\|\Phi_A(t,0)\| \le 2C2^{-\frac{t}{T}}$ for every $A \in \mathcal{S}$ and $t \ge 0$. This concludes the proof of the lemma. □

We can now deduce from Fenichel's lemma the equality between $\lambda(\mathcal{S})$ and $\hat{\lambda}(\mathcal{S})$.

Proposition 1.26 *If $\mathcal{S}$ long-time approximates $\mathcal{S}^\star$ (and, in particular, if $\mathcal{S}$ is weak-$\star$ compact), then $\lambda(\mathcal{S}) = \hat{\lambda}(\mathcal{S})$.*

Proof Recall that $\hat{\lambda}(\mathcal{S}) \le \lambda(\mathcal{S})$. We next prove the reverse inequality. For every $\mu > \hat{\lambda}(\mathcal{S})$, denote by $\mathcal{S}_\mu$ the set of signals of the form $A(\cdot) - \mu \mathrm{Id}_n$ with $A \in \mathcal{S}$. Then $\lambda(\mathcal{S}_\mu) = \lambda(\mathcal{S}) - \mu$ and $\hat{\lambda}(\mathcal{S}_\mu) = \hat{\lambda}(\mathcal{S}) - \mu < 0$ by (1.8). One deduces from Item 4 of Proposition 1.20 that every trajectory of $(\Sigma_{\mathcal{S}_\mu})$ converges to the origin. Since S_μ long-time approximate $S_\mu^\star$, it follows that $(\Sigma_{S_\mu^\star})$ is attractive. Hence $\lambda(\mathcal{S}_\mu^\star) < 0$ according to Fenichel's lemma, and then $\lambda(\mathcal{S}) = \lambda(\mathcal{S}^\star) < \mu$. The conclusion follows by letting μ tend to $\hat{\lambda}(\mathcal{S})$ from above. □

Remark 1.27 Proposition 1.26 implies the equality between $\lambda(\mathcal{S})$ and $\hat{\lambda}(\mathcal{S})$ when $\mathcal{S}$ is equal to $\mathcal{S}_{\mathrm{arb}}(\mathcal{M})$, $\mathcal{S}_{\mathrm{pc}}(\mathcal{M})$ or $\mathcal{S}_{\mathrm{cont}}(\mathcal{M})$ (cf. Proposition 1.11). In particular, $\lambda(\mathcal{M}) = \hat{\lambda}(\mathcal{M})$. For what concerns the case $\mathcal{S} = \mathcal{S}_{\mathrm{pe},T,\mu}(\mathcal{M})$, the weak-$\star$

compactness of $\mathcal{S}$ implies, again by Proposition 1.26, that $\lambda(\mathcal{S}) = \hat{\lambda}(\mathcal{S})$. In the cases $\mathcal{S} = \mathcal{S}_{\mathrm{dw},\tau}(\mathcal{M})$ or $\mathcal{S} = \mathcal{S}_{\mathrm{bv},T,\nu}$ we cannot directly apply the previous results. The equality $\lambda(\mathcal{S}) = \hat{\lambda}(\mathcal{S})$ however holds true, as it will be deduced by a different strategy in Chap. 6 (see Remark 6.19).

1.3.3 Related Exponential Rate Measures

Let us comment on two other measures of the asymptotic behavior of a switched system. The first one is the *Bohl exponent*, which is defined by

$$\lambda_B(\mathcal{S}) = \sup_{A\in\mathcal{S}} \limsup_{s,t\to\infty} \frac{\log \|\Phi_A(t+s,t)\|}{s}.$$

Directly from the definitions, one gets that

$$\hat{\lambda}(\mathcal{S}) \le \lambda_B(\mathcal{S}) \le \lambda(\mathcal{S}),$$

and, if $\mathcal{S}$ is not weak-$\star$ closed, the previous inequalities may be strict. Indeed, for $k \in \mathbb{N}$, define $a_k(\cdot)$ on $[0, k+k^2]$ by $a_k(t) = 0$ if $t \in [0,k)$ and $a_k(t) = -1$ if $t \in [k, k+k^2]$. Let now $A(\cdot)$ be the function defined on $\mathbb{R}$ equal to 1 on $(-\infty, 0)$ and equal to the concatenation of the a_k's on $[0,+\infty)$. Finally consider the class $\mathcal{S}$ of signals of the form $A(\cdot + h)$, $h \in \mathbb{R}$. Then one has $\hat{\lambda}(\mathcal{S}) = -1$, $\lambda_B(\mathcal{S}) = 0$, and $\lambda(\mathcal{S}) = 1$.

The second measure is the *joint spectral exponent* $\mu(\mathcal{S})$, which is defined by

$$\mu(\mathcal{S}) = \limsup_{t\to\infty} \sup_{A\in\mathcal{S}} \frac{\log(\rho(\Phi_A(t,0)))}{t},$$

where $\rho(M)$ denotes the spectral radius of M. It is immediate to see that Eqs. (1.7)–(1.9) hold true if one replaces $\lambda(\mathcal{S})$ by $\mu(\mathcal{S})$.

Moreover, since $\rho(M) \le \|M\|$ for every $M \in M_n(\mathbb{R})$, one immediately gets that

$$\mu(\mathcal{S}) \le \lambda(\mathcal{S}),$$

with strict inequality in general. The equality between $\mu(\mathcal{S})$ and $\lambda(\mathcal{S})$ will be obtained in the case $\mathcal{S} = \mathcal{S}_{\mathrm{arb}}(\mathcal{M})$ in Sect. 4.3.1 and for more general classes of signals in Sect. 6.1.3.

1.4 Block Reduction

In this section, we introduce the concepts of *invariant subspace* and *maximal invariant flag*, the latter being a generalization of Jordan blocks in the context of switched linear systems.

Definition 1.28 We say that a subspace V of $\mathbb{R}^n$ is *invariant* with respect to a set $\mathcal{M} \subset M_n(\mathbb{R})$ if and only if it is invariant with respect to every matrix $M \in \mathcal{M}$, i.e., $Mx \in V$ for every $x \in V$. A set $\mathcal{M} \subset M_n(\mathbb{R})$ is said to be *irreducible* if its only invariant subspaces are $\{0\}$ and $\mathbb{R}^n$. Otherwise it is said to be *reducible*.

We have the following useful results of the previous definitions.

Lemma 1.29 *A subspace V of $\mathbb{R}^n$ is invariant with respect to $\mathcal{M}$ if and only if it is invariant with respect to every matrix $\Phi_A(t, 0)$ for $t \geq 0$ and $A \in \mathcal{S}_{\mathrm{arb}}(\mathcal{M})$. Moreover, if $\mathcal{M}$ is irreducible, then for every $x \in \mathbb{R}^n \setminus \{0\}$ and every $\tau > 0$,*

$$\mathrm{span}\{\Phi_A(t, 0)x \mid t \in [0, \tau],\ A \in \mathcal{S}_{\mathrm{arb}}(\mathcal{M})\} = \mathbb{R}^n.$$

Proof Notice that V is invariant with respect to $\mathcal{M}$ if and only if each linear vector field $x \mapsto Mx$ with $M \in \mathcal{M}$ is tangent to V. The latter is equivalent to the property that every trajectory of $(\Sigma_{\mathcal{M}})$ with initial condition in V stays in V, which concludes the proof of the first part of the lemma.

Assume now that $\mathcal{M}$ is irreducible and set $V_\tau = \mathrm{span}\{\Phi_A(t, 0)x \mid t \in [0, \tau],\ A \in \mathcal{S}_{\mathrm{arb}}(\mathcal{M})\}$ for $\tau > 0$. Then $\tau \mapsto \dim V_\tau$ is nondecreasing and takes values in $\{1, \ldots, n\}$. Since it has a finite number of discontinuities, there exists $\bar{\tau} > 0$ such that V_τ is constantly equal to a linear subspace $V \subset \mathbb{R}^n$ on $(0, \bar{\tau})$. Then V is invariant with respect to $\{\Phi_A(t, 0) \mid t \in [0, \bar{\tau}),\ A \in \mathcal{S}_{\mathrm{arb}}(\mathcal{M})\}$, and hence with respect to every matrix in $\mathcal{M}$. Hence, $V = \mathbb{R}^n$ since $\mathcal{M}$ is irreducible, and then $V_\tau = \mathbb{R}^n$ for every $\tau > 0$. □

The following definition and subsequent result are useful to relate the properties of a reducible switched system to those of lower dimensional irreducible switched systems.

Definition 1.30 We say that

$$\{0\} = E_0 \subsetneq E_1 \subsetneq \cdots \subsetneq E_{r-1} \subsetneq E_r = \mathbb{R}^n \tag{1.12}$$

is an *invariant flag of length r for $\mathcal{M} \subset M_n(\mathbb{R})$* if every E_i is a subspace of $\mathbb{R}^n$ which is invariant with respect to $\mathcal{M}$. An invariant flag is said to be *maximal* if, for every $i = 1, \ldots, r$, there exists no subspace V such that $E_{i-1} \subsetneq V \subsetneq E_i$ and V is invariant with respect to $\mathcal{M}$. An invariant flag is said to be *trivial* if $r = 1$.

Proposition 1.31 *Consider a maximal invariant flag $E_1, \dots, E_r$ for $\mathcal{M}$. Set $n_i = \dim(E_i)$ for $i = 1, \dots, r$. Then the integer r does not depend on the maximal flag and there exists $P \in \mathrm{GL}(n)$ such that for every matrix $M \in \mathcal{M}$ one has*

$$PMP^{-1} = \begin{pmatrix} M_{11} & M_{12} & \cdots & & \\ 0 & M_{22} & M_{23} & \cdots & \\ 0 & 0 & M_{33} & M_{34} & \cdots \\ \vdots & \ddots & \ddots & \ddots & \\ 0 & \cdots & \cdots & 0 & M_{rr} \end{pmatrix}, \tag{1.13}$$

where each M_{ij} is a $(n_i - n_{i-1}) \times (n_j - n_{j-1})$ matrix. Define, for $i = 1, \dots, r$, $\mathcal{M}_i := \{M_{ii} \mid M \in \mathcal{M}\}$. Then $\mathcal{M}_1, \dots, \mathcal{M}_r$ are irreducible and unique up to reordering, i.e., if $\mathcal{M}_1, \dots, \mathcal{M}_r$ and $\mathcal{N}_1, \dots, \mathcal{N}_r$ are obtained as above, then there exist a permutation σ of $\{1, \dots, r\}$ and $P_1, \dots, P_r$ invertible matrices so that $\mathcal{N}_i = P_i \mathcal{M}_{\sigma(i)} P_i^{-1}$ for $i = 1, \dots, r$. The sets $\mathcal{M}_1, \dots, \mathcal{M}_r$ are called the irreducible blocks *of $\mathcal{M}$.*

Proof The reduction into the block form (1.13) simply follows from the definition of invariant flag. As for the uniqueness of the sets $\mathcal{M}_i$ (up to reordering), it is a consequence of the Jordan–Hölder theorem (cf. Theorem A.28) applied to $\mathbb{R}^n$ seen as a module over the ring of matrices generated by $\mathcal{M}$. □

It is useful to introduce the following notation. Let $P \in \mathrm{GL}(n)$ be chosen so that (1.13) holds for every $M \in \mathcal{M}$. Then, for $1 \leq i \leq j \leq r$, we denote by

$$\pi_{ij} : M_n(\mathbb{R}) \to M_{(n_i - n_{i-1}) \times (n_j - n_{j-1})}(\mathbb{R}) \tag{1.14}$$

the linear application mapping each M into the corresponding cropped matrix M_{ij}.

Given a class of signals $\mathcal{S}$, one deduces from Proposition 1.31 the existence of r families of signals

$$\mathcal{S}_i := \{A_{ii}(\cdot) \mid A(\cdot) \in \mathcal{S}\}, \qquad i = 1, \dots, r,$$

where $A_{ii}(t) = \pi_{ii}(A(t))$ for every $t \geq 0$ and $i = 1, \dots, r$, and π_{ii} is defined as in (1.14). Notice that $\mathcal{S}_1, \dots, \mathcal{S}_r$ are uniquely defined up to permutations and change of variables. We refer to $(\Sigma_{\mathcal{S}_1}), \dots, (\Sigma_{\mathcal{S}_r})$ as the *irreducible subsystems of* $(\Sigma_{\mathcal{S}})$.

One has the following result relating $\lambda(\mathcal{S})$ and the maximal Lyapunov exponents $\lambda(\mathcal{S}_i)$, $i = 1, \dots, r$.

Proposition 1.32 *Let $\mathcal{S}$ be a class of signals. Then one has $\lambda(\mathcal{S}) = \max_{i=1,\dots,r} \lambda(\mathcal{S}_i)$.*

Proof Thanks to (1.7) one can assume that every matrix $M \in \mathcal{M}$ has the block form given by (1.13). The inequality $\lambda(\mathcal{S}_i) \leq \lambda(\mathcal{S})$ for $i = 1, \dots, r$ follows from the fact that each solution of (Σ_{S_i}) is the projection on $\mathbb{R}^{n_i-n_{i-1}}$ of a solution of (Σ_S) with initial condition in $\mathrm{span}\{e_{n_{i-1}+1},\ldots,e_{n_i}\}$.

As for the equality $\max_{i=1,\ldots,r}\lambda(\mathcal{S}_i)=\lambda(\mathcal{S})$, we argue by induction on r. The case $r=1$ is trivial. Assume that the equality holds true for some positive integer r and consider a switched system with maximal invariant flag $(E_1,\ldots,E_{r+1})$ of length $r+1$. Write the state x of the system as $(x_1,\ldots,x_{r+1})$, according to the invariant flag decomposition, and set $y=(x_2,\ldots,x_{r+1})$. Let us consider for y the dynamics given by the linear switched system $\dot y(t)=\bar A(t)y(t)$ where $\bar A(\cdot)$ is any signal obtained from an element $A(\cdot)$ of $\mathcal{S}$ by deleting its first n_1 rows and its first n_1 columns.

Applying the induction hypothesis to the dynamics of y, one deduces from Item 3 of Proposition 1.20 that for every $\varepsilon>0$, there exists $C(\varepsilon)$ such that for all $t\geq 0$, $\|y(t)\|\leq C(\varepsilon)e^{t(\mu+\varepsilon)}\|y(0)\|$, where $\mu=\max_{i=2,\ldots,r+1}\lambda(\mathcal{S}_i)$.

Item 3 of Proposition 1.20 also yields that, for every $\varepsilon>0$ there exists $C_1(\varepsilon)$ such that for all $t\geq s\geq 0$, $\|\Phi_{A_{11}}(t,s)\|\leq C_1(\varepsilon)e^{(t-s)(\lambda(\mathcal{S}_1)+\varepsilon)}$.

Using the variation of constants formula (Lemma A.6) applied to the dynamics of x_1, one finally gets that there exists $K>0$ such that for every $\varepsilon>0$, for all $t\geq 0$,

$$\begin{aligned}\|x_1(t)\| &\leq C_1(\varepsilon)e^{t(\lambda(\mathcal{S}_1)+\varepsilon)}\|x_1(0)\|\\ &\quad + KC_1(\varepsilon)C(\varepsilon)\int_0^t e^{(t-s)(\lambda(\mathcal{S}_1)+\varepsilon)}e^{s(\mu+\varepsilon)}\|y(0)\|ds.\end{aligned}$$

It is immediate to deduce that for every $\varepsilon>0$, there exists $C'(\varepsilon)$ such that

$$\|x(t)\|\leq(1+t)C'(\varepsilon)e^{t(\max(\mu,\lambda(\mathcal{S}_1))+\varepsilon)}\|x(0)\|$$

for every $t\geq 0$. One deduces that $\lambda(\mathcal{S})\leq\max(\mu,\lambda(\mathcal{S}_1))+\varepsilon$. Since ε is arbitrary, this concludes the inductive step. □

As an application of the previous result, we have the following.

Corollary 1.33 *Let $\mathcal{M}$ be compact. Assume that there exists an invertible matrix $R\in M_n(\mathbb{C})$ such that for every $M\in\mathcal{M}$, the matrix RMR^{-1} is upper-triangular (not necessarily with real entries). Then $(\Sigma_{\mathcal{M}})$ is uniformly exponentially stable if and only if $\mathcal{M}$ is made of Hurwitz matrices.*

Proof With every matrix $Z=(z_{jk})_{j,k=1}^n\in M_n(\mathbb{C})$, let us associate a matrix $C(Z)=(c_{jk}(Z))_{j,k=1}^n$ in $M_{2n}(\mathbb{R})$, where each $c_{jk}(Z)$ represents the 2×2 block $\begin{pmatrix}\Re(z_{jk}) & -\Im(z_{jk})\\ \Im(z_{jk}) & \Re(z_{jk})\end{pmatrix}$.

Let $\tilde{\mathcal{M}}=\{C(RMR^{-1})\mid M\in\mathcal{M}\}$. By assumption, the elements in $\tilde{\mathcal{M}}$ are block upper-triangular with commuting blocks of size 2×2. For $j=1,\ldots,n$, let $\tilde{\mathcal{M}}_j$ be the subset of $M_2(\mathbb{R})$ equal to $\{c_{jj}(RMR^{-1})\mid M\in\mathcal{M}\}$. Since every $\tilde{\mathcal{M}}_j$ is made of commuting matrices and is compact, it is immediate to see that $\max_{j=1,\dots,n} \lambda(\tilde{\mathcal{M}}_j)$ is the maximum of the real parts of the eigenvalues of the matrices in $\mathcal{M}$.

Moreover, since (1.7) trivially extends to invertible elements of $M_n(\mathbb{C})$ one has $\lambda(\mathcal{M}) = \lambda(\tilde{\mathcal{M}})$. Applying Proposition 1.32 to $\mathcal{S}_{\text{arb}}(\tilde{\mathcal{M}})$, one gets that $\lambda(\mathcal{M}) = \max_{j=1,\dots,n} \lambda(\tilde{\mathcal{M}}_j)$ and hence the conclusion. □

1.5 Lifted and Projected Switched Systems

Given a class of switching signals $\mathcal{S}$, we can associate with $\mathcal{S}$, in addition to the system $(\Sigma_{\mathcal{S}})$, some other useful switched systems.

Definition 1.34 The *lifted switched system* $(\Sigma_{\mathcal{S}}^{\text{lift}})$ is the linear system on $M_n(\mathbb{R})$ given by

$$\dot{M}(t) = A(t)M(t), \qquad t \geq 0, \quad A(\cdot) \in \mathcal{S}, \quad M \in M_n(\mathbb{R}). \qquad (\Sigma_{\mathcal{S}}^{\text{lift}})$$

Notice that the fundamental matrix $\Phi_A(\cdot, 0)$ associated with the signal $A \in \mathcal{S}$ is the trajectory of $(\Sigma_{\mathcal{S}}^{\text{lift}})$ starting from the identity matrix. Due to the underlying linear structure, $(\Sigma_{\mathcal{S}}^{\text{lift}})$ can be seen as a switched system on $\mathbb{R}^{n^2}$ associated with a class of switching signals $\mathcal{S}^{\text{lift}}$ with values in $M_{n^2}(\mathbb{R})$.

Let $\mathbb{RP}^{n-1}$ be the $(n-1)$-dimensional projective space, which can be identified with the set of all one-dimensional subspaces of $\mathbb{R}^n$. Let $\Pi \colon \mathbb{R}^n \setminus \{0\} \to \mathbb{RP}^{n-1}$ denote the canonical projection onto $\mathbb{RP}^{n-1}$, which associates with $x \neq 0$ the one-dimensional space span$\{x\}$. A nonsingular $n \times n$ matrix M acts on $\mathbb{RP}^{n-1}$ by sending any one-dimensional subspace of $\mathbb{R}^n$ to its image by M, so that $M\Pi(x) := \Pi(Mx)$.

Given a matrix $M \in M_n(\mathbb{R})$, we denote by $\Pi_* M$ the vector field on $\mathbb{RP}^{n-1}$ obtained by projection of the vector field $x \mapsto Mx$ onto $T\mathbb{RP}^{n-1}$, i.e., for every $q = \Pi(x) \in \mathbb{RP}^{n-1}$ with $x \in \mathbb{R}^n \setminus \{0\}$,

$$\Pi_* M(q) = d\Pi_x(Mx),$$

where $d\Pi_x \colon T_x\mathbb{R}^n \to T_q\mathbb{RP}^{n-1}$ denotes the differential of Π at x. Note that $\Pi_* M$ is well-defined since $d\Pi_x(Mx)$ only depends on $q = \Pi(x)$.

Let us introduce the *projected switched system*

$$\dot{q}(t) = \Pi_* A(t)(q(t)), \qquad t \geq 0, \quad A \in \mathcal{S}, \quad q \in \mathbb{RP}^{n-1}. \qquad (\Pi\Sigma_{\mathcal{S}})$$

Since $\mathbb{RP}^{n-1}$ is not a linear space, such a system does not enter in the framework of switched linear systems discussed in the previous sections. However, we still use the expression *switched system* to stress that it represents a family of non-autonomous dynamical systems parameterized by the signal $A \in \mathcal{S}$.

Given a nontrivial trajectory $t \mapsto x(t)$ of $(\Sigma_{\mathcal{S}})$ associated with some $A \in \mathcal{S}$, $t \mapsto \Pi(\Phi_A(t,0)x(0)) = q(t)$ is the trajectory of $(\Pi\Sigma_{\mathcal{S}})$ associated with A and starting from $\Pi(x(0))$. One can then decompose $(\Sigma_{\mathcal{S}})$ into the cascade of a switched system on the compact $(n-1)$-dimensional manifold $\mathbb{RP}^{n-1}$ and the one-dimensional dynamics for $r(t) = \|x(t)\|$. More precisely, every nontrivial trajectory $t \mapsto x(t)$ of $(\Sigma_{\mathcal{S}})$ can be described, up to sign, by the function $t \mapsto (q(t), r(t))$, where $q(\cdot)$ is a trajectory of $(\Pi\Sigma_{\mathcal{S}})$ and

$$\frac{\dot{r}(t)}{r(t)} = \langle q(t), A(t)q(t)\rangle, \quad t \geq 0.$$

Here, for $q \in \mathbb{RP}^{n-1}$ and $M \in M_n(\mathbb{R})$, we write

$$\langle q, Mq\rangle = \frac{x^\top Mx}{\|x\|^2} \qquad \text{for any } x \in \Pi^{-1}(q).$$

Notice that we can rewrite $\hat{\lambda}(\mathcal{S})$ as

$$\hat{\lambda}(\mathcal{S}) = \sup\left\{\limsup_{t\to\infty}\frac{1}{t}\int_0^t \langle q(s), A(s)q(s)\rangle ds \mid A \in \mathcal{S},\ q \text{ solution of } (\Pi\Sigma_{\mathcal{S}})\right\}.$$

One advantage of the previous framework lies in the fact that the study of the asymptotic behavior of $(\Sigma_{\mathcal{S}})$ is now reduced to the study of a switched system on a compact manifold.

1.6 Notes and References

Switched linear systems have a relatively recent but extremely active history (see for instance [Lib03, Jun09] and references therein), due to the fact that they are used in several research areas ranging from number theory to network security management, from combinatorics on words to signal processing, etc. In control theory, switched linear systems can also be seen as particular instances of hybrid systems (see [GST12]) while they can be also used to model certain types of time-varying parametric uncertainties [BCK95].

In the control literature, the set of modes is usually finite and every mode represents an admissible dynamics of the system. The classes of signals in that case are subsets of piecewise constant functions taking values in that set of modes. In this monograph, sets of modes are simply assumed to be bounded subsets of matrices. As for the classes of signals given in Sect. 1.1.1, these are the most common ones used in the literature (see for instance [Lib03]), with the possible exception of the class of BV (bounded variation) signals, which can be seen as an extension of the class of signals satisfying the average dwell-time constraint allowing for piecewise continuous signals. Note that the Zeno phenomenon (infinite countable accumulation of discontinuities for a switching signal over a finite interval of time) is not excluded here as soon as one considers the class $\mathcal{S}_{\rm arb}$. As for the notion of *dwell-time*, it appears in the literature with other names, such as *refractory period* in [CP14] and it has been introduced in the context of switched systems in [Mor93]. Regarding *persistence of excitation*, it is widely used in adaptive control, cf. [ABJ+86]. Let us stress that, although switched systems have been defined in the vector space $\mathbb{R}^n$, most of the results of this monograph readily extend to switched systems defined in $\mathbb{C}^n$.

The concept of *long-time approximation* corresponds to a property established in [ISW03, Theorem 1], where nonlinear extensions of Corollary 1.4 and Proposition 1.11 are derived. Our proof of Proposition 1.11 provides a simplified argument exploiting the linear framework.

The uniform exponential rate has been introduced in the discrete-time setting in [RS60] under the name *joint spectral radius* in 1960 and the subject regained interest in the nineties with no interruption since then. For a discussion on different spectral exponents for linear flows in a general abstract setting, see [CK14]. As for Fenichel's lemma, it was first given in [Fen74] in the context of dynamical systems. For linear (or, more generally, homogeneous) switched systems and under arbitrary switching the result has been rediscovered later in [Ang99, ADLS09]. The Bohl exponent was first considered in [Boh13] and the joint spectral exponent has been introduced in [DL92] in the discrete-time setting under the name of *generalized spectral radius*. Section 1.4 gathers classical results (see, for instance, [Wir02] and [SWM+07]) while Corollary 1.33 is taken from [LHM99], where it is observed that the upper-triangularizability appearing in the statement of Corollary 1.33 corresponds to the solvability of the Lie algebra generated by $\mathcal{M}$. More general stability results based on Lie-algebraic methods can be found in [AL01, ABL12] (see also [Lib23] for a historical survey about these methods). The last section on lifted and projected switched systems is directly inspired by [CK14, Chapter 4] and it will be used later on in the text, cf. Sect. 6.2.

Chapter 2
Stability Analysis for Two-Dimensional Switched Systems

Starting from this chapter we address the problem of assessing the stability of a switched linear system. For most switched linear systems, the analysis of the stability properties represents an extremely complex problem. This is true also in the case of arbitrary switchings, that is, for switched systems of the form $(\Sigma_{\mathscr{M}})$, for most sets of matrices $\mathscr{M}$. The case in which $\mathscr{M}$ is formed by only two matrices in $M_2(\mathbb{R})$ is, somehow, an exception to this rule. Indeed, although the interaction between two stable planar dynamics may still produce an unstable behaviour, the stability analysis in this case may be reduced, for the most part, to the analysis of explicitly computable special trajectories. This leads to a complete classification of all possible asymptotic behaviors in terms of finitely many quantities explicitly depending on the coefficients of the modes through elementary functions. Let us stress that the methods used in this chapter are peculiar to systems evolving on the plane. In higher dimensions and even with only two modes, it is unknown if a complete classification of the stability properties by means of elementary functions of the matrix coefficients is feasible.

2.1 Stability Conditions for Two-Dimensional Switched Systems

We consider switched linear systems $(\Sigma_{\mathscr{M}})$ with $n = 2$ and $\mathscr{M} = \{M_1, M_2\}$. According to Corollary 1.8 and Proposition 1.14, $(\Sigma_{\mathscr{M}})$ shares the same stability properties as $(\Sigma_{\mathrm{co}(\mathscr{M})})$, where $\mathrm{co}(\mathscr{M})$ is the segment $[\![M_1, M_2]\!]$ of matrices between M_1 and M_2. Clearly, if one mode is unstable then the switched system is unstable as well, while if one of the two matrices possesses a zero eigenvalue or imaginary eigenvalues then uniform exponential stability is impossible, although the system may still be uniformly bounded (see Remark 2.4 below).

Y. Chitour et al., *Dynamics and Stability of Continuous-Time Switched Linear Systems*, PNLDE Subseries in Control 105,
https://doi.org/10.1007/978-3-031-82982-6_2

Hence we consider here the case in which each 2×2 matrix M_i is Hurwitz, whose study provides a panoramic view of the various distinct stability properties of $(\Sigma_{\mathcal{M}})$. The objective of this chapter is to characterize such stability properties in terms of coordinate-invariant parameters depending on M_1 and M_2 and the results are gathered in Theorem 2.1 given next. For that purpose, we associate with $\mathcal{M}$ several useful quantities, invariant with respect to a linear change of coordinates. The discriminant of a matrix $X \in M_2(\mathbb{R})$ is defined as

$$\delta_X = \operatorname{tr}(X)^2 - 4\det(X)\,.$$

We then define the following quantities associated with $\mathcal{M}$:

$$\tau_i = \begin{cases} \frac{\operatorname{tr}(M_i)}{\sqrt{|\delta_{M_i}|}} & \text{if } \delta_{M_i} \neq 0\,, \\ \frac{\operatorname{tr}(M_i)}{\sqrt{|\delta_{M_{3-i}}|}} & \text{if } \delta_{M_i} = 0 \text{ and } \delta_{M_{3-i}} \neq 0\,, \\ \frac{\operatorname{tr}(M_i)}{2} & \text{if } \delta_{M_1} = \delta_{M_2} = 0\,, \end{cases}$$

$$\kappa = \frac{2\tau_1\tau_2}{\operatorname{tr}(M_1)\operatorname{tr}(M_2)}\Big(\operatorname{tr}(M_1M_2) - \frac{1}{2}\operatorname{tr}(M_1)\operatorname{tr}(M_2)\Big)\,.$$

We will also make use of the following object relating two matrices $X, Y \in M_2(\mathbb{R})$

$$\Gamma(X, Y) = \frac{1}{2}(\operatorname{tr}(X)\operatorname{tr}(Y) - \operatorname{tr}(XY)).$$

Note that, as a straightforward consequence of the definitions and the fact that M_1, M_2 are supposed to be Hurwitz, $\tau_i < 0$ for $i = 1, 2$ and $\tau_i < -1$ whenever $\delta_{M_i} > 0$. Moreover, $\operatorname{sign}\Gamma(M_1, M_2) = \operatorname{sign}(\tau_1\tau_2 - \kappa)$. We further define Δ as the discriminant of the quadratic form $x \mapsto \det(M_1x, M_2x)$, that is,

$$\begin{aligned}\Delta &= 4(\Gamma(M_1, M_2)^2 - \Gamma(M_1, M_1)\Gamma(M_2, M_2)) \\ &= 4(\Gamma(M_1, M_2)^2 - \det(M_1)\det(M_2)),\end{aligned}$$

and, if $\Delta > 0$,

$$\mathcal{R} = \frac{2\Gamma(M_1, M_2) + \sqrt{\Delta}}{2\sqrt{\det(M_1)\det(M_2)}}e^{\tau_1 t_1 + \tau_2 t_2},$$

$$\text{where } t_i = \begin{cases} \frac{\pi}{2} - \arctan\frac{\operatorname{tr}(M_1)\operatorname{tr}(M_2)(\kappa\tau_i + \tau_{3-i})}{2\tau_1\tau_2\sqrt{\Delta}} & \text{if } \delta_{M_i} < 0, \\ \operatorname{arctanh}\frac{2\tau_1\tau_2\sqrt{\Delta}}{\operatorname{tr}(M_1)\operatorname{tr}(M_2)(\kappa\tau_i - \tau_{3-i})} & \text{if } \delta_{M_i} > 0, \\ \frac{\sqrt{\Delta}}{(\operatorname{tr}(M_1M_2) - \operatorname{tr}(M_1)\operatorname{tr}(M_2)/2)\tau_i} & \text{if } \delta_{M_i} = 0, \end{cases} \quad \text{for } i = 1, 2.$$

One also needs the following stability concept: $(\Sigma_{\mathcal{M}})$ is said to be *quadratically stable* if there exists a symmetric and positive-definite 2×2 matrix P such that

$$M_i^\top P + PM_i \prec 0, \tag{2.1}$$

for $i = 1, 2$, that is, if the function $V(x) = x^\top Px$ is a quadratic Lyapunov function for both $\dot{x} = M_1 x$ and $\dot{x} = M_2 x$ (we refer to Chap. 3 for a comprehensive discussion on Lyapunov functions). Quadratic stability is a stronger stability property than uniform exponential stability. In the following we will refer to the function V as to a common (quadratic) Lyapunov function for $(\Sigma_{\mathcal{M}})$. If (2.1) is replaced by the corresponding nonstrict inequality, we say that V is a nonstrict common (quadratic) Lyapunov function for $(\Sigma_{\mathcal{M}})$.

The following result, which is the main result of the chapter, provides a complete characterization of the stability properties of two-dimensional switched linear systems.

Theorem 2.1 *Let* $\mathcal{M} = \{M_1, M_2\}$, *where* M_1 *and* M_2 *are* 2×2 *Hurwitz matrices. Then the following holds true.*

(S₁) *If* $\mathrm{tr}(M_1 M_2) > 2\sqrt{\det(M_1)\det(M_2)} + \mathrm{tr}(M_1)\mathrm{tr}(M_2)$, *then* $(\Sigma_{\mathcal{M}})$ *is unbounded;*

(S₂) *If* $\mathrm{tr}(M_1 M_2) = 2\sqrt{\det(M_1)\det(M_2)} + \mathrm{tr}(M_1)\mathrm{tr}(M_2)$, *then* $(\Sigma_{\mathcal{M}})$ *is uniformly bounded but not uniformly exponentially stable;*

(S₃) $(\Sigma_{\mathcal{M}})$ *is quadratically stable if and only if*

$$-2\sqrt{\det(M_1)\det(M_2)} < \mathrm{tr}(M_1 M_2) < 2\sqrt{\det(M_1)\det(M_2)} + \mathrm{tr}(M_1)\mathrm{tr}(M_2); \tag{2.2}$$

(S₄) *If* $\mathrm{tr}(M_1 M_2) \le -2\sqrt{\det(M_1)\det(M_2)}$, *then* $(\Sigma_{\mathcal{M}})$ *is uniformly exponentially stable (but not quadratically stable) if* $\mathcal{R} < 1$, *uniformly bounded (but not uniformly exponentially stable) if* $\mathcal{R} = 1$, *and unbounded if* $\mathcal{R} > 1$.

Remark 2.2 We can easily interpret the conditions of Theorem 2.1 in terms of the uniform exponential rate $\lambda(\mathcal{M})$ defined in Chap. 1. In the cases **(S₁)**, **(S₂)**, and **(S₃)** we have $\lambda(\mathcal{M}) > 0$, $\lambda(\mathcal{M}) = 0$, and $\lambda(\mathcal{M}) < 0$, respectively. In the case **(S₄)** we have $\lambda(\mathcal{M}) > 0$, $\lambda(\mathcal{M}) = 0$, and $\lambda(\mathcal{M}) < 0$ if $\mathcal{R} > 1$, $\mathcal{R} = 1$, and $\mathcal{R} < 1$, respectively.

Remark 2.3 The stability conditions of Theorem 2.1 and the function $\mathcal{R}$ are preserved by linear coordinate transformations and by rescalings of the type $(M_1, M_2) \to (M_1/\alpha_1, M_2/\alpha_2)$ with α_1, α_2 positive. This is consistent with the fact that the stability properties of $(\Sigma_{\mathcal{M}})$ are preserved by linear coordinate transformations and do not depend on the parameterization of the integral curves of $x \mapsto M_1 x$ and $x \mapsto M_2 x$.

Remark 2.4 Theorem 2.1 may be extended to the degenerate case in which at least one among M_1 and M_2 gives rise to a stable dynamics without being Hurwitz.

The theorem remains almost unchanged with the exception that, assuming (2.2), the system admits now a nonstrict quadratic Lyapunov function, but it is not quadratically stable. Also, the expression $\mathcal{R}$ is not meaningful if one among M_1 and M_2 has a zero eigenvalue. The correct expression and, more generally, the extension of Theorem 2.1 to the degenerate case, may be obtained either by mimicking the arguments developed in this chapter or by seeing the matrices M_1, M_2 as limits of Hurwitz matrices for which we may apply Theorem 2.1. The latter approach relies on the fact that the invariant quantities involved in Theorem 2.1 and the trajectories of the system evolve continuously with respect to perturbations of the system.

As an example of application of the conditions in Theorem 2.1, let $\mathcal{M}$ be the subset of $M_2(\mathbb{R})$ made of the two matrices

$$M_1 = \begin{pmatrix} -0.1 & -1 \\ 1 & -0.1 \end{pmatrix}, \qquad M_2 = \begin{pmatrix} -0.1 & -a \\ 1/a & -0.1 \end{pmatrix}. \tag{2.3}$$

Note that both M_1, M_2 have complex nonreal eigenvalues, i.e., $\mathrm{sign}(\delta_{M_1}) = \mathrm{sign}(\delta_{M_2}) = -1$. Let us discuss the stability properties of $(\Sigma_{\mathcal{M}})$ in terms of the parameter $a \in \mathbb{R} \setminus \{0\}$.

A simple computation shows that the condition in $(\mathbf{S}_1)$ of Theorem 2.1 is satisfied if and only if $a + \frac{1}{a} < -2.04$, which corresponds to the fact that $a < -\bar{a}_1$ or a belongs to an open interval $(-\bar{a}_2, 0)$, where $\bar{a}_1 \approx 1.221$ and $\bar{a}_2 = \frac{1}{\bar{a}_1} \approx 0.819$. In this case $(\Sigma_{\mathcal{M}})$ is unbounded. The case $a = -1.5$ is depicted in Fig. 2.1a.

The condition in $(\mathbf{S}_2)$ is satisfied if and only if $a + \frac{1}{a} = -2.04$, that is, when $a \in \{-\bar{a}_1, -\bar{a}_2\}$. In this case $(\Sigma_{\mathcal{M}})$ is uniformly bounded, but it is not uniformly exponentially stable. The case $a = -\bar{a}_1$ is depicted in Fig. 2.1b.

Concerning $(\mathbf{S}_3)$, condition (2.2) is satisfied if and only if $a + \frac{1}{a} \in (-2.04, 2.04)$. This happens if $a \in (-\bar{a}_1, -\bar{a}_2)$ or $a \in (\bar{a}_2, \bar{a}_1)$. In this case $(\Sigma_{\mathcal{M}})$ is quadratically stable. Figure 2.2a and b represents the cases in which $a = -1$ and $a = 1.1$

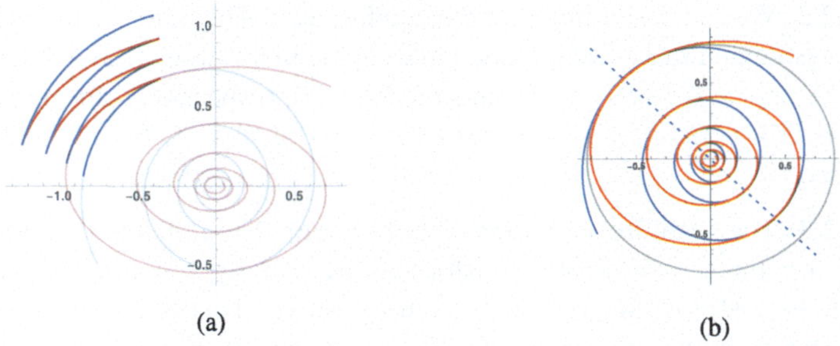

Fig. 2.1 Cases $(\mathbf{S}_1)$ and $(\mathbf{S}_2)$ with M_1, M_2 as in (2.3). Trajectories of M_1 are depicted in blue, trajectories of M_2 are depicted in red. (**a**) $a = -1.5$: $(\Sigma_{\mathcal{M}})$ is unbounded. (**b**) $a = -\bar{a}_1$: $(\Sigma_{\mathcal{M}})$ is uniformly bounded. The vectors M_1x, M_2x are parallel only along the dashed line. A level set of a nonstrict quadratic Lyapunov function is also depicted in gray

(a) (b)

Fig. 2.2 Case ($\mathbf{S}_3$) in Theorem 2.1, with M_1, M_2 as in (2.3). Trajectories of M_1 are depicted in blue, trajectories of M_2 are depicted in red. (**a**) $a = -1$: $(\Sigma_{\mathcal{M}})$ is quadratically stable. (**b**) $a = 1.1$: $(\Sigma_{\mathcal{M}})$ is quadratically stable

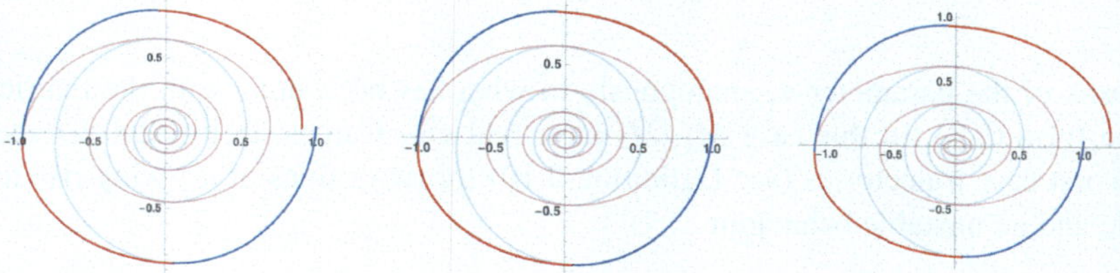

(a) (b) (c)

Fig. 2.3 Possible situations occurring in Case ($\mathbf{S}_4$) of Theorem 2.1, with M_1, M_2 as in (2.3). Trajectories of M_1 are depicted in blue, trajectories of M_2 are depicted in red. Bold curves correspond to worst case trajectories (see Definition 2.10). (**a**) $a = 1.3$: $(\Sigma_{\mathcal{M}})$ is uniformly exponentially stable. (**b**) $a = \bar{a}_4$: $(\Sigma_{\mathcal{M}})$ is uniformly bounded. (**c**) $a = 1.5$: $(\Sigma_{\mathcal{M}})$ is unbounded

(note that in the first case the trajectories corresponding to the two modes rotate in opposite sense, while in the second case they both rotate clockwise).

Finally, the condition in ($\mathbf{S}_4$) corresponds to the case in which $a \in (0, \bar{a}_2)$ or $a > \bar{a}_1$. In this case one needs to compute the value of $\mathcal{R}$ to establish the stability properties of the system. It turns out that $(\Sigma_{\mathcal{M}})$ is unbounded if a belongs to $(0, \bar{a}_3)$ or if it is larger than $\bar{a}_4$, where $\bar{a}_3 \approx 0.7315$ and $\bar{a}_4 = \frac{1}{\bar{a}_3} \approx 1.367$. $(\Sigma_{\mathcal{M}})$ is uniformly bounded, but not uniformly exponentially stable if $a \in \{\bar{a}_3, \bar{a}_4\}$ and it is uniformly exponentially stable (but not quadratically stable) if a belongs to $(\bar{a}_3, \bar{a}_2)$ or to $(\bar{a}_1, \bar{a}_4)$. Figure 2.3a–c corresponds to the cases $a = 1.3$, $a = \bar{a}_4$, and $a = 1.5$, respectively.

Note that similar examples may be found with one or both matrices possessing real eigenvalues. For instance, it is possible to describe the stability properties in the case

$$M_1 = \begin{pmatrix} -1.1 & 1 \\ 1 & -1.1 \end{pmatrix}, \qquad M_2 = \begin{pmatrix} -1.1 & a \\ 1/a & -1.1 \end{pmatrix} \tag{2.4}$$

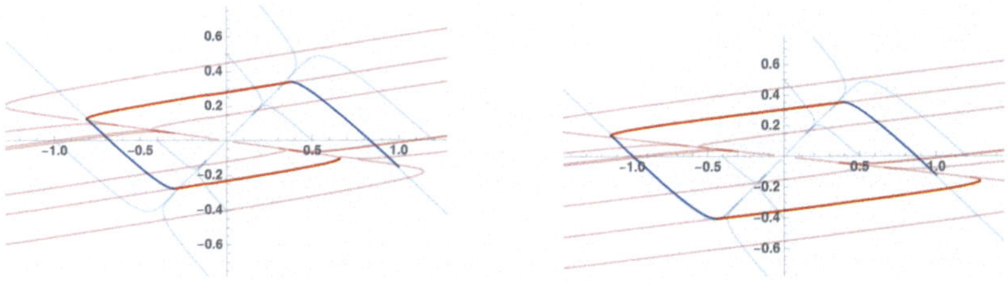

(a) (b)

Fig. 2.4 Case $(\mathbf{S}_4)$ with M_1, M_2 as in (2.4). Trajectories of M_1 are depicted in blue, trajectories of M_2 are depicted in red. Bold curves correspond to worst case trajectories. (**a**) $a = 9$: $(\Sigma_{\mathcal{M}})$ is uniformly exponentially stable. (**b**) $a = 7$: $(\Sigma_{\mathcal{M}})$ is unbounded

in terms of the parameter a, analogously to what has been done with the matrices in the form (2.3). In this case M_1, M_2 have real eigenvalues. In Fig. 2.4 we show the worst case trajectories (see Definition 2.10) for two values of a giving rise to a stable and an unstable behaviour.

2.2 Proof of Theorem 2.1

In order to prove Theorem 2.1, a preliminary step, motivated by Remark 2.3, consists in transforming the matrices M_1, M_2 into suitable normal forms, obtained by a linear change of coordinates and by rescaling by positive constants.

Lemma 2.5 *Up to a linear change of coordinates and a rescaling according to Remark 2.3, one of the following cases occurs:*

1. *if* $\det([M_1, M_2]) < 0$, *one has*

$$M_1 = \begin{pmatrix} \tau_1 & 1 \\ \operatorname{sign}(\delta_{M_1}) & \tau_1 \end{pmatrix} \tag{2.5}$$

   and there exists $F \in \mathbb{R}$, $|F| \geq 1$, *such that* $F + \frac{\operatorname{sign}(\delta_{M_1}\delta_{M_2})}{F} = 2\kappa$ *and*

$$M_2 = \begin{pmatrix} \tau_2 & \operatorname{sign}(\delta_{M_2})/F \\ F & \tau_2 \end{pmatrix}; \tag{2.6}$$

2. *if* $\det([M_1, M_2]) > 0$ *then* $\delta_{M_i} > 0$ *for* $i = 1, 2$, $\kappa \in (-1, 1)$, M_1 *is given by* (2.5) *and* M_2 *has the form*

$$M_2 = \begin{pmatrix} \tau_2 + \sqrt{1-\kappa^2} & \kappa \\ \kappa & \tau_2 - \sqrt{1-\kappa^2} \end{pmatrix}; \tag{2.7}$$

3. *if* $\det([M_1, M_2]) = 0$ *then either* $M_1 = -\mathrm{Id}_2$ *and* M_2 *is in Jordan form, or* M_1, M_2 *have the form (2.5)-(2.6) with* $|F| = 1$ *and* $\mathrm{sign}(\delta_{M_1}\delta_{M_2}) = 1$ *(in both cases* M_1 *and* M_2 *commute), or* M_1, M_2 *are in upper triangular form.*

Proof A direct computation shows that when M_1, M_2 are in the form (2.5)–(2.6) then

$$\det([M_1, M_2]) = -\left(F - \frac{\mathrm{sign}(\delta_{M_1}\delta_{M_2})}{F}\right)^2 \leq 0.$$

Assume now that $\delta_{M_1} < 0$. Then, up to a rescaling according to Remark 2.3 and a linear change of coordinates, one can put M_1 in the form (2.5). It is easy to see that M_2 can be put in the form (2.6) (while keeping M_1 unchanged) by applying a rotation and, possibly, a further rescaling.

Assume now that $\delta_{M_1} = 0$. Imposing $\det([M_1, M_2]) \neq 0$ we obtain that, up to a rescaling and a linear change of coordinates, M_1 coincides with the Jordan block (2.5) and we write $M_2 = \begin{pmatrix} a_1 & a_2 \\ a_3 & a_4 \end{pmatrix}$ for some a_1, a_2, a_3, a_4 in $\mathbb{R}$. We assume $a_3 \neq 0$ since otherwise $\det([M_1, M_2]) = 0$. Using the linear transformation $T = \begin{pmatrix} 1 & \frac{a_1-a_4}{2a_3} \\ 0 & 1 \end{pmatrix}$, one gets

$$T^{-1}M_1T = M_1 \quad \text{and} \quad T^{-1}M_2T = \begin{pmatrix} \frac{a_1+a_4}{2} & \frac{\delta_{M_2}}{4a_3} \\ a_3 & \frac{a_1+a_4}{2} \end{pmatrix}.$$

If $\delta_{M_2} = 0$ then it is easy to see that $T^{-1}M_2T$ is in the form (2.6). Otherwise, one can easily check that the normal form (2.6) is obtained by multiplying $T^{-1}M_2T$ by a factor $2/\sqrt{|\delta_{M_2}|}$.

Consider now the case $\delta_{M_1} > 0$. In this case we can diagonalize M_1 so that

$$M_1 = \begin{pmatrix} \lambda_1 & 0 \\ 0 & \lambda_2 \end{pmatrix}, \quad M_2 = \begin{pmatrix} a_1 & a_2 \\ a_3 & a_4 \end{pmatrix}, \qquad \lambda_i, a_j \in \mathbb{R} \text{ for } i = 1, 2,\ j = 1, 2, 3, 4.$$

If $\det([M_1, M_2]) = a_2a_3(\lambda_1 - \lambda_2)^2 < 0$, one gets, after using the linear transformation $U_1 = \begin{pmatrix} -\sqrt{-a_2/a_3} & \sqrt{-a_2/a_3} \\ 1 & 1 \end{pmatrix}$ that $\frac{2}{\sqrt{|\delta_{A_1}|}}U_1^{-1}M_1U_1$ and $\frac{2}{\sqrt{|\delta_{A_2}|}}U_1^{-1}M_2U_1$ have the forms (2.5) and (2.6), respectively. The inequality $|F| \geq 1$ is obtained up to exchanging the coordinates.

If $\det([M_1, M_2]) = a_2a_3(\lambda_1 - \lambda_2)^2 > 0$ then $\delta_{M_2} = (a_1 - a_4)^2 + 4a_2a_3 > 0$ and it is straightforward to see that $|\kappa| < 1$. Up to reversing the sign of one of the basis vectors and/or exchanging the coordinates, we may assume that $\lambda_2 > \lambda_1$ and $a_3 > 0$. Then it is easy to check that the normal forms (2.5) and (2.7) are obtained as $\frac{2}{\sqrt{|\delta_{A_1}|}}U_2^{-1}M_1U_2$ and $\frac{2}{\sqrt{|\delta_{A_2}|}}U_2^{-1}M_2U_2$, respectively, where $U_2 = \begin{pmatrix} \sqrt{a_2/a_3} & -\sqrt{a_2/a_3} \\ 1 & 1 \end{pmatrix}$.

Concerning the case $\det([M_1, M_2]) = 0$, the desired result follows easily from the arguments above. This concludes the proof of the lemma. □

We split the proof of Theorem 2.1 in two parts. We first provide a separate proof for Case ($\mathbf{S}_3$), while the rest of the proof will be given afterwards.

Proposition 2.6 *The switched linear system* $(\Sigma_{\mathcal{M}})$ *is quadratically stable if and only if* (2.2) *holds true.*

Proof Let us first assume that M_1, M_2 are in the form (2.5)–(2.6). Then the existence of a common quadratic Lyapunov function for the switched system is equivalent to the existence of a matrix $P = \left(\begin{smallmatrix} 1 & p \\ p & q \end{smallmatrix}\right)$ such that $P \succ 0$ and $Q_i = M_i^\top P + PM_i \prec 0$ for $i = 1, 2$. On the half-line $\ell = \{(p, q) \in \mathbb{R}^2 \mid p = 0,\ q > 0\}$ one has that $P \succ 0$ and that the diagonal coefficients of the matrices Q_i are negative. Consider now the case in which $\delta_{M_i} \neq 0$ for $i = 1, 2$. It is easy to check that the sets of parameters $(p, q) \in \mathbb{R}^2$ such that $\det(Q_i) > 0$ are enclosed by two ellipses on the half-plane defined by $q > 0$ (see Fig. 2.5). Moreover each ellipse has one of the two axes lying on the p-axis. Thus the two sets of parameters have nonempty intersection if and only if their intersection on the half-line ℓ is nonempty. By what precedes, this is equivalent to the existence of a common quadratic Lyapunov function. Letting $\alpha_1 < \beta_1$ and $\alpha_2 < \beta_2$ such that $(0, \alpha_1)$,$(0, \beta_1)$ and $(0, \alpha_2)$,$(0, \beta_2)$ are the zeros on ℓ of $\det(Q_1)$ and $\det(Q_2)$, respectively, then the two sets intersect if and only if the two conditions $\alpha_1 < \beta_2$ and $\alpha_2 < \beta_1$ hold. Since both $\alpha_1 > \beta_2$ and $\alpha_2 > \beta_1$ cannot hold simultaneously, one deduces that the two inequalities above are equivalent to the condition $(\beta_2 - \alpha_1)(\beta_1 - \alpha_2) > 0$. An explicit computation shows that

$$(\beta_2 - \alpha_1)(\beta_1 - \alpha_2)$$
$$= \frac{\left(2\sqrt{\det(M_1)\det(M_2)} + \frac{1}{2}\mathrm{tr}(M_1)\mathrm{tr}(M_2)\right)^2 - \left(\mathrm{tr}(M_1M_2) - \frac{1}{2}\mathrm{tr}(M_1)\mathrm{tr}(M_2)\right)^2}{F^2}.$$

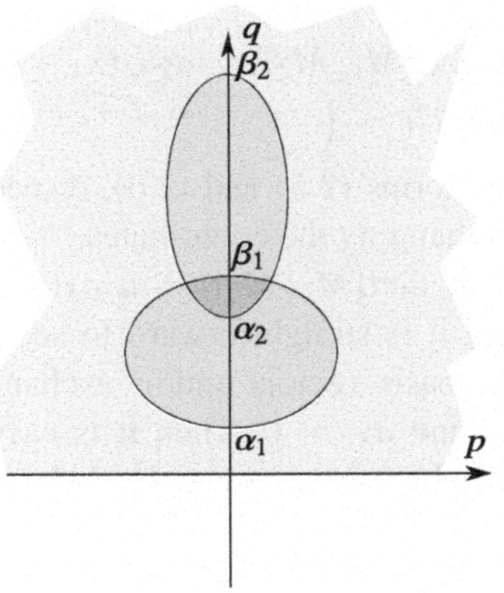

Fig. 2.5 Quadratic Lyapunov function in the space of parameters (p, q). The parameters in the darkest gray area, the intersection of the two ellipses, are such that $\det(Q_i) > 0,\ i = 1, 2$. The parameters obtained by further intersecting with the q-axis correspond to common quadratic Lyapunov functions

The right-hand side is positive if and only if (2.2) is satisfied, hence the lemma is proved for matrices in the form (2.5)–(2.6) with $\delta_{M_i} \neq 0$ for $i = 1, 2$. If only one among the δ_{M_i} is equal to zero then, up to exchanging M_1, M_2, we can assume that $\delta_{M_1} \neq 0$, and one proceeds exactly as in the previous case (with the difference that the parameters $(p, q) \in \mathbb{R}^2$ such that $\det(Q_2) > 0$ are those enclosed by the two lines defined by $q = \alpha_2 = 0$ and $q = \beta_2 = \frac{4\tau_2^2}{F^2}$). If $\delta_{M_1} = \delta_{M_2} = 0$, the set of parameters $(p, q) \in \mathbb{R}^2$ such that $\det(Q_1) > 0$ is the half-plane defined by $q > \alpha_1 = \frac{1}{4\tau_1^2}$. Similarly as before, the condition $\beta_2 > \alpha_1$ is equivalent to the existence of a common quadratic Lyapunov function and can be easily seen to be equivalent to (2.2).

For the normal forms considered in Item 2 of Lemma 2.5 one can directly check, on the one hand, that the square of the Euclidean norm is always a common quadratic Lyapunov function for the system and, on the other hand, that (2.2) is satisfied. In fact, in this case τ_1, τ_2 are both smaller than -1 and $\kappa \in (-1, 1)$, so that $0 < \mathrm{tr}(M_1 M_2) = 2\tau_1\tau_2 + 2\kappa < 4\tau_1\tau_2 = \mathrm{tr}(M_1)\mathrm{tr}(M_2)$.

It finally remains to consider the case in which $\det([M_1, M_2]) = 0$ and the matrices M_1, M_2 are in upper triangular form. In this case it is easy to see that the quadratic function $V(x) = x^\top P x$ with $P = \begin{pmatrix} 1 & 0 \\ 0 & q \end{pmatrix}$ is a common Lyapunov function for $(\Sigma_{\mathcal{M}})$ when q is large enough and that, at the same time, (2.2) is satisfied. □

Remark 2.7 It follows from Proposition 2.6 and its proof that if $\det([M_1, M_2]) \geq 0$ then $(\Sigma_{\mathcal{M}})$ is quadratically stable.

Remark 2.8 As follows from the proof of Proposition 2.6, if (2.2) holds true and the matrices M_1, M_2 are in the normal forms of Lemma 2.5, then the quadratic Lyapunov function for $(\Sigma_{\mathcal{M}})$ can be assumed of the form $x^\top D x$ with D diagonal.

Before completing the proof of Theorem 2.1 we provide intermediate results in the case in which M_1 and M_2 are in the form (2.5)–(2.6), allowing us to introduce the crucial notion of *worst case trajectory* and to prove Case (**S**$_4$). First, observe that the assumption in (**S**$_4$) implies that $\mathrm{tr}(M_1 M_2) < 0$ and, since $\mathrm{tr}(M_1 M_2) = F + \frac{\mathrm{sign}(\delta_{M_1}\delta_{M_2})}{F} + 2\tau_1\tau_2 = 2(\kappa + \tau_1\tau_2)$ and $|F| > 1$, one necessarily has that $\kappa < 0$ and $F < -1$.

Under the same assumption in (**S**$_4$), one has that $\Delta > 0$ and the set $\mathcal{Z}$ of points where the vector fields $M_1 x$ and $M_2 x$ are parallel, that is

$$\mathcal{Z} = \{x \in \mathbb{R}^2 \mid \det(M_1 x, M_2 x) = 0\},$$

is nonempty and consists of two straight lines passing through the origin. This is due to the fact that the discriminant of the quadratic form is equal to Δ.

In the following we say that a vector $v \in \mathbb{R}^2$ *points clockwise* at $x \in \mathbb{R}^2$ if the angle between x and v, measured in clockwise sense from x, is in the open interval $(0, \pi)$. This is equivalent to saying that $\mathrm{sign}(\det(v, x)) = 1$.

Lemma 2.9 *Let* M_1, M_2 *be in the form* (2.5)–(2.6) *and satisfy the assumption in* $(\mathbf{S}_4)$. *Then, for any nonzero* $x \in \mathbb{R}^2$, *at least one among the vectors* M_1x *and* M_2x *points clockwise at* x. *Moreover,* M_1x *and* M_2x *point in the same direction if* $x \in \mathcal{Z}$.

Proof Let $x \in \mathbb{R}^2$ have coordinates (x_1, x_2). Let us consider the quantities $\sigma_j = \mathrm{sign}(\det(M_jx, x))$ for $j = 1, 2$. By a direct computation and using the fact that F is negative, one has that $\sigma_1 = \mathrm{sign}(m^2 - \mathrm{sign}(\delta_{M_1}))$ and $\sigma_2 = \mathrm{sign}(F^2 - m^2\mathrm{sign}(\delta_{M_2}))$, where m is the slope of the line spanned by x, i.e., $x_2 = mx_1$. It is straightforward to show that if both σ_1 and σ_2 were non-positive then $F^2 \leq 1$, leading to a contradiction. This proves the first part of the lemma.

Let now $x \in \mathcal{Z}$. Then there exists $\alpha \in \mathbb{R}$ such that $M_2x = \alpha M_1x$ and, in particular, α is an eigenvalue of $M_1^{-1}M_2$. Note that the two eigenvalues of $M_1^{-1}M_2$ have the same sign, since $\det(M_1^{-1}M_2) > 0$. Moreover $M_1^{-1} = (\mathrm{tr}(M_1)\mathrm{Id}_2 - M_1)/\det(M_1)$, which implies that

$$\mathrm{tr}(M_1^{-1}M_2) = \mathrm{tr}\big((\mathrm{tr}(M_1)M_2 - M_1M_2)/\det(M_1)\big) = 2\Gamma(M_1, M_2)/\det(M_1) > 0,$$

so that both eigenvalues of $M_1^{-1}M_2$ are positive, which concludes the proof. □

The previous lemma allows us to introduce a useful tool for checking the stability of $(\Sigma_{\mathcal{M}})$ under the conditions in $(\mathbf{S}_4)$.

Definition 2.10 Assume that M_1, M_2 are in the form (2.5)–(2.6) and satisfy the assumption in Case $(\mathbf{S}_4)$. A *worst case trajectory* is a nonzero trajectory of the switched system that follows at almost every time the vector field forming the smallest angle measured in clockwise sense from the outward radial direction.

Lemma 2.9 guarantees the existence of a worst case trajectory starting from any nonzero initial point in $\mathbb{R}^2$. Recall that, given a planar trajectory $\gamma(\cdot)$ and its expression in polar coordinates $(r(\cdot), \theta(\cdot))$, the angular velocity is obtained as

$$\dot\theta(t) = \frac{1}{r^2(t)}\det(\gamma(t), \dot\gamma(t)). \tag{2.8}$$

We then deduce from Lemma 2.9 that a worst case trajectory has uniformly negative angular velocity. It is then clear that such a trajectory rotates clockwise infinitely many times around the origin and it is the concatenation of integral curves of $x \mapsto M_2x$, in the regions where $\det(M_1x, M_2x) > 0$, and integral curves of $x \mapsto M_1x$, wherever $\det(M_1x, M_2x) < 0$. Switches occur exactly at the intersection points with the two lines forming $\mathcal{Z}$. Also, at a point x_0 for which $\det(M_1x_0, M_2x_0) > 0$ the vector M_1x_0 points at the right of the (oriented) integral curve of $x \mapsto M_2x$, while the vector M_2x_0 points at the right of the (oriented) integral curve of $x \mapsto M_1x$ if $\det(M_1x_0, M_2x_0) < 0$ (see Fig. 2.6). Intuitively, worst case trajectories are related to the stability properties of $(\Sigma_{\mathcal{M}})$, since they "trap" trajectories of $(\Sigma_{\mathcal{M}})$. In particular a worst case trajectory converging to the origin forces the system to be attractive (and hence uniformly exponentially stable). This fact is formalized as follows.

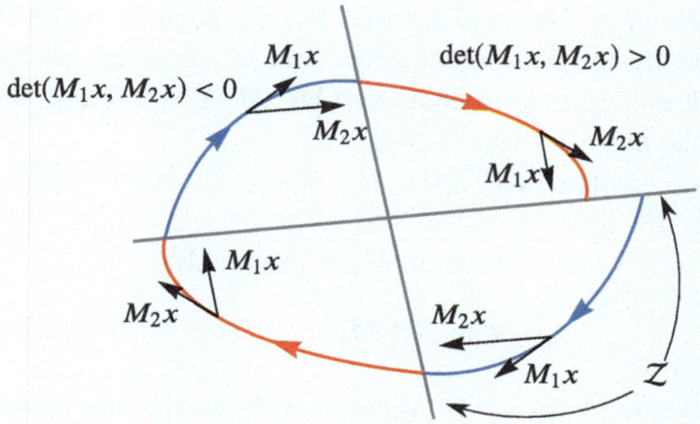

Fig. 2.6 Worst case trajectory in the case of an exponentially stable system

Lemma 2.11 *Assume that M_1, M_2 are in the form (2.5)–(2.6) and satisfy the assumption in Case ($\mathbf{S}_4$). Let η be the ratio between the modulus of the point obtained after a complete turn along a worst case trajectory and the modulus of its initial point. If $\eta > 1$, then $(\Sigma_{\mathcal{M}})$ is unbounded; if $\eta = 1$, then the worst case trajectory is periodic, $(\Sigma_{\mathcal{M}})$ is uniformly bounded and not uniformly exponentially stable; if $\eta < 1$ then $(\Sigma_{\mathcal{M}})$ is uniformly exponentially stable.*

Proof If $\eta \geq 1$ then the trajectory is either periodic or goes to infinity. Hence $(\Sigma_{\mathcal{M}})$ is not uniformly exponentially stable and it is unbounded in the case $\eta > 1$, since the norm of the worst case trajectory tends to infinity exponentially.

Consider now the case $\eta < 1$. Denote by $\gamma\colon [0, \tau] \to \mathbb{R}^2$ a worst case trajectory of $(\Sigma_{\mathcal{M}})$ making a complete turn at time τ and such that $\gamma(\tau) = \eta\gamma(0)$. Denote by Ω the region enclosed by the curve γ and the segment joining $\gamma(0)$ and $\gamma(\tau)$. Since γ is a worst case trajectory, Ω is invariant for $(\Sigma_{\mathcal{M}})$ (see Fig. 2.6). Hence, $(\Sigma_{\mathcal{M}})$ is uniformly bounded.

Let us consider the set of matrices $\mathcal{M}^\varepsilon = \mathcal{M} + \varepsilon \mathrm{Id}_2$, where $\varepsilon > 0$, and notice that the elements of $\mathcal{M}^\varepsilon$ are Hurwitz for ε small enough. Since $\operatorname{tr}(M_1M_2)$ and $\sqrt{\det(M_1)\det(M_2)}$ depend continuously on the matrices M_1, M_2, we deduce that for ε small enough $(\Sigma_{\mathcal{M}^\varepsilon})$ falls either in Case ($\mathbf{S}_3$) or Case ($\mathbf{S}_4$). The first case is actually impossible, since the quadratic stability of $(\Sigma_{\mathcal{M}^\varepsilon})$ would imply the quadratic stability of $(\Sigma_{\mathcal{M}})$, contradicting Proposition 2.6. We can therefore assume that $(\Sigma_{\mathcal{M}^\varepsilon})$ falls in Case ($\mathbf{S}_4$). Let $\gamma_\varepsilon\colon [0, \tau_\varepsilon] \to \mathbb{R}^2$ be a worst case trajectory of $(\Sigma_{\mathcal{M}_\varepsilon})$ starting at $\gamma(0)$ and making a complete turn at time τ_ε, and let $\eta_\varepsilon > 0$ be such that $\gamma_\varepsilon(\tau_\varepsilon) = \eta_\varepsilon\gamma_\varepsilon(0)$. Notice that τ_ε is upper bounded by some positive τ_*, since the angular velocity associated with γ_ε is bounded from above by a negative constant independent of ε (cf. Lemma 2.9 and Eq. (2.8)). Since the curve $t \mapsto e^{-\varepsilon t}\gamma_\varepsilon(t)$ is a trajectory of $(\Sigma_{\mathcal{M}})$ starting at $\gamma(0)$ and making a complete turn in time τ_ε, one gets that $\eta_\varepsilon e^{-\varepsilon\tau_\varepsilon} \leq \eta$ and hence, $\eta_\varepsilon \leq \eta e^{\varepsilon\tau_*}$, yielding for ε small enough that $\eta_\varepsilon < 1$. Then $(\Sigma_{\mathcal{M}^\varepsilon})$ is uniformly bounded and the lemma is proved by noticing, by (1.8) in Proposition 1.20, that $\lambda(\mathcal{M}) = \lambda(\mathcal{M}^\varepsilon) - \varepsilon \leq -\varepsilon < 0$. The conclusion follows by Item 5 in Proposition 1.20. □

We are ready to finish the proof of Theorem 2.1.

Proof of Theorem 2.1 As ($\mathbf{S}_3$) has been shown in Proposition 2.6, we just need to study the three other cases, and in particular, by Remark 2.7, we assume without loss of generality that $\det([M_1, M_2]) < 0$. Thus, from now on, we take M_1, M_2 in the normal forms (2.5)–(2.6).

Assume that $\Gamma(M_1, M_2) \leq -\sqrt{\det(M_1)\det(M_2)}$ and let

$$\begin{aligned}\phi(\sigma) &= \det(\sigma M_1 + (1-\sigma)M_2)\\ &= \sigma^2\det(M_1) + 2\sigma(1-\sigma)\Gamma(M_1, M_2) + (1-\sigma)^2\det(M_2).\end{aligned}$$

A straightforward computation shows that the minimum of $\sigma \mapsto \phi(\sigma)$ on $[0, 1]$ is obtained at

$$\sigma_0 = \frac{\det(M_2) - \Gamma(M_1, M_2)}{\det(M_1) + \det(M_2) - 2\Gamma(M_1, M_2)} \in (0, 1)$$

and that

$$\phi(\sigma_0) = \frac{-\Delta}{4(\det(M_1) + \det(M_2) - \Gamma(M_1, M_2))} \leq 0.$$

In particular, in Case ($\mathbf{S}_1$) we have $\phi(\sigma_0) < 0$ and therefore the matrix $\sigma_0 M_1 + (1-\sigma_0)M_2 \in [\![M_1, M_2]\!]$ has a positive real eigenvalue, so that the system is unbounded.

Similarly, in Case ($\mathbf{S}_2$) we have $\phi(\sigma_0) = 0$, so that the system cannot be uniformly exponentially stable. In addition, we claim that it admits a nonstrict common quadratic Lyapunov function. Indeed, consider a sequence $\mathcal{M}_k = \{M_1^{(k)}, M_2^{(k)}\}$ of pairs of matrices converging to $\mathcal{M} = \{M_1, M_2\}$ and satisfying ($\mathbf{S}_3$). By Remark 2.8 there exists a sequence of common quadratic Lyapunov functions $V_k(x) = x^\top P_k x$ for $(\Sigma_{\mathcal{M}_k})$, with P_k diagonal and of trace one converging to a diagonal positive semi-definite matrix P^* with trace one satisfying $Q_i = M_i^\top P^* + P^* M_i \preceq 0$ for $i = 1, 2$. One can easily show that $P^* \succ 0$, since otherwise $\det(Q_1) = -1$. Thus $V^*(x) = x^\top P^* x$ is a nonstrict common quadratic Lyapunov function for $(\Sigma_{\mathcal{M}})$.

Let us now prove ($\mathbf{S}_4$). We use the fact that any worst case trajectory is the concatenation of integral curves of $x \mapsto M_2 x$, in the regions where $\det(M_1 x, M_2 x) > 0$, and integral curves of $x \mapsto M_1 x$, wherever $\det(M_1 x, M_2 x) < 0$. Setting $x = (1, m)^\top$ we have

$$\begin{aligned}&\det(M_1 x, M_2 x)\\ &\quad = (\tau_2 - \tau_1 \operatorname{sign}(\delta_{M_2})/F)\, m^2 + (F - \operatorname{sign}(\delta_{M_1}\delta_{M_2})/F)\, m\\ &\qquad + (\tau_1 F - \tau_2 \operatorname{sign}(\delta_{M_1})).\end{aligned}$$

Note that the discriminant of the previous quadratic polynomial in m coincides with Δ and its zeroes are given by

$$m_{\pm} = \frac{-F + \text{sign}(\delta_{M_1}\delta_{M_2})/F \pm \sqrt{\Delta}}{2(\tau_2 - \tau_1 \text{sign}(\delta_{M_2})/F)}$$

whenever $\tau_2 - \tau_1 \text{sign}(\delta_{M_2})/F \neq 0$. If $\tau_2 - \tau_1 \text{sign}(\delta_{M_2})/F = 0$ we set $m_- = \tau_1$ and $m_+ = \infty$. Since any worst case trajectory rotates clockwise, it coincides with an integral curve of $x \mapsto M_2 x$ after crossing the line spanned by $w_- = (1, m_-)^\top$ and, vice versa, it coincides with an integral curve of $x \mapsto M_1 x$ after a crossing with the line spanned by $w_+ = (1, m_+)^\top$ (corresponding to the vertical axis if $m_+ = \infty$).

By symmetry reasons, in order to determine if the system is uniformly exponentially stable or not, it is enough to compute, for a worst case trajectory, the ratio $\mathcal{R}$ between the modulus of the point obtained after a half-turn and the modulus of the initial point. If t_1 and t_2 are the time lengths of the arcs of the worst case trajectory during the half-turn, then

$$\mathcal{R} = \tilde{\mathcal{R}} e^{\tau_1 t_1 + \tau_2 t_2},$$

where $\tilde{\mathcal{R}}$ is the ratio obtained after a half-turn by following the integral curve corresponding to

$$\tilde{M}_1 = \begin{pmatrix} 0 & 1 \\ \text{sign}(\delta_{M_1}) & 0 \end{pmatrix}$$

for time t_1 and then the integral curve of

$$\tilde{M}_2 = \begin{pmatrix} 0 & \text{sign}(\delta_{M_2})/F \\ F & 0 \end{pmatrix}$$

for time t_2. In order to compute $\tilde{\mathcal{R}}$, let us observe that $V_1(z) = z_2^2 - \text{sign}(\delta_{M_1}) z_1^2$ and $V_2(z) = F^2 z_1^2 - \text{sign}(\delta_{M_2}) z_2^2$ are quadratic first integrals for the differential equations $\dot{z} = \tilde{M}_1 z$ and $\dot{z} = \tilde{M}_2 z$, respectively. Moreover, the curve defining $\tilde{\mathcal{R}}$ corresponds to the integration of $x \mapsto \tilde{M}_1 x$ from w_+ to $\sqrt{\frac{V_1(w_+)}{V_1(w_-)}} w_-$ or $-\sqrt{\frac{V_1(w_+)}{V_1(w_-)}} w_-$ and then of $x \mapsto \tilde{M}_2 x$ up to the point $-\sqrt{\frac{V_2(w_-)}{V_2(w_+)} \cdot \frac{V_1(w_+)}{V_1(w_-)}} w_+$. By replacing the values $w_\pm$ obtained above and after simplifications, one gets

$$\tilde{\mathcal{R}} = \sqrt{\frac{V_2(w_-)}{V_2(w_+)} \cdot \frac{V_1(w_+)}{V_1(w_-)}} = \frac{2\Gamma(M_1, M_2) + \sqrt{\Delta}}{2\sqrt{\det(M_1)\det(M_2)}}.$$

It remains to compute the switching times. To compute the time length for the arc of trajectory of the first mode M_1, let us consider the corresponding dynamics of the slope $m = x_2/x_1$, where $x = (x_1, x_2)^\top$. It is easy to see that this is given by

$$\dot{m}(t) = \text{sign}(\delta_{M_1}) - m(t)^2.$$

As a consequence, the time length of the arc between the lines of directions w_+ and w_- is equal to

$$\int_{m_-}^{+\infty} \frac{1}{m^2 - \text{sign}(\delta_{M_1})} dm + \int_{-\infty}^{m_+} \frac{1}{m^2 - \text{sign}(\delta_{M_1})} dm$$

if $\tau_2 - \tau_1 \text{sign}(\delta_{M_2})/F \leq 0$ and is equal to

$$\int_{m_-}^{m_+} \frac{1}{m^2 - \text{sign}(\delta_{M_1})} dm$$

if $\tau_2 - \tau_1 \text{sign}(\delta_{M_2})/F > 0$. By an explicit integration and after simplification, one gets (independently of the sign of $\tau_2 - \tau_1 \text{sign}(\delta_{M_2})/F$) that $t_1 = \frac{\pi}{2} - \arctan \frac{2(\kappa\tau_1 + \tau_2)}{\sqrt{\Delta}}$ if $\text{sign}(\delta_{M_1}) = -1$, $t_1 = \text{arctanh} \frac{\sqrt{\Delta}}{2(\kappa\tau_1 - \tau_2)}$ if $\text{sign}(\delta_{M_1}) = 1$, and $t_1 = \frac{\sqrt{\Delta}}{2\kappa\tau_1}$ if $\text{sign}(\delta_{M_1}) = 0$.

By symmetry reasons (in particular the transformation $y_1 = x_2$, $y_2 = Fx_1$ exchanges the roles of M_1 and M_2) the time length t_2 of the arc of the mode M_2 is computed by means of the same formula, when one exchanges the values of δ_{M_1} and τ_1 with δ_{M_2} and τ_2, respectively. This complete the proof in the case in which M_1, M_2 are in normal form.

For a general pair of matrices (M_1, M_2), the formula for t_i in Theorem 2.1 is obtained by taking into account the rescaling of $\sqrt{\Delta}$ when passing from M_1, M_2 to their corresponding normal forms. □

2.3 Notes and References

Concerning the problem of characterizing global uniform asymptotic stability for switched linear systems in the planar case, some necessary and sufficient conditions have been obtained in [Pya71], where the problem stems in the context of absolute stability; the result is achieved via a variational approach [Pya70] characterizing the "most unstable" trajectories of the system, analogous to the worst case trajectories studied in this chapter (see also the later works [Bar89, PR96] and, for an overview on the variational approach in the context of switched systems, [Mar06]). Necessary and sufficient conditions in the more general case of planar switched systems with $\mathcal{M} = \{M_1, \ldots, M_m\}$ (actually, in the case in which the system is driven by m homogeneous vector fields) can be found in [Fil80]: the conditions therein are expressed via inequalities involving integrals of elementary functions and are obtained by studying the representation of the switched system in polar coordinates. The approach followed in this monograph is the one of [Bos02, BB08, BBM09]: the aim is to provide necessary and sufficient conditions characterizing the various stability properties of planar switched systems with $\mathcal{M} = \{M_1, M_2\}$ in terms of elementary functions of the coefficients of the matrices. We remark that the paper [Bos02] is mainly motivated by the results in [AL01], dealing with higher dimensional switched systems of the form $(\Sigma_{\mathcal{M}'})$ with $\mathcal{M}' = \{M_1', M_2'\} \subset M_n(\mathbb{R})$: in [AL01] it is shown that if the Lie algebra generated by $\mathcal{M}'$ and the identity matrix has dimension at most four, then one can construct a pair $\mathcal{M} = \{M_1, M_2\} \subset M_2(\mathbb{R})$ such that $(\Sigma_{\mathcal{M}'})$ is uniformly exponentially stable if and only if $(\Sigma_{\mathcal{M}})$ is.

Concerning Condition $(\mathbf{S}_3)$ in Theorem 2.1, which characterizes quadratic stability for planar switched systems, equivalent conditions are derived e.g in [SN99, SN02]; see also Theorem 3.14 in Chap. 3 below.

We conclude this section by mentioning some works in which the approach and the context bear some similarities with those of this chapter. Concerning stability of planar switched nonlinear systems with two available modes, an approach inspired from [Bos02] is followed in [BCS06], where some sufficient conditions and some necessary conditions are provided based on properties of the set where the two vector fields are collinear. In [BBMZ14], motivated by the results of [BBM09] concerning the planar switched system $(\Sigma_{\mathcal{M}})$, $\mathcal{M} = \{M_1, M_2\}$, the authors perform a stability analysis assuming that the switchings between the two planar dynamics are regulated by a Markov process. The situation appears to be far more complex compared to the deterministic case, and a general characterization of the stability properties of the system appears to be out of reach. Under the assumption that there exists a non-Hurwitz convex combination of the matrices M_1, M_2, the authors prove that the asymptotic behaviour is almost surely characterized by an exponential rate only depending on the jump rate (the explicit computation of such an exponential rate is in general a difficult problem, in [BBMZ14] it is provided for a couple of examples). In particular, the switched system is almost surely stable for sufficiently small jump rates and almost surely unstable for sufficiently large jump rates.

Chapter 3
Direct and Converse Lyapunov Techniques

One of the most popular concepts in the stability analysis of dynamical systems is the notion of Lyapunov function. For physical models or natural processes, and in presence of dissipative phenomena, the role of Lyapunov function is naturally played by energy or storage functions. Proving the existence of such functions allows one to assess the stability of a dynamical system without a precise knowledge of the corresponding trajectories. In the context of switched systems, when dealing with stability notions that are uniform with respect to the family $\mathcal{S}$ of switching laws, a fundamental role is played by the concept of common Lyapunov function. In this chapter we first introduce this notion and we discuss its use in the stability analysis of switched linear systems. We then focus on the analysis of switched linear systems admitting common quadratic Lyapunov functions, referred to as *quadratically stable* switched systems. In particular we provide necessary and/or sufficient conditions for a switched linear system to be quadratically stable and, in the case of planar switched systems with only two modes, we provide simple algebraic characterizations of quadratic stability. We then discuss converse Lyapunov results, showing that the existence of a common Lyapunov function is not only a sufficient, but also a necessary condition for stability of switched linear systems. We also put forward the notion of *universal* classes of Lyapunov functions, i.e., families $\mathcal{F}$ of candidate Lyapunov functions such that for any bounded set of matrices $\mathcal{M} \subset M_n(\mathbb{R})$ the corresponding switched system $(\Sigma_{\mathcal{M}})$ admits a common Lyapunov function in $\mathcal{F}$. In particular, we present necessary and/or sufficient conditions for a class $\mathcal{F}$ to be universal.

3.1 Common Lyapunov Functions

Common Lyapunov functions, defined here below in a form adapted to our setting, may be used to obtain both theoretical results and numerical tests for stability.

Y. Chitour et al., *Dynamics and Stability of Continuous-Time Switched Linear Systems*, PNLDE Subseries in Control 105,
https://doi.org/10.1007/978-3-031-82982-6_3

Definition 3.1 We say that a continuous function $V : \mathbb{R}^n \to [0, +\infty)$ is a *nonstrict common Lyapunov function* for (Σ_S) if it is *positive-definite*, that is, $V(0) = 0$ and $V(x) > 0$ for every $x \neq 0$, and V is non-increasing along trajectories of (Σ_S). If, moreover, V is decreasing along nonzero trajectories of (Σ_S), we say that V is a *common Lyapunov function* for (Σ_S).

The preceding definition immediately leads to the following remark, for which we need to recall a standard definition: given $\alpha > 0$, a function $V : \mathbb{R}^n \to \mathbb{R}$ is said to be *absolutely homogeneous of degree* α if $V(\lambda x) = |\lambda|^\alpha V(x)$ for every $x \in \mathbb{R}^n$ and $\lambda \in \mathbb{R}$. If $\alpha = 1$ we simply say that V is *absolutely homogeneous*.

Remark 3.2 If $\varphi : [0, +\infty) \to [0, +\infty)$ is continuous, strictly increasing, and such that $\varphi(0) = 0$ and V is a (possibly nonstrict) common Lyapunov function, then $\varphi \circ V$ is also a common Lyapunov function. In particular, the positive multiple of a common Lyapunov function is a common Lyapunov function and if there exists an absolutely homogeneous common Lyapunov function, then for every $\alpha > 0$ there exists an absolutely homogeneous common Lyapunov function of degree α.

We have the following adaptation of the classical (direct) *Lyapunov theorem*, also referred to as *second Lyapunov method*.

Theorem 3.3 *A switched linear system (Σ_S) admitting a nonstrict common Lyapunov function is uniformly bounded and $\lambda(\mathcal{S}) \leq 0$. If $\mathcal{S}$ is weak-$\star$ compact and (Σ_S) admits a common Lyapunov function then (Σ_S) is uniformly exponentially stable.*

Proof Let $m > 0$ be the minimum of the nonstrict common Lyapunov function V over the unit sphere S^{n-1}. By continuity of V, there exists $\delta \in (0, 1)$ so that $V(x) \leq m/2$ as soon as $\|x\| \leq \delta$. Therefore, every trajectory of (Σ_S) starting in the ball $B(0, \delta)$ does not exit the unit ball $B(0, 1)$. By linearity of (Σ_S), it follows that (Σ_S) is uniformly bounded. The inequality $\lambda(\mathcal{S}) \leq 0$ follows immediately from the definition given in (1.6).

An argument for the second part of the statement goes as follows. We claim that every trajectory of (Σ_S) starting in the ball $B(0, \delta)$ converges to the origin. Fix $x \in B(0, \delta)$ and a signal $A \in \mathcal{S}$. Let l be the limit as t tends to infinity of the strictly decreasing function $t \mapsto V(\phi_A(t, 0)x)$. For every increasing sequence of times t_n tending to infinity, the corresponding sequence of points $\phi_A(t_n, 0)x$ remains in $B(0, 1)$ and hence admits a limit point $\bar{x}$ in $B(0, 1)$. Let $A^\star \in \mathcal{S}$ be a limit point for the weak-$\star$ topology of the sequence of signals $A(\cdot + t_n)$. By Proposition 1.3, for every $s > 0$, $V(\phi_{A^\star}(s, 0)\bar{x})$ is the limit, as n tends to infinity, of $V(\phi_A(s + t_n, 0)x)$, which is equal to l. Therefore V is constant along the trajectory $\phi_{A^\star}(\cdot, 0)\bar{x}$, implying that $\bar{x} = 0$. We have proved the claim. By linearity of (Σ_S), one deduces that (Σ_S) is attractive and one concludes by Fenichel's lemma (Lemma 1.25). □

Remark 3.4 In Definition 3.1 we do not require Lyapunov functions to be proper, i.e., such that the corresponding sublevel sets are compact. This assumption is not needed in the proof of Theorem 3.3. This is due to the fact that the existence of a Lyapunov function in the sense of Definition 3.1 implies local stability of $(\Sigma_{\mathcal{M}})$

and hence, by linearity of the system, its global stability. Notice also that, given a possibly non-proper Lyapunov function V for $(\Sigma_{\mathcal{M}})$, any Minkowski functional associated with a sublevel set of V, e.g., the function $\tilde{V}(x) = \min\{r \mid x \in r\mathcal{D}\}$ where $\mathcal{D} = \{x \in \mathbb{R}^n \mid V(x) \leq 1\}$, is a proper Lyapunov function for $(\Sigma_{\mathcal{M}})$.

In order to explicitly verify the monotonicity of V along the trajectories of the system, some differentiability assumption on V is usually imposed. Furthermore, the constraints defining the family $\mathcal{S}$ of switching laws may hamper the possibility of characterizing easily this monotonicity property. This is why, in this chapter, we mostly focus on families of (candidate) Lyapunov functions satisfying some regularity conditions, e.g., Lipschitz or piecewise smooth, and we limit our study to switched systems of the form $(\Sigma_{\mathcal{M}})$, that is, to arbitrary switching.

In order to verify that a function V that is C^1 on $\mathbb{R}^n \setminus \{0\}$ is a common Lyapunov function, a standard test consists in checking that

$$\nabla V(x)^\top Mx < 0 \qquad \forall x \in \mathbb{R}^n \setminus \{0\},\ \forall M \in \mathcal{M}. \tag{3.1}$$

Note that such a condition only depends on $\mathcal{M}$, yielding a stability test for $(\Sigma_{\mathcal{M}})$.

As we will see in Proposition 3.18, the continuous differentiability (outside the origin) of the common Lyapunov function, which is fundamental in order to check (3.1), is not a restrictive assumption when checking the uniform exponential stability of a switched linear system. On the other hand, a switched system $(\Sigma_{\mathcal{M}})$ possessing the weaker property of uniform boundedness does not always admit a C^1 nonstrict common Lyapunov function. An example of such a system will be discussed later in Example 4.10. Furthermore, even in case of uniform exponential stability, it may be useful to provide a criterion to check the existence of a non-differentiable common Lyapunov function, e.g., piecewise linear or piecewise quadratic. For these reasons we will introduce a criterion which generalizes the dissipation condition $\nabla V(x)^\top Mx < 0$ and characterizes the family of (possibly nonstrict) common Lyapunov functions in a nonsmooth setting. We will make use of a preliminary result expressing the variation of a convex function V along a trajectory in terms of the subdifferential of V. We recall that the subdifferential $\partial V(x)$ at a point $x \in \mathbb{R}^n$ is defined as

$$\partial V(x) = \{\ell \in \mathbb{R}^n \mid \ell^\top (y - x) \leq V(y) - V(x),\ \forall y \in \mathbb{R}^n\}. \tag{3.2}$$

Lemma 3.5 *Let $V : \mathbb{R}^n \to \mathbb{R}$ be a convex function and $\varphi : I \to \mathbb{R}^n$ be an absolutely continuous function, with $I \subseteq \mathbb{R}$ an open interval. Then $V \circ \varphi$ is absolutely continuous and the equality*

$$\frac{d}{dt} V(\varphi(t)) = \ell^\top \dot{\varphi}(t), \quad \forall \ell \in \partial V(\varphi(t)),$$

holds true at almost every $t \in I$.

Proof As V is convex, V is locally Lipschitz continuous (Item 2 of Proposition A.15) and the composition $V \circ \varphi$ is absolutely continuous. Hence for almost every $t \in I$ the derivatives of both φ and $V \circ \varphi$ are well-defined. By definition of subdifferential, for every $t, s \in I$ and $\ell \in \partial V(\varphi(t))$, we have

$$\ell^\top(\varphi(s) - \varphi(t)) \leq V(\varphi(s)) - V(\varphi(t)).$$

We deduce that

$$\frac{d}{dt}V(\varphi(t)) = \lim_{s \to t^+} \frac{V(\varphi(s)) - V(\varphi(t))}{s - t} \geq \ell^\top \lim_{s \to t^+} \frac{\varphi(s) - \varphi(t)}{s - t} = \ell^\top \dot{\varphi}(t)$$

holds true for almost every $t \in I$ and for every $\ell \in \partial V(\varphi(t))$. Similarly, taking the limit as $s \to t^-$, we obtain that $\frac{d}{dt}V(\varphi(t)) \leq \ell^\top \dot{\varphi}(t)$ for almost every $t \in I$ and for every $\ell \in \partial V(\varphi(t))$. This concludes the proof of the lemma. □

We next state a result providing a test extending that given in (3.1), which is used for checking the nonincreasing character of a candidate Lyapunov function along trajectories of a switched linear system.

Proposition 3.6 *Let* $V : \mathbb{R}^n \to [0, +\infty)$ *be a positive-definite convex function. Then* V *is a nonstrict common Lyapunov function for* $(\Sigma_{\mathcal{M}})$ *if and only if*

$$\ell^\top Mx \leq 0, \qquad \forall x \in \mathbb{R}^n \setminus \{0\},\ \forall \ell \in \partial V(x),\ \forall M \in \mathcal{M}. \tag{3.3}$$

Moreover, if in the condition (3.3) *the inequality is replaced by a strict inequality, then* V *is a common Lyapunov function for* $(\Sigma_{\mathcal{M}})$.

Proof The second part of the proposition and the *if* implication in the first part directly follow from Lemma 3.5. We are left to show that if V is a nonstrict common Lyapunov function for $(\Sigma_{\mathcal{M}})$, then the inequality (3.3) holds true. By contradiction, suppose that there exist $x \in \mathbb{R}^n$, $\ell \in \partial V(x)$, and $M \in \mathcal{M}$ such that $\ell^\top Mx > 0$. By Item 5 of Proposition A.15, one may find a differentiability point y of V such that y is arbitrarily close to x and $\nabla V(y)^\top My > 0$, that is, V is increasing at $t = 0$ along the trajectory $t \mapsto e^{tM}y$, leading to a contradiction. □

3.2 Common Quadratic Lyapunov Functions

In view of Theorem 3.3, in order to ensure the uniform exponential stability of a switched system with compact set of modes, it is enough to show the existence of a common Lyapunov function. One of the simplest and reasonable classes of functions where to seek a candidate Lyapunov function is that of positive quadratic functions. We have the following definition.

Definition 3.7 We say that a function $V: \mathbb{R}^n \to [0, +\infty)$ is a *common quadratic Lyapunov function* for $(\Sigma_{\mathcal{M}})$ if there exist P, Q positive-definite such that $V(x) = x^\top P x$ for $x \in \mathbb{R}^n$ and $M^\top P + PM \preceq -Q$ for every $M \in \mathcal{M}$. If $(\Sigma_{\mathcal{M}})$ admits a common quadratic Lyapunov function then $(\Sigma_{\mathcal{M}})$ is said to be *quadratically stable*.

When the set of modes is finite or polytopic, the existence of a quadratic common Lyapunov function is associated with the existence of a solution P of linear matrix inequalities of the type $M^\top P + PM \prec 0$ for finitely many $M \in \mathcal{M}$. Such linear matrix inequalities may be efficiently investigated by nowadays classical software tools. We recall that each exponentially stable unswitched linear system is quadratically stable. While quadratically stable switched systems are uniformly exponentially stable by Theorem 3.3, the converse is false in general, as shown by the example in (2.3) with $a \in (\bar{a}_3, \bar{a}_2) \cup (\bar{a}_1, \bar{a}_4)$ (see p. 29). However, the numerical test for quadratic stability can be adapted for checking the existence of Lyapunov functions in the broader class of sums of squares (i.e., homogeneous polynomials of even degree that can be decomposed into finite sums of squares of polynomials). The existence of such a polynomial Lyapunov function is equivalent to the uniform exponential stability of $(\Sigma_{\mathcal{M}})$, as explained in the next section (see Corollary 3.22 below).

Showing quadratic stability by numerical techniques is a relatively easy and well understood task. However, the issue of characterizing algebraically the family of quadratically stable switched linear systems is in general more complicated and results are known only for specific cases.

For instance, as regards switched linear systems whose modes are upper-triangular, quadratic stability is equivalent to the fact that every matrix is Hurwitz, as shown in the next proposition, which strengthens Corollary 1.33.

Proposition 3.8 *Let $\mathcal{M}$ be compact. Assume that there exists an invertible matrix $R \in M_n(\mathbb{C})$ such that for every $M \in \mathcal{M}$, the matrix RMR^{-1} is upper-triangular (not necessarily with real entries). Then $(\Sigma_{\mathcal{M}})$ is quadratically stable if and only if $\mathcal{M}$ is made of Hurwitz matrices.*

Proof We only need to prove the "if" part, the other implication being trivial. We claim that given a compact subset $\mathcal{M}_c$ of $\mathcal{M}_n(\mathbb{C})$ made of Hurwitz upper-triangular matrices, there exists a real diagonal matrix $P \succ 0$ such that $Z^*P + PZ$ is a negative-definite Hermitian matrix for every $Z \in \mathcal{M}_c$, where Z^* denotes the conjugate transpose of the matrix Z. Recall that $Q \in \mathcal{M}_n(\mathbb{C})$ is a negative-definite Hermitian matrix if $Q^* = Q$ and $x^*Qx < 0$ for every $x \in \mathbb{C}^n \setminus \{0\}$.

We proceed by induction on n. For $n = 1$ the result is obvious since $z^* + z = \bar{z} + z = 2\Re(z)$ is negative for every $z \in \mathcal{M}_c$. Assume that the result holds true up to dimension n and that $\mathcal{M}_c \subset M_{n+1}(\mathbb{C})$. Then the set

$$\mathcal{M}_c' = \left\{ Z \in M_n(\mathbb{C}) \mid \begin{pmatrix} Z & * \\ 0 & z \end{pmatrix} \in \mathcal{M}_c \right\}$$

is made of $n \times n$ Hurwitz matrices. By the inductive assumption, there exists a real diagonal matrix $P' \succ 0$ such that $Z^*P' + P'Z$ is negative-definite for every $Z \in \mathcal{M}_c'$. Let us now prove that for $p > 0$ large enough the matrix $P = \begin{pmatrix} P' & 0 \\ 0 & p \end{pmatrix}$ is such that $Y^*P + PY$ is negative-definite for every $Y \in \mathcal{M}_c$. Indeed, if $Y = \begin{pmatrix} Z & w \\ 0 & z \end{pmatrix}$, then

$$Y^*P + PY = \begin{pmatrix} Z^*P' + P'Z & P'w \\ w^*P' & 2\Re(z)p \end{pmatrix}$$

and, by the induction hypothesis and thanks to Sylvester's criterion (see Theorem A.27), the latter is negative-definite if the determinant of $-Y^*P - PY$ is positive for every $Y \in \mathcal{M}_c$. Since

$$\det(-Y^*P - PY) = -2\Re(z)p\det(-Z^*P' - P'Z) + \delta(Y),$$

where $\delta(Y)$ depends continuously on Y and is independent of p. By compactness of $\mathcal{M}_c$, p can be chosen large enough so that $\det(-Y^*P - PY) > 0$ for every $Y \in \mathcal{M}_c$, and the claim follows.

Let us now apply the claim to $\mathcal{M}_c = R\mathcal{M}R^{-1}$ and consider the corresponding matrix P. Let Q_r be the real part of the positive-definite Hermitian matrix $Q = R^*PR$. Notice that Q_r is symmetric.

Then, for every nonzero $x \in \mathbb{R}^n$ and every $M \in \mathcal{M}$, one has

$$x^TQ_rx = \mathrm{Re}(x^TQx) = x^TQx > 0$$

$$x^T(M^TQ_r + Q_rM)x = \mathrm{Re}(x^T(M^TQ + QM)x) = x^T(M^TQ + QM)x < 0.$$

Hence, Q_r is also positive-definite and the symmetric matrix $M^TQ_r + Q_rM$ is negative-definite for every $M \in \mathcal{M}$. This concludes the proof that $(\Sigma_{\mathcal{M}})$ is quadratically stable, since $\mathcal{M}$ is compact. □

We next derive a simple sufficient condition for the non-existence of a common quadratic Lyapunov function (Proposition 3.10). The result is based on the following technical lemma.

Lemma 3.9 *Let $M_1, M_2 \in M_n(\mathbb{R})$ be nonsingular. Then the set $[\![M_1, M_2]\!]$ contains a singular matrix if and only if the matrix product $M_1M_2^{-1}$ have a negative real eigenvalue.*

Proof For every $\xi \neq 0$, one has

$$\det(\xi M_1 + (1-\xi)M_2) = \xi^n \det(M_2)\det\left(M_1M_2^{-1} + \frac{1-\xi}{\xi}\mathrm{Id}_n\right),$$

hence the conclusion. □

Proposition 3.10 *Assume either that* $\mathrm{co}(\mathcal{M})$ *contains a singular matrix or, otherwise, that there exist* $M_1, M_2 \in \mathrm{co}(\mathcal{M})$ *such that one of the following conditions is satisfied: (i) one among* $[\![M_1, M_2]\!]$ *and* $[\![M_1, M_2^{-1}]\!]$ *contains a non-Hurwitz matrix, (ii) one among* M_1M_2 *and* $M_1M_2^{-1}$ *possesses a negative real eigenvalue. Then* $(\Sigma_{\mathcal{M}})$ *is not quadratically stable.*

Proof If $(\Sigma_{\mathcal{M}})$ is quadratically stable with common quadratic Lyapunov function $x \mapsto x^\top Px$ then every $M \in \mathrm{co}(\mathcal{M})$ satisfies $M^\top P + PM \prec 0$, and hence it is Hurwitz (in particular, nonsingular). Moreover, for every $M \in \mathrm{co}(\mathcal{M})$,

$$(M^{-1})^\top P + PM^{-1} = (M^{-1})^\top (PM + M^\top P)M^{-1} \prec 0.$$

As a consequence, given $M_1, M_2 \in \mathrm{co}(\mathcal{M})$, the matrices in $[\![M_1, M_2]\!]$ and $[\![M_1, M_2^{-1}]\!]$ are also Hurwitz. Applying Lemma 3.9 to the pairs $\{M_1, M_2\}$ and $\{M_1, M_2^{-1}\}$ yields the last part of the proposition. □

Let us now state a necessary and sufficient nonstrict linear matrix inequality condition for the existence of a common quadratic Lyapunov function when the set of modes is finite.

Theorem 3.11 *Let* $\mathcal{M} = \{M_1, \dots, M_k\}$*. Then* $(\Sigma_{\mathcal{M}})$ *is not quadratically stable if and only if there exist* $k + 1$ *positive semi-definite* $n \times n$ *matrices* $Q_0, \dots, Q_k$*, not all zero, such that*

$$Q_0 = \sum_{i=1}^{k} (M_i Q_i + Q_i M_i^\top).$$

Proof For every positive integer h, let $\mathrm{Sym}_h(\mathbb{R})$ be the space of $h \times h$ symmetric matrices. Since the bilinear map $(X, Y) \mapsto \mathrm{tr}(YX)$ is a scalar product on $\mathrm{Sym}_h(\mathbb{R})$, every linear functional on $\mathrm{Sym}_h(\mathbb{R})$ is of the form $X \mapsto \mathrm{tr}(QX)$ for some $Q \in \mathrm{Sym}_h(\mathbb{R})$.

Define now the linear mapping $F\colon \mathrm{Sym}_n(\mathbb{R}) \to \mathrm{Sym}_{n(k+1)}(\mathbb{R})$ by

$$F(P) = \begin{pmatrix} P & 0 & \dots & 0 \\ 0 & -PM_1 - M_1^\top P & \dots & 0 \\ \vdots & & \ddots & \vdots \\ 0 & \dots & 0 & -PM_k - M_k^\top P \end{pmatrix}.$$

Since $\mathcal{M}$ is finite, it holds that $(\Sigma_{\mathcal{M}})$ is not quadratically stable if and only if the linear matrix inequality

$$F(P) \succ 0, \qquad P \in \mathrm{Sym}_n(\mathbb{R}), \tag{3.4}$$

is unfeasible. This is equivalent to saying that the convex cone of positive-definite matrices and the image of the linear map F are disjoint in $\mathrm{Sym}_{n(k+1)}(\mathbb{R})$. Hence, by the hyperplane separation theorem (Theorem A.14), this is equivalent to the existence of $Q \in \mathrm{Sym}_{n(k+1)}(\mathbb{R})$ such that $\mathrm{tr}(QF(P)) \leq 0$ for every $P \in \mathrm{Sym}_n(\mathbb{R})$ and $\mathrm{tr}(QZ) > 0$ for every positive-definite matrix $Z \in \mathrm{Sym}_{n(k+1)}(\mathbb{R})$. We claim that $\mathrm{tr}(QZ) > 0$ for every $Z \succ 0$ if and only if Q is nonzero and positive semi-definite. Indeed, let $v_1, \dots, v_{n(k+1)}$ be an orthonormal basis of eigenvectors of Q with corresponding eigenvalues $q_1, \dots, q_{n(k+1)}$. For every $Z \succ 0$ one has that $\mathrm{tr}(QZ) = \sum_{j=1}^{n(k+1)} q_j v_j^\top Z v_j$. The latter is clearly positive as soon as Q is nonzero and positive semi-definite. Conversely, assume that $\mathrm{tr}(QZ) > 0$ for every $Z \succ 0$. Considering Z of the form $Z = \sum_{j=1}^{n(k+1)} z_j v_j v_j^\top$ with $z_1, \dots, z_{n(k+1)} > 0$, one has that $\mathrm{tr}(QZ) = \sum_{j=1}^{n(k+1)} q_j z_j > 0$. If one of the scalars $q_1, \dots, q_{n(k+1)}$ is negative (say q_1), one gets a contradiction by choosing $z_1 = 1$ and $z_j = \varepsilon, 2 \leq j \leq n(k+1)$ for $\varepsilon > 0$ small enough. Moreover one of the q_i must be positive otherwise $Q = 0$, which is again not possible. Hence the claim.

Equation (3.4) is then unfeasible if and only if there exists a nonzero $Q \succeq 0$ such that $\mathrm{tr}(QF(P)) \leq 0$ for every $P \in \mathrm{Sym}_n(\mathbb{R})$, which can be rewritten as

$$\mathrm{tr}(Q_0 P - \sum_{i=1}^{k}(Q_i P M_i + Q_i M_i^\top P)) = \mathrm{tr}\big((Q_0 - \sum_{i=1}^{k}(M_i Q_i + Q_i M_i^\top))P\big) \leq 0,$$

where, for $i = 0, \dots, k$, Q_i are the diagonal blocks of Q. Note that $Q_0, \dots, Q_k$ are necessarily positive semi-definite (one can see that by choosing appropriate Z) and at least one of them is positive-definite since otherwise $\mathrm{tr} Q = 0$, implying that all the eigenvalues of Q are zero. The result follows since $P \in \mathrm{Sym}_n(\mathbb{R})$ is arbitrary.

□

Characterizations of quadratic stability exist in the case where $\mathcal{M}$ consists of two matrices $M_1, M_2 \in M_n(\mathbb{R})$ and under some additional assumptions on M_1, M_2. For instance, whenever the difference $M_1 - M_2$ has rank equal to one, quadratic stability can be characterized as in the following theorem, given here without proof.

Theorem 3.12 ([KN06, SMOC04]) *Let* $\mathcal{M} = \{M, M + bc^\top\}$ *where* $M \in M_n(\mathbb{R})$ *and* b, c *are two nonzero vectors of* $\mathbb{R}^n$. *Then the following statements are equivalent.*

1. *The switched system* $(\Sigma_{\mathcal{M}})$ *is quadratically stable.*
2. *The matrix* $M(M + bc^\top)$ *has no negative real eigenvalues.*
3. *For every nonzero* $v, w \in \mathbb{R}^n$, *one has that*

$$Mvv^\top + vv^\top M^\top + (M + bc^\top)ww^\top + ww^\top(M + bc^\top)^\top \neq 0.$$

4. *For every* $\omega \in \mathbb{R}$ *one has* $1 + \mathrm{Re}(c^\top(i\omega \mathrm{Id}_n - M)^{-1}b) > 0$.

Remark 3.13 Notice that the implication $2 \implies 1$ provides a converse to Proposition 3.10 and the equivalence between Item 1 and Item 4 can be interpreted as a generalized Kalman–Yakubovich–Popov lemma.

The situation is much simpler for two-dimensional systems. In particular, in the special case $\mathcal{M} = \{M_1, M_2\} \subset M_2(\mathbb{R})$, a characterization of quadratic stability by means of easily checkable algebraic conditions may be stated as follows.

Theorem 3.14 *Consider a switched system* $(\Sigma_{\mathcal{M}})$ *where* $\mathcal{M} = \{M_1, M_2\} \subset M_2(\mathbb{R})$ *and assume that* M_1, M_2 *are Hurwitz matrices. The following conditions are equivalent.*

(C_1) $(\Sigma_{\mathcal{M}})$ *is quadratically stable,*
(C_2) *the matrices* M_1M_2 *and* $M_1M_2^{-1}$ *have no negative real eigenvalues,*
(C_3) *the intervals* $[\![M_1, M_2]\!]$ *and* $[\![M_1, M_2^{-1}]\!]$ *are made of nonsingular matrices,*
(C_4) *the intervals* $[\![M_1, M_2]\!]$ *and* $[\![M_1, M_2^{-1}]\!]$ *are made of Hurwitz matrices,*
(C_5) M_1, M_2 *satisfy the condition*

$$-2\sqrt{\det(M_1)\det(M_2)} < \operatorname{tr}(M_1M_2) < 2\sqrt{\det(M_1)\det(M_2)} + \operatorname{tr}(M_1)\operatorname{tr}(M_2).$$

Proof We have

$$(C_1) \Rightarrow (C_4) \Rightarrow (C_3) \Leftrightarrow (C_2),$$

where the second implication is obvious, the first one corresponds to Proposition 3.10, and the equivalence between (C_3) and (C_2) follows from Lemma 3.9. The equivalence between (C_2) and (C_5) comes from a straightforward analysis of the roots of the characteristic polynomials $\lambda^2 - \operatorname{tr}(M_1M_2)\lambda + \det(M_1M_2)$ and $\lambda^2 - \operatorname{tr}(M_1M_2^{-1})\lambda + \det(M_1M_2^{-1})$, and by replacing $M_2^{-1} = \frac{1}{\det(M_2)}(-M_2 + \operatorname{tr}(M_2)\mathrm{Id}_2)$ in $\operatorname{tr}(M_1M_2^{-1})$.

Finally, the equivalence between (C_5) and (C_1) has been established in Theorem 2.1, Case $(\mathbf{S}_3)$. □

We close this section by providing a technical result on the existence of a common quadratic Lyapunov function for a class of switched systems depending on parameters. The result actually says that, up to suitably choosing the value of such parameters, a common quadratic Lyapunov function exists. The result will be useful in Chap. 7 when studying the stabilizability of controlled switched systems.

Let us introduce the following notation: let e_n be the vector $(0,\dots,0,1)^\top \in \mathbb{R}^n$ and, for every $a=(a_1,\dots,a_{n-1})\in\mathbb{R}^{n-1}$, denote by D_a^n the $n\times n$ matrix having $a_1,\dots,a_{n-1}$ given by

$$D_a^n = \begin{pmatrix} 0 & a_1 & 0 & \cdots & 0 \\ \vdots & \ddots & \ddots & \ddots & \vdots \\ 0 & \cdots & 0 & a_{n-2} & 0 \\ 0 & \cdots & \cdots & 0 & a_{n-1} \\ 0 & \cdots & \cdots & \cdots & 0 \end{pmatrix}.$$

The following proposition holds true.

Proposition 3.15 *Consider $\underline{b}>0$ and, for $i=1,\dots,n-1$, let $0<\underline{a}_i\le\overline{a}_i$. Then, there exist $K\in\mathbb{R}^n$ and Q positive-definite such that $M^\top Q+QM\prec-\mathrm{Id}_n$ for every*

$$M\in\mathcal{M}_K := \{D_a^n - b\,e_nK^\top \mid b\ge\underline{b},\ a_i\in[\underline{a}_i,\overline{a}_i]\,\text{for}\ i=1,\dots,n-1\}.$$

Proof Reasoning by induction, we start with $n=1$ and the statement reduces to finding $k\in\mathbb{R}$ and $q>0$ such that $-2bkq\le-1$ for every $b\ge\underline{b}$. By taking $q=1$ and $k=1/\underline{b}$ we get the result.

Let now $n>1$ and assume that the property holds true with n replaced by $n-1$. Set $\tilde{e}_{n-1}=(0,\dots,0,1)^\top\in\mathbb{R}^{n-1}$ and $K=(k_nw^\top,k_n)^\top$ with $w\in\mathbb{R}^{n-1}$ and $k_n>0$ to be determined. Notice that

$$D_a^n - b\,e_nK^\top = \begin{pmatrix} D_{a'}^{n-1} & a_{n-1}\tilde{e}_{n-1} \\ -b\,k_nw^\top & -bk_n \end{pmatrix},$$

where $a'=(a_1,\dots,a_{n-2})$.

For every invertible $n\times n$ matrix L, by multiplying the matrix inequality $M^\top Q+QM\prec-\mathrm{Id}_n$ on the left by $L^\top$ and on the right by L, we get the equivalent inequality

$$(L^\top M^\top(L^\top)^{-1})L^\top QL + L^\top QL(L^{-1}ML)\prec -L^\top L.$$

Up to renaming Q the new variable $\frac{1}{\rho}L^\top QL$, where ρ is such that $L^\top L\preceq\rho\mathrm{Id}_n$, we are left to find Q positive-definite such that

$$(L^{-1}ML)^\top Q + Q(L^{-1}ML)\prec-\mathrm{Id}_n, \qquad \forall M\in\mathcal{M}_K. \tag{3.5}$$

Consider the $n\times n$ matrix L_w given by

$$L_w = \begin{pmatrix} \mathrm{Id}_{n-1} & 0 \\ -w^\top & 1 \end{pmatrix}.$$

For every $M = D_a^n - b\,e_n K^\top \in \mathscr{M}_K$, one gets that

$$L_w^{-1} M L_w = \begin{pmatrix} D_{a'}^{n-1} - a_{n-1}\tilde{e}_{n-1} w^\top & a_{n-1}\tilde{e}_{n-1} \\ w^\top (D_{a'}^{n-1} - a_{n-1}\tilde{e}_{n-1} w^\top) & -bk_n + a_{n-1} w^\top \tilde{e}_{n-1} \end{pmatrix}. \tag{3.6}$$

Applying the induction hypothesis for $n-1$ with $a = a'$ and $b = a_{n-1}$, we get that there exists $w \in \mathbb{R}^{n-1}$ and a positive-definite $(n-1) \times (n-1)$ matrix Q' (depending only on $\underline{a}_1, \ldots, \underline{a}_{n-1}$ and $\overline{a}_1, \ldots, \overline{a}_{n-2}$) so that

$$(D_{a'}^{n-1} - a_{n-1}\tilde{e}_{n-1} w^\top)^\top Q' + Q'(D_{a'}^{n-1} - a_{n-1}\tilde{e}_{n-1} w^\top) \prec -\mathrm{Id}_{n-1}. \tag{3.7}$$

We next show that the proposition is proved by taking $Q = \begin{pmatrix} 2Q' & 0 \\ 0 & 1 \end{pmatrix}$. To see that, set $y = (z, y_n) \in \mathbb{R}^n$ with $z \in \mathbb{R}^{n-1}$ and $y_n \in \mathbb{R}$ and let us develop

$$(z^\top \; y_n)\Big((L_w^{-1}(D_a^n - b\,e_n K^\top)L_w)^\top Q + Q L_w^{-1}(D_a^n - b\,e_n K^\top)L_w\Big)\begin{pmatrix} z \\ y_n \end{pmatrix}, \tag{3.8}$$

according to (3.6) and the definition of Q. Using (3.7) and the bounds on the entries of a imposed by the definition of $\mathscr{M}_K$, one obtains that (3.8) is lower than or equal to

$$-2\|z\|^2 + C(\bar{a})|y_n|\|z\| - y_n^2(2bk_n - C(\bar{a})),$$

where $C(\bar{a})$ is a positive constant only depending on the constants $\underline{a}_1, \ldots, \underline{a}_{n-1}$ and $\overline{a}_1, \ldots, \overline{a}_{n-1}$. Since $b \geq \underline{b} > 0$, the above quantity is lower than or equal to $-\frac{3}{2}\|y\|^2$ for k_n large enough depending on $C(\bar{a})$ and $\underline{b}$ only. This concludes the proof of (3.5), and therefore of the proposition. □

3.3 Converse Lyapunov Constructions and Universal Classes of Lyapunov Functions

Converse Lyapunov theorems state that existence of a common Lyapunov function is not just a sufficient condition for uniform exponential stability but it is also necessary. Moreover such results often provide useful structural information on the candidate Lyapunov functions. The following is a simple example of converse Lyapunov result concerning the uniform stability of switched linear systems.

Proposition 3.16 *Assume that $(\Sigma_{\mathscr{M}})$ is uniformly bounded. Then there exists a nonstrict common Lyapunov function V for $(\Sigma_{\mathscr{M}})$ which is convex and absolutely homogeneous.*

Proof Let us consider the map $V : \mathbb{R}^n \to \mathbb{R}_+$ defined as

$$V(x) = \sup_{t\geq 0,\ A\in \mathcal{S}_{\text{arb}}(\mathcal{M})} \|\Phi_A(t,0)x\|.$$

By definition of V and by convexity of the function $x \mapsto \|\Phi_A(t,0)x\|$ for any given $t \geq 0$ and $A \in \mathcal{S}_{\text{arb}}(\mathcal{M})$, we have

$$\begin{aligned}\|\Phi_A(t,0)(\xi x_1 + (1-\xi)x_2)\| &\leq \xi\|\Phi_A(t,0)x_1\| + (1-\xi)\|\Phi_A(t,0)x_2\| \\ &\leq \xi V(x_1) + (1-\xi)V(x_2)\end{aligned} \tag{3.9}$$

for every $\xi \in [0, 1]$ and $x_1, x_2 \in \mathbb{R}^n$. Taking the supremum on the left-hand side for $t \geq 0$ and $A \in \mathcal{S}_{\text{arb}}(\mathcal{M})$ we thus get $V(\xi x_1 + (1-\xi)x_2) \leq \xi V(x_1) + (1-\xi)V(x_2)$, showing the convexity of V. Also, convex functions are locally Lipschitz (Item 2 in Proposition A.15) and thus continuous. The absolute homogeneity of V is an immediate consequence of its definition. Finally, for every $B \in \mathcal{S}_{\text{arb}}(\mathcal{M})$ and $s \geq 0$,

$$\begin{aligned}V(\Phi_B(s,0)x) &= \sup_{\tau\geq 0,\ A\in \mathcal{S}_{\text{arb}}(\mathcal{M})} \|\Phi_A(\tau,0)\Phi_B(s,0)x\| \\ &= \sup_{t\geq s, A\in \mathcal{S}_{\text{arb}}(\mathcal{M})} \|\Phi_{B|_{[0,s]}*A}(t,0)x\| \\ &\leq \sup_{t\geq 0,\ A\in \mathcal{S}_{\text{arb}}(\mathcal{M})} \|\Phi_{B|_{[0,s]}*A}(t,0)x\| \\ &\leq \sup_{t\geq 0,\ A\in \mathcal{S}_{\text{arb}}(\mathcal{M})} \|\Phi_A(t,0)x\| = V(x),\end{aligned}$$

so that V is a nonstrict common Lyapunov function. □

We have presented in the previous section several conditions ensuring the existence or non-existence of a common quadratic Lyapunov function. However, as previously noticed, uniform exponential stability of a switched linear system does not imply that a common quadratic Lyapunov function exists. In other words, the class of quadratic forms is not *universal*, in the sense of the following definition.

Definition 3.17 A set $\mathcal{P}$ of functions from $\mathbb{R}^n$ to $\mathbb{R}$ is a *universal class of Lyapunov functions* if for every uniformly exponentially stable $(\Sigma_{\mathcal{M}})$ there exists a common Lyapunov function for $(\Sigma_{\mathcal{M}})$ in $\mathcal{P}$.

An equivalent formulation of the universality of a class $\mathcal{P}$ is that the converse Lyapunov theorem holds true within $\mathcal{P}$.

We show below that universal classes do exist. From the point of view of testing the stability of a switched linear system, the interest of a universal class depends on its numerical tractability. In particular, interesting candidate classes $\mathcal{P}$ are those parametric families of functions for which the property of being positive-definite and strictly decreasing along all admissible dynamics can be translated into numerically verifiable relations or inequalities, such as linear matrix inequalities. The result below presents a first example of universal class.

Proposition 3.18 *Let $\alpha \geq 1$ and $\mathcal{P}$ be the class of positive-definite, convex, absolutely homogeneous functions of degree α on $\mathbb{R}^n$ that are smooth on $\mathbb{R}^n \setminus \{0\}$. Then $\mathcal{P}$ is universal. Moreover, for every uniformly exponentially stable $(\Sigma_{\mathcal{M}})$ there exists a common Lyapunov function $V \in \mathcal{P}$ for $(\Sigma_{\mathcal{M}})$ such that $\nabla V(x)^\top Mx < -\|x\|^\alpha$ for every $x \in \mathbb{R}^n \setminus \{0\}$ and $M \in \mathcal{M}$.*

Proof Let us first consider the case $\alpha = 1$. Let $(\Sigma_{\mathcal{M}})$ be uniformly exponentially stable. We define $V_0 \colon \mathbb{R}^n \to [0, \infty)$ by

$$V_0(x) = \sup_{A \in \mathcal{S}_{\mathrm{arb}}(\mathcal{M})} \int_0^{+\infty} \|\Phi_A(t,0)x\| dt \,.$$

The function V_0 is well defined since there exist positive constants C, μ such that, for all $A \in \mathcal{S}_{\mathrm{arb}}(\mathcal{M})$, $t \geq 0$, and $x \in \mathbb{R}^n$,

$$\|\Phi_A(t,0)x\| \leq C e^{-\mu t} \|x\|.$$

Note that V_0 is absolutely homogeneous. We next show that V_0 is convex. By (3.9), for every $t \geq 0$ and $A \in \mathcal{S}_{\mathrm{arb}}(\mathcal{M})$, one has

$$\begin{aligned}
&\int_0^{+\infty} \|\Phi_A(t,0)(\xi x + (1-\xi)y)\| dt \\
&\qquad \leq \xi \int_0^{+\infty} \|\Phi_A(t,0)x\|\, dt + (1-\xi) \int_0^{+\infty} \|\Phi_A(t,0)y\|\, dt \\
&\qquad \leq \xi\, V_0(x) + (1-\xi)\, V_0(y).
\end{aligned}$$

By definition of $V_0(\xi\, x + (1-\xi)\, y)$, we then have

$$V_0(\xi\, x + (1-\xi)\, y) \leq \xi\, V_0(x) + (1-\xi)\, V_0(y), \quad \forall \xi \in [0,1], \quad \forall x, y \in \mathbb{R}^n.$$

We now prove that V_0 is decreasing along nontrivial trajectories. For $B \in \mathcal{S}_{\mathrm{arb}}(\mathcal{M})$, $x \neq 0$, and $h > 0$ we have

$$\begin{aligned}
V_0(\Phi_B(h,0)x) &= \sup_{A \in \mathcal{S}_{\mathrm{arb}}(\mathcal{M})} \int_0^{+\infty} \|\Phi_{B|_{[0,h]} * A}(s+h,0)x\| ds \\
&\leq \sup_{A \in \mathcal{S}_{\mathrm{arb}}(\mathcal{M})} \int_0^{+\infty} \|\Phi_A(s+h,0)x\| ds \\
&= \sup_{A \in \mathcal{S}_{\mathrm{arb}}(\mathcal{M})} \int_h^{+\infty} \|\Phi_A(s,0)x\| ds
\end{aligned}$$

$$
\begin{aligned}
&= \sup_{A\in\mathcal{S}_{\mathrm{arb}}(\mathcal{M})} \left(\int_0^{+\infty} \|\Phi_A(s,0)x\|ds - \int_0^h \|\Phi_A(s,0)x\|ds\right)\\
&\le \sup_{A\in\mathcal{S}_{\mathrm{arb}}(\mathcal{M})} \int_0^{+\infty} \|\Phi_A(s,0)x\|ds - \inf_{A\in\mathcal{S}_{\mathrm{arb}}(\mathcal{M})} \int_0^h \|\Phi_A(s,0)x\|ds\\
&= V_0(x) - \inf_{A\in\mathcal{S}_{\mathrm{arb}}(\mathcal{M})} \int_0^h \|\Phi_A(s,0)x\|ds\\
&\le V_0(x) - \|x\| \int_0^h e^{-s\kappa} ds,
\end{aligned}
$$

where the last inequality follows from Equation (1.1) and $\kappa = \sup_{M\in\mathcal{M}} \|M\| < \infty$ since $\mathcal{M}$ is bounded. Hence,

$$
\limsup_{h\to 0} \frac{V_0(\Phi_B(h,0)x) - V_0(x)}{h} \le -\|x\|. \tag{3.10}
$$

Let us now mollify V_0 in order to make it smooth outside the origin. For $x \in \mathbb{R}^n$, define

$$
V_1(x) = \int_{\mathrm{SO}(n)} f(R)\, V_0(Rx)\, dm(R),
$$

where $m(\cdot)$ denotes the Haar measure on the compact group $\mathrm{SO}(n)$ (cf. Theorem A.21) and $f\colon \mathrm{SO}(n) \to [0,+\infty)$ is a smooth function with support in a small neighborhood of the identity matrix and such that

$$
\int_{\mathrm{SO}(n)} f(R)\, dm(R) = 1. \tag{3.11}
$$

Notice that V_1 is absolutely homogeneous, convex, and positive outside the origin. Let us prove that V_1 is smooth outside the origin. Fix $x \in \mathbb{R}^n \setminus \{0\}$. Define $W(\nu, U) = V_1(\nu U x)$ for $\nu > 0$ and $U \in \mathrm{SO}(n)$. By the change of variable $R \mapsto RU$, one has

$$
W(\nu, U) = \nu \int_{\mathrm{SO}(n)} f(RU^\top) V_0(Rx)\, dm(R).
$$

It is immediate to see that W is smooth. Since, moreover, the differential of $(\nu, U) \mapsto \nu U x$ at $(1, \mathrm{Id}_n)$ is surjective, one deduces by the local inversion theorem that V_1 is smooth at x.

We are left to prove that V_1 is decreasing along nontrivial trajectories. Pick $x \neq 0$ and notice that

$$V_1(\Phi_B(h,0)x) - V_1(x) = \int_{\mathrm{SO}(n)} f(R)(V_0(R\Phi_B(h,0)x) - V_0(Rx))\, dm(R)$$

and

$$\begin{aligned}\frac{V_0(R\Phi_B(h,0)x) - V_0(Rx)}{h} &= \frac{V_0(R\Phi_B(h,0)x) - V_0(\Phi_B(h,0)Rx)}{h} \\ &\quad + \frac{V_0(\Phi_B(h,0)Rx) - V_0(Rx)}{h}.\end{aligned}$$

Since V_0 is a norm, one gets that

$$\begin{aligned}\left|\frac{V_0(R\Phi_B(h,0)x) - V_0(\Phi_B(h,0)Rx)}{h}\right| &\leq \frac{V_0(R\Phi_B(h,0)x - \Phi_B(h,0)Rx)}{h} \\ &\leq V_0\left((R - \mathrm{Id}_n)\frac{\Phi_B(h,0) - \mathrm{Id}_n}{h}x\right) \\ &\quad + V_0\left(\frac{\Phi_B(h,0) - \mathrm{Id}_n}{h}(R - \mathrm{Id}_n)x\right) \\ &\leq C\|R - \mathrm{Id}_n\|\,\|x\|,\end{aligned}$$

where C does not depend on x, R, and h small enough. Taking the support of f small enough, using (3.10) and the normalization (3.11) one deduces that

$$\limsup_{h\to 0} \frac{V_1(\Phi_B(h,0)x) - V_1(x)}{h} \leq -\frac{\|x\|}{2}.$$

The conclusion follows by taking $V = 2V_1$.

The general case $\alpha \geq 1$ follows by considering $C_\alpha V^\alpha$ for a suitable positive constant C_α. □

As a corollary, it is possible to deduce the upper-semicontinuity of the Lyapunov exponent, a property which was already noticed to hold true in Remark 1.23.

Corollary 3.19 *Let $(\Sigma_{\mathcal{M}})$ be uniformly exponentially stable. Then there exists $\varepsilon > 0$ such that, if a compact set $\mathcal{M}' \subset M_n(\mathbb{R})$ satisfies $d_H(\mathrm{co}(\mathcal{M}), \mathrm{co}(\mathcal{M}')) < \varepsilon$ then $(\Sigma_{\mathcal{M}'})$ is uniformly exponentially stable. In particular, the function $\mathcal{M} \mapsto \lambda(\mathcal{M})$ defined on the set of compact subsets of $M_n(\mathbb{R})$ endowed with the Hausdorff distance is upper-semicontinuous.*

Proof According to Remark 1.21, $(\Sigma_{\mathcal{M}})$ is uniformly exponentially stable if and only if $(\Sigma_{\mathrm{co}(\mathcal{M})})$ is. By Proposition 3.18, there exists a common absolutely homogeneous Lyapunov function V, continuously differentiable outside the origin, and such that $\nabla V(x)^\top Mx \leq -1$, for every $M \in \mathrm{co}(\mathcal{M})$ and $x \in S^{n-1}$. For every $M' \in \mathcal{M}'$ there exists, by assumption, $M \in \mathrm{co}(\mathcal{M})$ such that $\|M' - M\| < \varepsilon$ and we have

$$\nabla V(x)^\top M'x = \nabla V(x)^\top Mx + \nabla V(x)^\top (M' - M)x \leq -1 + \max_{x \in S^{n-1}} \|\nabla V(x)\|\varepsilon.$$

The result follows by taking, e.g., $\varepsilon = 1/(2 \max_{x \in S^{n-1}} \|\nabla V(x)\|)$.

As regards the upper-semicontinuity, it is enough to show that for every $\delta > 0$ and $\mathcal{M} \subset M_n(\mathbb{R})$ compact, there exists a neighborhood $\mathcal{U}$ of $\mathcal{M}$ such that $\lambda(\mathcal{M}') < \lambda(\mathcal{M}) + \delta$ for every $\mathcal{M}' \in \mathcal{U}$. Let $\kappa = \lambda(\mathcal{M}) + \delta$. Then $\lambda(\mathcal{M} - \kappa \mathrm{Id}_n) = -\delta < 0$ and, hence, by the first part of the statement, there exists $\varepsilon > 0$ such that $\lambda(\mathcal{M}' - \kappa \mathrm{Id}_n) < 0$ if $d_H(\mathrm{co}(\mathcal{M}), \mathrm{co}(\mathcal{M}')) < \varepsilon$. In particular, if $d_H(\mathrm{co}(\mathcal{M}), \mathrm{co}(\mathcal{M}')) < \varepsilon$ then one deduces that $\lambda(\mathcal{M}') < \kappa = \lambda(\mathcal{M}) + \delta$. To conclude the proof, we claim that $d_H(\mathcal{M}, \mathcal{M}') \geq d_H(\mathrm{co}(\mathcal{M}), \mathrm{co}(\mathcal{M}'))$. Indeed, notice that

$$\max_{M \in \mathcal{M}} d(M, \mathcal{M}') \geq \max_{M \in \mathcal{M}} d(M, \mathrm{co}(\mathcal{M}')) = \max_{M \in \mathrm{co}(\mathcal{M})} d(M, \mathrm{co}(\mathcal{M}')).$$

The inequality is immediate since $\mathcal{M}' \subset \mathrm{co}(\mathcal{M}')$, while the equality follows from the convexity of the function $M \mapsto d(M, \mathrm{co}(\mathcal{M}'))$. Similarly

$$\max_{M' \in \mathcal{M}'} d(M', \mathcal{M}) \geq \max_{M' \in \mathrm{co}(\mathcal{M}')} d(M', \mathrm{co}(\mathcal{M})).$$

The claim follows at once from the definition of the Hausdorff distance. □

Given a switched linear system $(\Sigma_{\mathcal{M}})$, the family $\mathcal{P}$ identified by Proposition 3.18 is too broad to admit a tractable parameterization, suitable for investigating numerically the existence of a common Lyapunov function. In order to identify other universal classes of Lyapunov functions, let us provide the following result, expressing the fact that density in the class of convex absolutely homogeneous functions (for the topology of uniform convergence on compact sets) is a sufficient condition for universality. The proof of this result exploits a semicontinuity property of the subdifferential that is specific to convex functions.

Proposition 3.20 *Let $\mathcal{P}$ be a subset of the family of convex absolutely homogeneous functions from $\mathbb{R}^n$ to $\mathbb{R}_+$. Assume that for every convex absolutely homogeneous function $V : \mathbb{R}^n \to \mathbb{R}_+$ and every $\delta > 0$ there exists a function W in $\mathcal{P}$ such that $|V(x) - W(x)| \leq \delta$ for every $x \in S^{n-1}$. Then $\mathcal{P}$ is a universal class of Lyapunov functions.*

Proof Let $(\Sigma_{\mathcal{M}})$ be uniformly exponentially stable. Let $V \in C^0(\mathbb{R}^n, \mathbb{R}_+) \cap C^\infty(\mathbb{R}^n \setminus \{0\}, \mathbb{R}_+)$ be the absolutely homogeneous common Lyapunov function provided by Proposition 3.18. In order to prove the proposition, it is enough to show that any convex absolutely homogeneous function close enough to V on S^{n-1} in uniform norm is itself a common Lyapunov function for $(\Sigma_{\mathcal{M}})$. First notice that any such function close enough to V is itself positive-definite.

Then, proceeding by contradiction, let us assume that there exists a sequence of convex absolutely homogeneous functions $(W_k)_{k\in\mathbb{N}}$ converging uniformly to V on S^{n-1} as k goes to infinity and such that each W_k is not strictly decreasing along at least one trajectory of the system. In particular the derivative of W_k along such a trajectory is nonnegative on a set of times of positive measure. By Lemma 3.5 and absolute homogeneity of W_k, we deduce that there exist $x_k \in S^{n-1}$ and $M_k \in \mathcal{M}$ such that $\ell_k^\top M_k x_k \geq 0$ for every $\ell_k \in \partial W_k(x_k)$. By compactness, we may assume that x_k tends to $\bar{x} \in S^{n-1}$ as k goes to infinity. Then, by Item 4 of Proposition A.15, ℓ_k converges to $\nabla V(\bar{x})$. As

$$|\nabla V(\bar{x})^\top M_k \bar{x} - \ell_k^\top M_k x_k| \leq |(\nabla V(\bar{x}) - \ell_k)^\top M_k \bar{x}| + |\ell_k^\top M_k(\bar{x} - x_k)|$$

and $\mathcal{M}$ is bounded, we obtain that

$$\limsup_{k\to\infty} \nabla V(\bar{x})^\top M_k \bar{x} = \limsup_{k\to\infty} \ell_k^\top M_k x_k \geq 0,$$

contradicting the fact that $\nabla V(\bar{x})^\top M_k \bar{x} \leq -1$.

□

Remark 3.21 Based on the general fact that a class $\mathcal{P}$ is universal if and only if the class $\cup_{\eta>0} \eta\mathcal{P}$ is universal, one can extend Proposition 3.20 by weakening the approximation assumption on $\mathcal{P}$ as follows: for every convex absolutely homogeneous function $V\colon \mathbb{R}^n \to \mathbb{R}_+$, there exist $\eta > 0$ and $W \in \mathcal{P}$ such that $|V(x) - \eta W(x)| < 1$ for every $x \in S^{n-1}$. Indeed, if $\eta > 0$ and $W_\delta \in \mathcal{P}$ are such that $|\eta W_\delta(x) - \frac{1}{\delta} V(x)| < 1$ for $x \in S^{n-1}$ then $W := \eta\delta W_\delta$ satisfies $|V(x) - W(x)| \leq \delta$ for $x \in S^{n-1}$.

As an application of the previous result, two notable examples of universal classes of Lyapunov functions are provided below.

Corollary 3.22 *The family of polyhedral functions*

$$\left\{ \max_{k=1,\dots,N} |\ell_k^\top x| \mid N \in \mathbb{N},\ \ell_1, \dots, \ell_N \in \mathbb{R}^n \right\}$$

and that of homogeneous sums of squares

$$\left\{ \sum_{k=1}^{N} (\ell_k^\top x)^{2d} \mid d, N \in \mathbb{N},\ \ell_1, \dots, \ell_N \in \mathbb{R}^n \right\}$$

are universal classes of Lyapunov functions.

Proof Let V be a convex absolutely homogeneous function. Let $(x_i)_{i\in\mathbb{N}}$ be a dense sequence in S^{n-1}, and $\ell_i \in \partial V(x_i)$. We consider the increasing sequence of absolutely homogeneous functions defined by

$$W_i(x) = \max_{j=1,\ldots,i} |\ell_j^\top x|, \qquad \forall x \in \mathbb{R}^n.$$

Observe that each W_i is convex and $W_i(x) \le V(x)$ for every $x \in \mathbb{R}^n$. Indeed

$$\begin{aligned}
|\ell_j^\top x| &= \max\{\ell_j^\top x, \ell_j^\top(-x)\} \\
&= \max\{\ell_j^\top(x - x_j), \ell_j^\top(-x - x_j)\} + \ell_j^\top x_j \\
&\le V(x) - V(x_j) + \ell_j^\top x_j \\
&= V(x),
\end{aligned}$$

for every positive integer j, by definition of subgradient and since $V(x) = V(-x)$, and $V(x_j) = \ell_j^\top x_j$. We deduce that $W_i(x_k) = V(x_k)$ for every $k \in \mathbb{N}$ and $i \ge k$, hence $\lim_{i\to\infty} W_i(x_k) = V(x_k)$ and we can apply Item 7 of Proposition A.15 to conclude that the sequence of functions W_i converges to V uniformly on S^{n-1}. By applying Proposition 3.20 we get that the family of polyhedral functions is a universal class of Lyapunov functions.

For every positive integer i, let us now consider the absolutely homogeneous functions defined by

$$Z_i(x) = \left(\sum_{j=1}^{i} |\ell_j^\top x|^{2i}\right)^{\frac{1}{2i}}, \qquad \forall x \in \mathbb{R}^n.$$

The function Z_i is convex since it is the composition of the $2i$-norm on $\mathbb{R}^i$, i.e., $\|y\|_{2i} = \left(\sum_{j=1}^{i} y_j^{2i}\right)^{\frac{1}{2i}}$ for $y \in \mathbb{R}^i$, with the linear function from $\mathbb{R}^n$ to $\mathbb{R}^i$ mapping x to $(\ell_1^\top x, \ldots, \ell_i^\top x)^\top$. Moreover it is immediate to see that $W_i(x) \le Z_i(x) \le i^{\frac{1}{2i}} W_i(x)$, and since

$$\begin{aligned}
|Z_i(x) - V(x)| &\le |Z_i(x) - W_i(x)| + |W_i(x) - V(x)| \\
&\le (i^{\frac{1}{2i}} - 1) W_i(x) + |W_i(x) - V(x)|
\end{aligned}$$

we have that Z_i tends to V uniformly on S^{n-1} as i goes to infinity. By applying again Proposition 3.20, it follows that the family of homogeneous sums of squares is a universal class of Lyapunov functions. □

Remark 3.23 According to Remark 3.2, the first part of Corollary 3.22 remains valid if one replaces the piecewise linear functions $\max_{k=1,\ldots,N} |\ell_k^\top x|$ with the functions $\left(\max_{k=1,\dots,N}|\ell_k^\top x|\right)^q = \max_{k=1,\dots,N}|\ell_k^\top x|^q$, for any given $q > 1$. In particular, for $q = 2$, we have that the piecewise quadratic functions form a universal class of Lyapunov functions.

Next, we provide some restrictions on the classes of functions which may be universal. For this purpose, we will make use of the following technical result.

Lemma 3.24 *Let $\mathcal{M}_1, \mathcal{M}_2$ be bounded subsets of $M_n(\mathbb{R})$ and assume that $(\Sigma_{\mathcal{M}_1})$ is uniformly bounded. For $\nu > 0$, denote by $\mathcal{M}_2^\nu$ the set of matrices of the form $M - \nu\mathrm{Id}_n$ for $M \in \mathcal{M}_2$ and set $\mathcal{M} = \mathcal{M}_1 \cup \mathcal{M}_2^\nu$. Then, the switched system $(\Sigma_{\mathcal{M}})$ is uniformly bounded for $\nu > 0$ large enough.*

Proof Proposition 3.16 guarantees the existence of a convex nonstrict common Lyapunov function V for $(\Sigma_{\mathcal{M}_1})$, absolutely homogeneous. Intuitively speaking, the lemma follows from the fact that, for ν large enough, the vectors Mx, with $M \in \mathcal{M}_2^\nu$ and $x \in \mathbb{R}^n$, point towards the interior of the sublevel set $V^{-1}([0, V(x)])$. Let us formalize this idea. Since $\ell^\top x = V(x)$ whenever $\ell \in \partial V(x)$, by boundedness of $\mathcal{M}_2$ and of $\cup_{x\in V^{-1}(1)} \partial V(x)$ (Item 6 of Proposition A.15), for $\nu > 0$ large enough one has $\ell^\top(M - \nu\mathrm{Id}_n)x = \ell^\top Mx - \nu\ell^\top x = \ell^\top Mx - \nu < 0$ for any $M \in \mathcal{M}_2$, $x \in V^{-1}(1)$, and $\ell \in \partial V(x)$. The result is then an immediate consequence of Proposition 3.6. □

Theorem 3.25 *Let $n \geq 2$ and $\mathcal{P}$ be a compact subset of the space of continuous functions from $\mathbb{R}^n$ to $\mathbb{R}$ that are analytic on $\mathbb{R}^n \setminus \{0\}$, endowed with the topology of uniform convergence on bounded sets. Assume that $\mathcal{P}$ does not contain the zero function. Then $\mathcal{P}$ cannot be a universal class of Lyapunov functions.*

Proof We start by showing the theorem in the case $n = 2$. We proceed by contradiction, hence assuming that every uniformly exponentially stable switched system in the case where $\mathcal{M}$ consists of two matrices in $M_2(\mathbb{R})$ admits a common Lyapunov function in $\mathcal{P}$. We consider a switched system corresponding to $\mathcal{M}^0 = \{M_1^0, M_2^0\} \subset M_2(\mathbb{R})$, where M_1^0, M_2^0 are Hurwitz, the corresponding trajectories rotate clockwise around the origin, the system is uniformly bounded, but not attractive, and starting from every initial nonzero condition there exists a unique periodic trajectory, with four switches per period. The existence of such a system is obtained in Theorem 2.1, where it corresponds to the case $(\mathbf{S}_4)$, $\mathcal{R} = 1$. In particular, there exist $t_1, t_2 > 0$ such that $e^{t_1 M_1^0} e^{t_2 M_2^0}$ has an eigenvalue equal to -1, corresponding to an eigenvector x_0. Set $T = t_1 + t_2$ and consider the switched systems associated with $\mathcal{M}^\varepsilon = \{M_1^\varepsilon, M_2^\varepsilon\} \subset M_2(\mathbb{R})$, where $M_i^\varepsilon = M_i^0 - \varepsilon\mathrm{Id}_2$ for $i = 1, 2$. For $\varepsilon \geq 0$ we consider the T-periodic switching sequence $A^\varepsilon(\cdot)$ which takes values M_1^ε for $t \in [0, t_1)$ and M_2^ε for $t \in [t_1, T)$.

Since $\lambda(\mathcal{M}^\varepsilon) = -\varepsilon$, system $(\Sigma_{\mathcal{M}^\varepsilon})$ is uniformly exponentially stable for $\varepsilon > 0$. Hence, by assumption, it admits a common Lyapunov function $V_\varepsilon(\cdot)$ in $\mathcal{P}$. Since the latter is compact, there exists a sequence $(\varepsilon_k)_{k\in\mathbb{N}}$ converging to zero such that $(V_{\varepsilon_k})_{k\in\mathbb{N}}$ converges to some $\bar{V} \in \mathcal{P}$. Moreover, for every $t \geq 0$,

$$\bar{V}(\Phi_{A^0}(t, 0)x_0) = \lim_{k\to\infty} V_{\varepsilon_k}(\Phi_{A^{\varepsilon_k}}(t, 0)x_0) \leq \lim_{k\to\infty} V_{\varepsilon_k}(x_0) = \bar{V}(x_0).$$

Since $\bar{V}(\Phi_{A^0}(2T,0)x_0) = \bar{V}(x_0)$ we deduce that $\bar{V}$ is constant along the trajectory $\Phi_{A^0}(\cdot,0)x_0$. The function $t \mapsto \bar{V}(e^{tM_1^0}x_0)$ is analytic for $x_0 \neq 0$, being the composition of analytic functions, and it is constantly equal to $\bar{V}(x_0)$ for $t \in [0,t_1]$. By analyticity, $t \mapsto \bar{V}(e^{tM_1^0}x_0)$ is constant for all $t > 0$ and therefore it must be identically equal to zero since $\bar{V}(\lim_{t\to\infty} e^{tM_1^0}x_0) = \bar{V}(0) = 0$. Since every nonzero point of $\mathbb{R}^2$ may be written as $\mu e^{tM_1^0}x_0$ for some positive μ and t, we deduce that $\bar{V}$ must be identically zero, contradicting the assumptions on $\mathcal{P}$.

We are left to prove the result for $n > 2$. For this purpose we consider $\mathcal{M}_1 = \{\bar{M}_1^0, \bar{M}_2^0\}$ with

$$\bar{M}_i^0 = \begin{pmatrix} M_i^0 & 0 \\ 0 & -\mathrm{Id}_{n-2} \end{pmatrix},$$

where the matrices M_i^0 are defined as above. Let $\lambda > 0$ and $\mathcal{M}_2^\lambda$ be given by Lemma 3.24 with $\mathcal{M}_2 = \{M \in \mathfrak{so}(n) \mid \|M\| \leq 1\}$, where $\mathfrak{so}(n)$ denotes the space of skew-symmetric $n \times n$ matrices. Define $\bar{\mathcal{M}}^0 = \mathcal{M}_1 \cup \mathcal{M}_2^\lambda$ and, for $\varepsilon > 0$, consider the switched system corresponding to $\bar{\mathcal{M}}^\varepsilon = \bar{\mathcal{M}}^0 - \varepsilon \mathrm{Id}_n$. It is clear that $(\Sigma_{\bar{\mathcal{M}}^\varepsilon})$ is uniformly exponentially stable for every $\varepsilon > 0$.

Letting $\Pi_{1,2}$ be the (x_1,x_2) plane, i.e., $\Pi_{1,2} = \{x \in \mathbb{R}^n \mid x_3 = \cdots = x_n = 0\}$, we notice that, starting from every $\bar{x} \in \Pi_{1,2}$, there exists a periodic trajectory of $(\Sigma_{\bar{\mathcal{M}}^0})$ lying on $\Pi_{1,2}$. The restrictions of functions in $\mathcal{P}$ to $\Pi_{1,2}$ form a compact set of analytic functions on this plane. As in the case $n = 2$, we prove by contradiction that $\mathcal{P}$ is not universal. Assume that there exists a sequence $(V_{\varepsilon_k}(\cdot))_{k\in\mathbb{N}}$ of Lyapunov functions in $\mathcal{P}$ for $(\Sigma_{\bar{\mathcal{M}}^{\varepsilon_k}})$ converging to $\bar{V} \in \mathcal{P}$. We can show as before that $\bar{V}$ is equal to zero on $\Pi_{1,2}$. Because of the choice of $\mathcal{M}_2$ and by construction of $\bar{\mathcal{M}}^0$, every one-dimensional linear subspace of $\mathbb{R}^n$ may be reached in finite time from $\Pi_{1,2}$ via a trajectory of $(\Sigma_{\bar{\mathcal{M}}^0})$. Since $\bar{V}$ is non-increasing along such a trajectory, we deduce that $\bar{V} \equiv 0$ on $\mathbb{R}^n$, obtaining a contradiction. □

Remark 3.26 The assumption that the zero function is not in $\mathcal{P}$ cannot be removed from the hypotheses of Theorem 3.25. Indeed, consider the subset of polynomial functions made of the zero polynomial and, for every $N \geq 0$, the polynomials of degree N whose maximum of the absolute value of the coefficients is upper bounded by a positive constant c_N, where the latter is chosen in such a way that the maximum on the ball $B(0,N)$ of the polynomial is less than or equal to $1/(N+1)$. Since the class $\mathcal{P}$ contains a multiple of any polynomial, it is universal by Corollary 3.22. It is also compact since, given a sequence in $\mathcal{P}$, it has a subsequence with either degree going to infinity or constant degree. In the former case, the subsequence converges to zero for the topology of uniform convergence on bounded sets, while in the latter one the coefficients are uniformly bounded and hence the sequence admits a further converging subsequence.

As a consequence of Theorem 3.25 we obtain the following corollary which provides, in particular, a partial counterpart to Corollary 3.22 for homogeneous polynomial functions. Namely, we recover that, if we impose a uniform bound on the degree, such functions do not form a universal class of Lyapunov functions.

Corollary 3.27 *Let $n \geq 2$ and $\mathcal{P}$ be a finite-dimensional vector subspace of the space of continuous functions from $\mathbb{R}^n$ to $\mathbb{R}$ that are analytic on $\mathbb{R}^n \setminus \{0\}$. Then $\mathcal{P}$ is not universal. In particular, for every positive integer m, the set of polynomial functions of degree at most m from $\mathbb{R}^n$ to $\mathbb{R}$ is not a universal class of Lyapunov functions.*

Proof Let $\{f_1, \ldots, f_N\}$ be a basis of $\mathcal{P}$. The linear map $\varphi\colon \mathbb{R}^N \to \mathcal{P}$ defined as $\varphi(\alpha) = \sum_{i=1}^N \alpha_i f_i$ is continuous when $\mathcal{P}$ is endowed with the topology of uniform convergence on compact sets. Indeed, on every compact set $K \subset \mathbb{R}^n$ and for every $\alpha^1, \alpha^2 \in \mathbb{R}^N$, one has

$$\|\varphi(\alpha^1)|_K - \varphi(\alpha^2)|_K\|_\infty \leq \|\alpha^1 - \alpha^2\| \max_{i=1,\ldots,N} \|f_i|_K\|_\infty.$$

Hence the image of the unit sphere via the map φ is a compact set $\bar{\mathcal{P}} \subset \mathcal{P}$. Furthermore, $\bar{\mathcal{P}}$ does not contain the zero function because $\{f_1, \ldots, f_N\}$ is a linearly independent subset of $\mathcal{P}$. Applying Theorem 3.25 we obtain that $\bar{\mathcal{P}}$ is not universal. As each element of $\mathcal{P}$ is a scalar multiple of an element of $\bar{\mathcal{P}}$ we deduce that $\mathcal{P}$ is not universal either, concluding the proof of the first part of the corollary. Concerning the second part, it is enough to observe that the set of polynomial functions of degree at most m from $\mathbb{R}^n$ to $\mathbb{R}$ is a finite-dimensional vector space of analytic functions.

□

Example 3.28 Let $n = 2$, fix a positive integer m, and consider the vector space $\mathcal{P}$ of functions from $\mathbb{R}^2$ to $\mathbb{R}$ defined, in polar coordinates, by

$$(r, \theta) \mapsto r a_0 + r \sum_{i=1}^{M} (a_i \cos(2i\theta) + b_i \sin(2i\theta)),$$

where $a_0, \ldots, a_m, b_1, \ldots, b_m$ are in $\mathbb{R}$. Each function in $\mathcal{P}$ is analytic outside the origin of $\mathbb{R}^2$ and absolutely homogeneous. By Corollary 3.27 the class $\mathcal{P}$ is not universal.

Non-universality can be further proved for families of functions involving maxima and minima within a finite collection of polynomials, such as the class of polyhedral functions V of the form

$$\begin{aligned} V(x) &= \max\{|\ell_1^\top x|, \ldots, |\ell_N^\top x|\} \\ &= \max\{\ell_1^\top x, \ldots, \ell_N^\top x, -\ell_1^\top x, \ldots, -\ell_N^\top x\}, \end{aligned}$$

with $\ell_1, \ldots, \ell_N \in \mathbb{R}^n$, where N is fixed. This partial counterpart to Corollary 3.22 is a consequence of the following more general result.

Proposition 3.29 *Let $n \geq 2$ and $\mathcal{P}_d^n$ be the family of polynomial functions in $\mathbb{R}^n$ of degree at most d and l be a positive integer. Consider the family*

$$\mathcal{P}_{d,l}^n = \{V \in C(\mathbb{R}^n, \mathbb{R}) \mid \exists V_1, \ldots, V_l \in \mathcal{P}_d^n \text{ s.t. } V(x) \in \{V_1(x), \ldots, V_l(x)\},\ \forall x \in \mathbb{R}^n\}.$$

Then, $\mathcal{P}_{d,l}^n$ is not universal.

Proof We first claim that if $\mathcal{P}_{d,l}^n$ is universal, the same is true for $\mathcal{P}_{d,l}^2$. Indeed, for every $\mathcal{M} \subset M_2(\mathbb{R})$ such that $(\Sigma_{\mathcal{M}})$ is uniformly exponentially stable, consider $\hat{\mathcal{M}} \subset M_n(\mathbb{R})$ given by

$$\hat{\mathcal{M}} = \left\{ \begin{pmatrix} M & 0 \\ 0 & -\mathrm{Id}_{n-2} \end{pmatrix} \mid M \in \mathcal{M} \right\}.$$

If $\hat{V} \in \mathcal{P}_{d,l}^n$ is a common Lyapunov function for $(\Sigma_{\hat{\mathcal{M}}})$, then $V : \mathbb{R}^2 \ni x \mapsto \hat{V}(x, 0)$ is a common Lyapunov function for $(\Sigma_{\mathcal{M}})$ and $V \in \mathcal{P}_{d,l}^2$.

We are left to prove that $\mathcal{P}_{d,l}^2$ is not universal. Consider the switched systems $(\Sigma_{\mathcal{M}^\varepsilon})$ introduced in the proof of Theorem 3.25, which are uniformly exponentially stable for $\varepsilon > 0$, and only uniformly bounded for $\varepsilon = 0$. Assume by contradiction that $\mathcal{P}_{d,l}^2$ is universal and, in particular, that for every $\varepsilon > 0$ there exists a common Lyapunov function $V^\varepsilon \in \mathcal{P}_{d,l}^2$ for $(\Sigma_{\mathcal{M}^\varepsilon})$. By definition of $\mathcal{P}_{d,l}^2$, for every $\varepsilon > 0$ there exist l polynomials $P_1^\varepsilon, \ldots, P_l^\varepsilon$ of degree at most d such that $V^\varepsilon(x) \in \{P_1^\varepsilon(x), \ldots, P_l^\varepsilon(x)\}$. Given $j, k \in \{1, \ldots, l\}$, we investigate the set of zeroes of the polynomial Q_{jk}^ε defined as the homogeneous polynomial corresponding to the terms of maximal degree of $P_j^\varepsilon - P_k^\varepsilon$. For this purpose, recall that every homogeneous polynomial Q of positive degree m may be factorized as $Q = \prod_{k=1}^m (\alpha_k x_1 + \beta_k x_2)$, where $\alpha_k, \beta_k \in \mathbb{C}$ for $k = 1, \ldots, m$, so that its zeroes correspond to the intersection of the unit circle S^1 with at most m lines through the origin. Hence, it follows that either $Q_{jk}^\varepsilon \equiv 0$ (i.e., $P_j^\varepsilon \equiv P_k^\varepsilon$) or Q_{jk}^ε vanishes at most $2d$ times on the unit circle. Moreover, for every $\varepsilon > 0$, the integer $N = 2d\binom{l-1}{2} + 1$ is a strict upper bound for the total number of zeroes of Q_{jk}^ε for $j, k \in \{1, \ldots, l\}$. Partitioning the circle into N arcs $C_1, \ldots, C_N$ of equal length, for every $\varepsilon > 0$ there exists an arc C_{n_ε} which contains no zero of the nontrivial polynomials Q_{jk}^ε in its interior. Denote by $\mathcal{A}_{n_\varepsilon}$ the closed middle third of C_{n_ε}.

We next claim that for every $\varepsilon > 0$ there exists $\nu_\varepsilon > 0$ large enough such that the restriction of V^ε to the dilated arc $\nu_\varepsilon \mathcal{A}_{n_\varepsilon}$ coincides with the restriction to the same arc of one of the polynomials $P_1^\varepsilon, \ldots, P_l^\varepsilon$. By definition of the function V^ε and taking into account its continuity, it is enough to prove that, for every $\varepsilon > 0$ there exists $\nu_\varepsilon > 0$ large enough such that, in the interior of the arc $\nu_\varepsilon \mathcal{A}_{n_\varepsilon}$, one has for every $j, k \in \{1, \ldots, l\}$ that either $P_j^\varepsilon \equiv P_k^\varepsilon$ or $P_j^\varepsilon - P_k^\varepsilon$ is never vanishing. To see that, it is enough to prove that if Q_{jk}^ε does not vanish on $\mathcal{A}_{n_\varepsilon}$ then $P_j^\varepsilon - P_k^\varepsilon$ does not vanish on $\nu_\varepsilon \mathcal{A}_{n_\varepsilon}$ for $\nu_\varepsilon > 0$ large enough independent of j, k. In that case, one has, for $\nu > 0$ large enough, $x \in S^1$ and $j, k \in \{1, \ldots, l\}$,

$$(P_j^\varepsilon - P_k^\varepsilon)(\nu x) = \nu^{d'} \left(Q_{jk}^\varepsilon(x) + o(1) \right),$$

where d' is the positive degree of Q_{jk}^ε and $o(1)$ is a function of x and ν tending to zero as ν tends to infinity uniformly with respect to $x \in S^1$ and $j, k \in \{1, \ldots, l\}$. This concludes the proof of the claim.

Since the arcs $\mathcal{A}_1, \ldots, \mathcal{A}_N$ do not depend on ε, there exist one of them, denoted by $\mathcal{A}$, and sequences $(\varepsilon_m)_{m \in \mathbb{N}}$, $(\nu_m)_{m \in \mathbb{N}}$ in $(0, +\infty)$, and $(V_m)_{m \in \mathbb{N}}$ in $\mathcal{P}_d^2$ such that $\lim_{m \to \infty} \varepsilon_m = 0$ and $V^{\varepsilon_m} = V_m$ on $\nu \mathcal{A}$ for every $m \in \mathbb{N}$ and every $\nu \geq \nu_m$. Let $\hat{V}_m \in \mathcal{P}_d^2$ be the homogeneous term of maximal degree of V_m. Notice that

$$\hat{V}_m(x) = \lim_{\nu \to +\infty} \nu^{-d_m} V_m(\nu x), \qquad \forall x \in \mathbb{R}^2,$$

where d_m denotes the degree of V_m. Up to normalizing V^{ε_m}, we may assume that the maximum of the moduli of the coefficients of the polynomial $\hat{V}_j$ is equal to one. Thus, up to extracting a subsequence, $\hat{V}_m$ converges uniformly on compact sets to some nonzero $\bar{V} \in \mathcal{P}_d^2$.

Similarly to the proof of Theorem 3.25, we can construct a periodic trajectory $t \mapsto \Phi_{A^0}(t, 0)\bar{x}$ starting at $\bar{x}$ in the interior of the arc $\mathcal{A}$, with $A^0(\cdot)$ piecewise constant taking values in $\mathcal{M}^0$. Consider the switching laws $A^\varepsilon(\cdot) = A^0(\cdot) - \varepsilon \mathrm{Id}_2$ taking values in $\mathcal{M}^\varepsilon$. For every $t \geq 0$ such that $\Phi_{A^0}(t, 0)\bar{x} \in \mathcal{A}$ and for every $m \in \mathbb{N}$, we have

$$\begin{aligned}
\hat{V}_m(\Phi_{A^{\varepsilon_m}}(t, 0)\bar{x}) &= \lim_{\nu \to +\infty} \nu^{-d_m} V_m(\nu \Phi_{A^{\varepsilon_m}}(t, 0)\bar{x}) \\
&= \lim_{\nu \to +\infty} \nu^{-d_m} V^{\varepsilon_m}(\Phi_{A^{\varepsilon_m}}(t, 0)\nu\bar{x}) \\
&\leq \lim_{\nu \to +\infty} \nu^{-d_m} V^{\varepsilon_m}(\nu\bar{x}) = \lim_{\nu \to +\infty} \nu^{-d_m} V_m(\nu\bar{x}) = \hat{V}_m(\bar{x}),
\end{aligned}$$

and therefore

$$\bar{V}(\Phi_{A^0}(t, 0)\bar{x}) = \lim_{m \to \infty} \hat{V}_m(\Phi_{A^{\varepsilon_m}}(t, 0)\bar{x}) \leq \lim_{m \to \infty} \hat{V}_m(\bar{x}) = \bar{V}(\bar{x}).$$

We then deduce that $t \mapsto \bar{V}(\Phi_{A^0}(t, 0)\bar{x}_0)$ is constant on $\{t \geq 0 \mid \Phi_{A^0}(t, 0)\bar{x}_0 \in \mathcal{A}\}$. Moreover $\Phi_{A^0}(t, 0) = e^{tM_i^0}$ for some $i = 1, 2$, for t small enough, and by repeating the argument in the proof of Theorem 3.25 we obtain that $\bar{V} \equiv 0$, yielding a contradiction. □

3.4 Notes and References

The notion of Lyapunov function dates back to the work of A.M. Lyapunov [Lya92]. Since then, it has been refined and adapted to various contexts. For instance, the extension to the case of differential inclusions is essentially straightforward (see e.g. [BR05]), and the usual concept of common Lyapunov function for a switched system may be seen as a particular case of it, since, by Filippov's Selection Lemma, the solutions of $(\Sigma_{\mathcal{M}})$ coincide with those of the differential inclusion $\dot{x} \in \mathcal{M}x$ when $\mathcal{M}$ is compact and convex (see, for instance, [AC84]). Furthermore, stability analysis via Lyapunov methods for (control) systems admitting parametric uncertainties has been an important topic in the control community since at least the seventies, see e.g. [PTS77].

In this book we introduce a general notion of common Lyapunov function V, by only assuming continuity of V and monotonicity along trajectories of the switched system $(\Sigma_{\mathcal{S}})$. Theorem 3.3 is then an adaptation of the classical second method of Lyapunov to our setting. Proposition 3.6 provides conditions characterizing the monotonicity property of convex Lyapunov functions for $(\Sigma_{\mathcal{M}})$ via the notion of subdifferential, similarly to what is done in, e.g., [BC99]. Although this chapter focuses on the case of arbitrary switching, let us mention that there exists another classical concept of Lyapunov function, namely that of *multiple Lyapunov function* introduced in [PD91] (see also [Bra98]), which is less conservative than that of common Lyapunov function when one considers restrained switching.

There is a vast literature concerning the characterization of quadratically stable switched linear system (see, e.g., [SWM+07] and references therein). It is well-known that not every exponentially stable switched system admits a quadratic common Lyapunov function [CL93, DM99]. However, it is always possible to express a common Lyapunov function belonging to the more general class of homogeneous sum of squares of polynomials as a quadratic common Lyapunov function for a system of higher dimension, cf. [CGTV09]. From a numerical point of view, this leads to less conservative linear matrix inequalities compared to the ones in Definition 3.7.

Proposition 3.8, stating that a switched system is quadratically stable whenever all modes are Hurwitz and can be put in triangular form by means of a common coordinate transformation, is classical and can be found, for instance, in [CLR97, MMK97, SN98].

Proposition 3.10, as well as a partial converse result obtained under additional assumptions, including the existence of a nonstrict quadratic common Lyapunov function, may be found in [SNM03]. The converse result applies in particular in the case $n = 2$ when only two modes are available. This case is considered in Theorem 3.14, where a series of well-known conditions equivalent to quadratic stability are presented, cf. [CL93, SN02, BBM09]. It is unknown, to the knowledge of the authors of this book, if the converse of Proposition 3.10 is true in general. In the case of positive switched linear systems with $n = 2$ and $\mathcal{M} = \{M_1, M_2\}$, it has been shown in [GSM07] that quadratic stability and exponential stability are equivalent, and correspond to the fact that every matrix in $[\![M_1, M_2^{-1}]\!]$ is Metzler–Hurwitz or, equivalently, that $M_1 M_2^{-1}$ has no negative eigenvalue. For more stability results via Lyapunov methods for positive systems, see [BCV+15].

The necessary and sufficient conditions for quadratic stability expressed in Theorem 3.11 are classical and can be found, e.g., in [BEGFB94]. Slightly different linear matrix inequality conditions may be found e.g. in [KP87]. Concerning Theorem 3.12, the equivalence of the given conditions has been shown in [SMOC04, KN06]. Proposition 3.15 is essentially a rephrasing of [GK94, Lemma 4.0], where the authors attribute the argument to W.P. Dayawansa (see also [CUB20]).

Concerning Proposition 3.16, we note that, unlike the linear case, in a nonlinear setting the sole uniform boundedness of the system does not guarantee in general the existence of a continuous nonstrict Lyapunov function. A counterexample is discussed for instance in [Kra63, p. 46]. The concept of universal class of Lyapunov function, in relation with converse Lyapunov results, has been introduced in [BM99]. The literature dealing with converse Lyapunov theorems, starting from the works by J.L. Massera and J. Kurzweil in the 1950s (see e.g. [Mas56, Kur56, Wil69, LSW96, HS11]) is quite rich. The results concerning the existence of smooth Lyapunov functions for nonlinear systems with global asymptotic stability properties require the development of rather sophisticated techniques. Concerning robust asymptotic stability with respect to a closed invariant set in presence of perturbation terms, converse Lyapunov theorems have been obtained in [LSW96]. In the context of switched systems (even in a nonlinear setting) such results establish the equivalence between the global uniform asymptotic stability and the existence of a smooth common Lyapunov function. For switched linear systems the construction of a common Lyapunov function is more direct and natural due, essentially, to the homogeneous nature of the system and the equivalence between asymptotic and exponential stability discussed in Sect. 1.3. The converse Lyapunov result stated in Proposition 3.18 can be found in a slightly different form in [Mei79]. Let us remark that converse Lyapunov results in a similar spirit than Proposition 3.18 have also been derived for linear switched systems evolving on infinite dimensional Banach spaces, see for instance [HS11]. Proposition 3.20 and the subsequent corollary formalize classical ideas for the construction of universal classes of Lyapunov functions (see, e.g. [MP86a, MP86b, BM99]). Finally, the non-existence results Theorem 3.25 and Corollary 3.27 can be found in [MCS23] and take inspiration from [MBC06, Theorem 2].

Chapter 4
Barabanov Norms

In this chapter we focus our attention on the case of unrestricted switching, i.e., on the properties of switched systems $(\Sigma_{\mathcal{M}})$, which correspond to the class $\mathcal{S}_{\mathrm{arb}}(\mathcal{M})$ of arbitrarily switching signals.

Recall that the largest exponential rate among all trajectories of $(\Sigma_{\mathcal{M}})$ is given by the quantity $\lambda(\mathcal{M})$, in such a way that the trajectories of $(\Sigma_{\mathcal{M}-\lambda(\mathcal{M})\mathrm{Id}_n})$ cannot diverge exponentially. It turns out, as we will see in this chapter, that when the trajectories of $(\Sigma_{\mathcal{M}-\lambda(\mathcal{M})\mathrm{Id}_n})$ do not diverge, it is possible to identify a norm which plays the role of nonstrict common Lyapunov function for $(\Sigma_{\mathcal{M}-\lambda(\mathcal{M})\mathrm{Id}_n})$. Such a norm is said to be *extremal* for $(\Sigma_{\mathcal{M}})$. Under suitable extra hypotheses on $\mathcal{M}$, we then prove that one can select an extremal norm for $(\Sigma_{\mathcal{M}})$, called *Barabanov norm*, with the additional property that for every possible choice of initial condition there exists a trajectory of $(\Sigma_{\mathcal{M}-\lambda(\mathcal{M})\mathrm{Id}_n})$ with constant norm, i.e., for which the norm of the corresponding trajectory of $(\Sigma_{\mathcal{M}})$ has exponential rate exactly equal to $\lambda(\mathcal{M})$. Any such trajectory of $(\Sigma_{\mathcal{M}})$ is said to be extremal.

Barabanov norms exist when $\mathcal{M}$ is irreducible and allow one to have an accurate description of the behavior of trajectories with fastest growth rate or least decay rate. Up to decomposing $\mathcal{M}$ in blocks corresponding to irreducible subsystems of $(\Sigma_{\mathcal{M}})$, Barabanov norms can be used to obtain fine asymptotic properties of any switched linear system with unrestricted switching, such as precise upper bounds on the polynomial degree of divergence of the trajectories of $(\Sigma_{\mathcal{M}-\lambda(\mathcal{M})\mathrm{Id}_n})$ in terms of so-called resonances between irreducible subsystems.

In general Barabanov norms are not unique, in the sense that a switched system can admit several Barabanov norms with different level sets. We provide conditions guaranteeing uniqueness of the Barabanov norm up to positive scalar multiplication. We also study the regularity of Barabanov norms and discuss under which conditions on $\mathcal{M}$ every Barabanov norm is strictly convex.

We conclude the chapter by providing qualitative properties of switched linear systems with unrestrained switching that can be obtained using Barabanov norms as a tool for analysis. In particular, we prove that $\lambda(\mathcal{M})$ coincides with the joint

Y. Chitour et al., *Dynamics and Stability of Continuous-Time Switched Linear Systems*, PNLDE Subseries in Control 105,
https://doi.org/10.1007/978-3-031-82982-6_4

spectral exponent associated with $\mathcal{S}_{\mathrm{arb}}(\mathcal{M})$ and that $\mathcal{M} \mapsto \lambda(\mathcal{M})$ is continuous on a suitable topological space of sets of modes.

4.1 Existence of Barabanov Norms

We start by giving the definitions of the fundamental concepts of extremal and Barabanov norms.

Definition 4.1 A norm v in $\mathbb{R}^n$ is said to be *extremal for* $\mathcal{M}$ if, for every trajectory $x(\cdot)$ of $(\Sigma_{\mathcal{M}})$, $v(x(t)) \le v(x(0))e^{\lambda(\mathcal{M})t}$, for every $t \ge 0$.

A norm v is said to be *Barabanov for* $\mathcal{M}$ if it is extremal for $\mathcal{M}$ and, in addition, for every $x_0 \in \mathbb{R}^n$, there exists a trajectory $x(\cdot)$ of $(\Sigma_{\mathcal{M}})$ starting from x_0 satisfying $v(x(t)) = v(x_0)e^{\lambda(\mathcal{M})t}$, for every $t \ge 0$. Any such trajectory with $x_0 \neq 0$ is called an *extremal trajectory for* v.

The following remarks are immediate consequences of the above definition.

Remark 4.2 A norm v is extremal for $\mathcal{M}$ if $v(e^{tM}x) \le e^{t\lambda(\mathcal{M})}v(x)$ for every $x \in \mathbb{R}^n$, $t \ge 0$, and $M \in \mathcal{M}$. In other words, a norm v is extremal if and only if the set $\{x \in \mathbb{R}^n \mid v(x) \le 1\}$ is forward invariant with respect to the flow of $M - \lambda(\mathcal{M})\mathrm{Id}_n$ for every $M \in \mathcal{M}$.

Remark 4.3 If v is an extremal (respectively, Barabanov) norm for $\mathcal{M}$, then, for every $\alpha \in \mathbb{R}$, v is also an extremal (respectively, Barabanov) norm for $\mathcal{M} + \alpha\mathrm{Id}_n$.

Observe that an extremal norm is a nonstrict common Lyapunov function for $(\Sigma_{\mathcal{M}-\lambda(\mathcal{M})\mathrm{Id}_n})$. In particular, if such a norm exists and if $\lambda(\mathcal{M}) = 0$, then $(\Sigma_{\mathcal{M}})$ is uniformly bounded.

The existence of extremal and Barabanov norms is not guaranteed even in the case where $\mathcal{M}$ is a singleton $\{M\}$. Consider for instance the case where $n = 2$ and $M = \left(\begin{smallmatrix} 0 & 1 \\ 0 & 0 \end{smallmatrix}\right)$. Then $\lambda(\mathcal{M}) = 0$ and, since $(\Sigma_{\mathcal{M}})$ has unbounded trajectories, $\mathcal{M}$ admits no extremal norm.

When $M = \left(\begin{smallmatrix} 0 & 0 \\ 0 & -1 \end{smallmatrix}\right)$, $\lambda(\mathcal{M})$ is still equal to zero and the Euclidean norm is clearly extremal, but $\mathcal{M}$ admits no Barabanov norm: indeed, the trajectory $t \mapsto (0, e^{-t})$ would otherwise be the extremal one starting from $(0, 1)$, which is not possible since it tends to zero as t goes to infinity.

An example of singleton $\mathcal{M} = \{M\}$ for which a Barabanov norm v exists is obtained by taking $M = \left(\begin{smallmatrix} 0 & 1 \\ -1 & 0 \end{smallmatrix}\right)$ and v equal to the Euclidean norm.

Remark 4.4 Assume that $\mathcal{M}$ admits an extremal norm v and denote by $\|\cdot\|_v$ the induced matrix norm. Then

$$\|\Phi_A(t+s, s)\|_v \le e^{\lambda(\mathcal{M})t}, \qquad s, t \ge 0, \quad A \in \mathcal{S}_{\mathrm{arb}}(\mathcal{M}). \tag{4.1}$$

As a consequence, the infimum appearing in Item 3 of Proposition 1.20 is actually reached, i.e.,

$$\lambda(\mathcal{M}) = \min\{\lambda \mid \exists C > 0 \text{ such that } \|\Phi_A(t+s,s)\| \leq Ce^{\lambda t}\ \forall A \in \mathcal{S}_{\text{arb}}(\mathcal{M})\ \forall t,s \geq 0\}. \tag{4.2}$$

If, moreover, v is a Barabanov norm, then for every initial condition x_0, there exists a trajectory of $(\Sigma_{\mathcal{M}-\lambda(\mathcal{M})\mathrm{Id}_n})$ lying on the *Barabanov sphere* $v^{-1}(v(x_0))$ and for every $A \in \mathcal{S}_{\text{arb}}(\mathcal{M})$ giving rise to an extremal trajectory one has

$$\|\Phi_A(t,0)\|_v = e^{\lambda(\mathcal{M})t}, \qquad t \geq 0. \tag{4.3}$$

Conversely, if $A \in \mathcal{S}_{\text{arb}}(\mathcal{M})$ satisfies (4.3), then it gives rise to an extremal trajectory. Indeed, taking $t = k \in \mathbb{N}$ in (4.3), we can select a sequence of initial conditions $x_k \in v^{-1}(1)$ such that $v(\Phi_A(k,0)x_k) = e^{\lambda(\mathcal{M})k}$ and hence $v(\Phi_A(\tau,0)x_k) = e^{\lambda(\mathcal{M})\tau}$ for every $\tau \in [0,k]$, by extremality of v. Any accumulation point of $(x_k)_{k\in\mathbb{N}}$ is therefore the initial condition for an extremal trajectory associated with the signal A, as follows by continuity of $x \mapsto \Phi_A(t,0)x$ for every fixed $t \geq 0$.

Definition 4.5 The set $\mathcal{M}$ is said to be *nondefective* if $(\Sigma_{\mathcal{M}-\lambda(\mathcal{M})\mathrm{Id}_n})$ is bounded, that is, if there exists $C > 0$ such that $\|\Phi_A(t,0)\| \leq Ce^{t\lambda(\mathcal{M})}$ for every $t \geq 0$ and every $A \in \mathcal{S}_{\text{arb}}(\mathcal{M})$. Otherwise, we say that $\mathcal{M}$ is *defective*.

Notice that $\mathcal{M}$ is nondefective if and only if the infimum in Item 3 of Proposition 1.20 can be replaced by a minimum as in (4.2). The following additional characterization holds true.

Lemma 4.6 *The set $\mathcal{M}$ admits an extremal norm if and only if it is nondefective.*

Proof Following Remark 4.4, if $\mathcal{M}$ admits an extremal norm then it is nondefective. In order to prove the converse, define

$$v(x) = \sup_{A\in\mathcal{S}_{\text{arb}}(\mathcal{M}),\ t\geq 0} e^{-t\lambda(\mathcal{M})}\|\Phi_A(t,0)x\|, \qquad x \in \mathbb{R}^n.$$

Then v is finite by the nondefectiveness assumption and positive-definite because $v(x) \geq \|x\|$ for every $x \in \mathbb{R}^n$. Let us show that v is a norm. The homogeneity is obvious and the subadditivity follows from the inequality

$$\|\Phi_A(t,0)(x+y)\| \leq \|\Phi_A(t,0)x\| + \|\Phi_A(t,0)y\|, \qquad \forall A \in \mathcal{S}_{\text{arb}}(\mathcal{M}),\ \forall t \geq 0.$$

In order to show that v is extremal, consider $B \in \mathcal{S}_{\rm arb}(\mathcal{M})$, $x_0 \in \mathbb{R}^n$, $t \geq 0$, and notice that

$$\begin{aligned} v(\Phi_B(t,0)x_0) &= \sup_{A\in\mathcal{S}_{\rm arb}(\mathcal{M}),\, s\geq 0} e^{-s\lambda(\mathcal{M})}\|\Phi_A(s,0)\Phi_B(t,0)x_0\| \\ &= e^{t\lambda(\mathcal{M})} \sup_{A\in\mathcal{S}_{\rm arb}(\mathcal{M}),\, s\geq 0} e^{-(t+s)\lambda(\mathcal{M})}\|\Phi_{B_{[0,t]}*A}(t+s,0)x_0\| \\ &\leq e^{t\lambda(\mathcal{M})} \sup_{\tilde{A}\in\mathcal{S}_{\rm arb}(\mathcal{M}),\, \tilde{s}\geq 0} e^{-\tilde{s}\lambda(\mathcal{M})}\|\Phi_{\tilde{A}}(\tilde{s},0)x_0\| \\ &= e^{t\lambda(\mathcal{M})}v(x_0). \end{aligned}$$

□

Let us associate with $\mathcal{M}$ the sets of matrices

$$\mathscr{S} = \cup_{t\geq 0}\mathscr{S}_t, \qquad \mathscr{S}_t = \{\Phi_A(t,0) \mid A \in \mathcal{S}_{\rm arb}(\mathcal{M})\}.$$

Notice that $\mathscr{S}$ is a semigroup by Eq. (1.2) and the closure under concatenation of $\mathcal{S}_{\rm arb}(\mathcal{M})$.

A preliminary result towards the existence of Barabanov norms is the following.

Lemma 4.7 *If $\mathcal{M}$ is irreducible then it is nondefective.*

Proof Up to considering $\mathcal{M} - \lambda(\mathcal{M})\mathrm{Id}_n$, we can assume without loss of generality that $\lambda(\mathcal{M}) = 0$.

Then

$$V := \{x \in \mathbb{R}^n \mid \sup_{M\in\mathscr{S}} \|Mx\| < \infty\}$$

is a subspace of $\mathbb{R}^n$ invariant with respect to every matrix in $\mathscr{S}$ and hence also with respect to those in $\mathcal{M}$ (see Lemma 1.29). We deduce from the irreducibility of $\mathcal{M}$ that either $V = \{0\}$ or $V = \mathbb{R}^n$.

Assume first that $V = \{0\}$. This means that for every $x \in \mathbb{R}^n$ there exists $t > 0$ and $M \in \mathscr{S}_t$ such that $\|Mx\| \geq 2\|x\|$. We claim that t can be taken uniformly bounded with respect to x. Indeed, assume by contradiction that there exist $(t_k)_{k\in\mathbb{N}}$ and $(x_k)_{k\in\mathbb{N}}$ such that $t_k \to \infty$ as $k \to \infty$ and $\|Mx_k\| < 2\|x_k\|$ for every $\tau \in [0, t_k]$ and $M \in \mathscr{S}_\tau$. Without loss of generality we can assume that $\|x_k\| = 1$ for every $k \in \mathbb{N}$ and that x_k converges to some unit vector x_∞ as $k \to \infty$. Notice now that, for every $M \in \mathscr{S}$, for every k large enough $\|Mx_k\| < 2\|x_k\|$. Hence $\|Mx_\infty\| \leq 2\|x_\infty\|$ for every $M \in \mathscr{S}$. This implies that x_∞ is in V, contradicting the assumption that $V = \{0\}$. Then there exists $\bar{t} > 0$ such that, for every $x \in \mathbb{R}^n$, one has $\|\Phi_A(t,0)x\| \geq 2\|x\|$ for some $t \in [0, \bar{t}]$ and $A \in \mathcal{S}_{\rm arb}(\mathcal{M})$. Since such times t are uniformly lower bounded by Proposition 1.1, one easily constructs by concatenation a signal $A_\infty \in \mathcal{S}_{\text{arb}}(\mathcal{M})$ such that

$$\limsup_{t\to\infty} \frac{\log \|\Phi_{A_\infty}(t,0)\|}{t} > 0.$$

This implies that $\lambda(\mathcal{M}) > 0$, which is a contradiction.

As a consequence, V must be equal to $\mathbb{R}^n$. In that case, we conclude the argument by contradiction. Since $\lambda(\mathcal{M}) = 0$, defectiveness implies that $\sup_{M\in\mathcal{S}} \|M\|$ is not finite. Then there would exist a sequence $(M_k)_{k\in\mathbb{N}}$ in $\mathcal{S}$ and a sequence of unit vectors $(x_k)_{k\in\mathbb{N}}$ in $\mathbb{R}^n$ so that $\|M_k x_k\|$ is equal to $\|M_k\|$ and tends to infinity as k tends to ∞. Up to a subsequence, one has that $\lim_{k\to\infty} x_k = x$, with $x \in \mathbb{R}^n$, and,

$$\|M_k x\| \geq \|M_k x_k\| - \|M_k(x_k - x)\| \geq \|M_k\|(1 - \|x_k - x\|) \geq \frac{\|M_k\|}{2},$$

where the last inequality holds true for k large enough. Hence $x \notin V$, which is a contradiction. □

We can now state the following fundamental result, which can be seen as a converse nonstrict Lyapunov result (for the system $(\Sigma_{\mathcal{M}-\lambda(\mathcal{M})\mathrm{Id}_n})$).

Theorem 4.8 *Let $\mathcal{M}$ be convex, irreducible, and compact. Then there exists a Barabanov norm for $\mathcal{M}$.*

Proof Notice that there exists a Barabanov norm for $\mathcal{M}$ if and only if there exists a Barabanov norm for $\mathcal{M} - \lambda(\mathcal{M})\mathrm{Id}_n$. Hence, we assume without loss of generality that $\lambda(\mathcal{M}) = 0$.

Define, for every $x \in \mathbb{R}^n$,

$$v(x) = \limsup_{t\to+\infty} \sup_{M\in\mathcal{S}_t} \|Mx\|. \tag{4.4}$$

By Theorem A.11 and Corollary 1.4, it follows that for every $t \geq 0$ $\mathcal{S}_t$ is compact, hence the supremum in the above equation is a maximum. Moreover, from Lemma 4.7, one gets that $\mathcal{M}$ is nondefective, implying the finiteness of v. Notice that, by the fact that $\mathcal{S}$ is a semigroup and by the definition of v,

$$v(Mx) \leq v(x), \qquad \forall M \in \mathcal{S}. \tag{4.5}$$

We claim that v is a norm. Homogeneity is obvious and subadditivity follows by noticing that

$$\begin{aligned} v(x+y) &\leq \limsup_{t\to+\infty} \max_{M\in\mathcal{S}_t} (\|Mx\| + \|My\|) \\ &\leq \limsup_{t\to+\infty} \max_{M\in\mathcal{S}_t} \|Mx\| + \limsup_{t\to\infty} \max_{M\in\mathcal{S}_t} \|My\| \end{aligned}$$

$$= v(x) + v(y). \tag{4.6}$$

Let us assume by contradiction that $v(x) = 0$ for some nonzero x. It follows from (4.5) that $v(Mx) = 0$ for every $M \in \mathcal{S}$. Since, moreover, as noticed in Lemma 1.29, the linear space generated by $\{Mx \mid M \in \mathcal{S}\}$ is equal to $\mathbb{R}^n$, then $v \equiv 0$ on $\mathbb{R}^n$. Hence $(\Sigma_{\mathcal{M}})$ is attractive and it follows from Fenichel's lemma (Lemma 1.25) that $\lambda(\mathcal{M}) < 0$, leading to a contradiction. This concludes the proof that v is a norm, which is extremal by (4.5).

We now claim that, for every $x \in \mathbb{R}^n$ and $t \geq 0$,

$$v(x) = \max_{M \in \mathcal{S}_t} v(Mx).$$

To see that, fix $x \in \mathbb{R}^n$ and $t \geq 0$ and notice that $v(x) = \lim_{k\to\infty} \|N_k M_k x\|$ where $M_k \in \mathcal{S}_t$ and $N_k \in \mathcal{S}_{t_k}$ for some increasing sequence of times t_k tending to infinity as k goes to infinity. By compactness of $\mathcal{S}_t$, the sequence of matrices $(M_k)_{k\in\mathbb{N}}$ converges (up to a subsequence) to some matrix $M \in \mathcal{S}_t$ and one has that

$$v(x) = \lim_{k\to\infty} \|N_k M x\| \leq v(Mx),$$

where the inequality follows from the definition of $v(Mx)$. One finally gets the claim by using (4.5).

Taking $t = 1$, one can associate with each $x \in \mathbb{R}^n$ a function $A_x \in L^\infty([0, 1], \mathcal{M})$ such that $v(x) = v(\Phi_{A_x}(1, 0)x)$. Concatenating a sequence of functions of this type, we end up with a signal $A \in \mathcal{S}_{\text{arb}}(\mathcal{M})$ such that $v(\Phi_A(k, 0)x) = v(x)$ for every $k \in \mathbb{N}$. We conclude the proof that v is a Barabanov norm by deducing from (4.5) that $v(\Phi_A(t, 0)x) = v(x)$ for every $t \geq 0$. □

We next provide examples where Barabanov norms are explicitly constructed via either closed formulas or level sets.

Example 4.9 Consider the matrices

$$M_1 = \begin{pmatrix} -1 & 0 \\ 0 & 0 \end{pmatrix}, \quad M_2 = \begin{pmatrix} 0 & 0 \\ 0 & -1 \end{pmatrix}, \quad M_3 = \begin{pmatrix} -\alpha & 1 \\ -1 & -\alpha \end{pmatrix}, \qquad \alpha > 0.$$

The trajectories corresponding to the switching law constantly equal to M_1 (respectively, M_2) are horizontal (respectively, vertical) and converge asymptotically to the vertical line (respectively, horizontal), while the trajectories corresponding to the switching law constantly equal to M_3 correspond to spirals converging asymptotically to the origin. For every $\beta > 0$, we consider the norm v_β on $\mathbb{R}^2$ given by $v_\beta(x) = \max\{|x_1|, \beta|x_2|\}$. Since M_1 and M_2 are singular and the Euclidean norm is a nonstrict common Lyapunov function for $(\Sigma_{\{M_1,M_2\}})$ and $(\Sigma_{\{M_1,M_2,M_3\}})$, it follows that $\lambda(\{M_1, M_2\}) = \lambda(\{M_1, M_2, M_3\}) = 0$.

Thanks to Remark 4.2, any v_β is extremal for $\{M_1, M_2\}$ and one checks that it is actually a Barabanov norm for this set. Note also that $\{M_1, M_2\}$ is neither convex

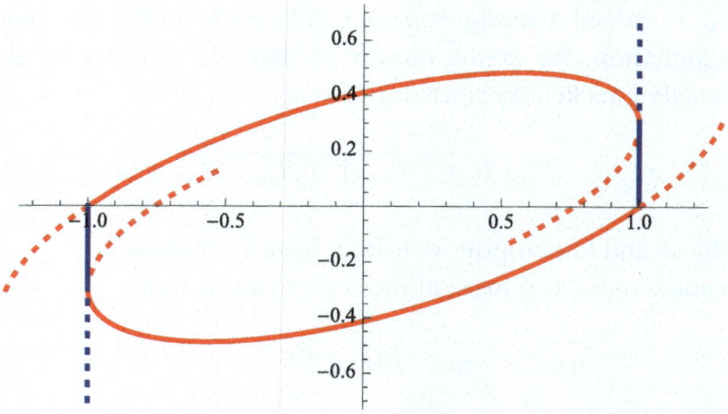

Fig. 4.1 Barabanov norm construction for the switched system of Example 4.10

nor irreducible. Moreover, one has for the same reasons that, if $\alpha \geq 1$, then v_β with $\beta \in [\frac{1}{\alpha}, \alpha]$ is a Barabanov norm for $\{M_1, M_2, M_3\}$. Moreover, the Barabanov norm for $\{M_1, M_2, M_3\}$ defined in (4.4) corresponds to the choice $\beta = 1$.

Example 4.10 Suppose that $\mathscr{M} = [\![M_1, M_2]\!]$ with

$$M_1 = \begin{pmatrix} 0 & 0 \\ 0 & -1 \end{pmatrix}, \qquad M_2 = \begin{pmatrix} \alpha & 3 \\ -0.6 & 0.7 \end{pmatrix},$$

where $\alpha \approx -0.8896$ is the supremum among all values α such that $\lambda(\mathscr{M}) = 0$. The latter condition is equivalent to the fact that the trajectory of $t \mapsto e^{tM_2}x_0$ with $x_0 = (-1, 0)^\top$ touches tangentially the line $x_1 = 1$. In this case, it is easy to see that the closed curve constructed in Fig. 4.1 by gluing together four trajectories of the system is the level set $\Gamma = v^{-1}(1)$ of some Barabanov norm v for $\mathscr{M}$.

Moreover, the extremal trajectories for v lying on Γ converge either to $(1, 0)^\top$ or to $(-1, 0)^\top$. On the other hand it is possible to construct trajectories of the system starting from Γ, turning around the origin an infinite number of times and staying arbitrarily close to Γ. One deduces that the Barabanov norm for $\mathscr{M}$ defined in (4.4) admits Γ as level set and is equal to ρv, where $\rho = \max\{\|x\| \mid x \in \Gamma\}$. Finally, note that the fact that Γ contains segments implies that v is not strictly convex.

Remark 4.11 Let us provide a useful yet different characterization of the Barabanov norm constructed in the proof of Theorem 4.8. Assume that $\lambda(\mathscr{M}) = 0$ and define the *limit semigroup* $\mathscr{S}_\infty \subset M_n(\mathbb{R})$ as

$$\mathscr{S}_\infty = \{M \mid \exists (A_k, t_k)_{k\in\mathbb{N}} \subset S_{\mathrm{arb}}(\mathscr{M}) \times [0, \infty) \text{ s.t. } t_k \to \infty \text{ and } \Phi_{A_k}(t_k, 0) \to M\}.$$

Notice that $\mathscr{S}_\infty$ is indeed a semigroup, as it follows from the fact that $\mathcal{S}_{\mathrm{arb}}(\mathcal{M})$ is closed by concatenation. We claim, moreover, that $\mathscr{S}_\infty$ is compact and nonempty. Closedness is easily checked by rewriting $\mathscr{S}_\infty$ as

$$\mathscr{S}_\infty = \cap_{s>0}\overline{\{\Phi_A(t,0) \mid A \in \mathcal{S}_{\mathrm{arb}}(\mathcal{M}),\ t > s\}},$$

while boundedness and nonemptiness follow from Lemma 4.7.

A simple compactness argument allows one to show that

$$v(x) = \max_{M \in \mathscr{S}_\infty} \|Mx\|, \qquad x \in \mathbb{R}^n, \tag{4.7}$$

where v is defined as in (4.4).

Remark 4.12 If $\mathcal{M}$ is irreducible then, by applying Theorem 4.8 to $\overline{co}(\mathcal{M})$, one deduces that there exists an extremal norm for $\mathcal{M}$.

We can strengthen Theorem 4.8 as follows.

Proposition 4.13 *Let $\mathcal{M}$ be compact and convex. Then the following three conditions are equivalent:*

(C_1) *There exists a Barabanov norm for $\mathcal{M}$;*
(C_2) *$\mathcal{M}$ is nondefective and for every proper invariant subspace V of $\mathbb{R}^n$ with respect to $\mathcal{M}$, $\lambda(\mathcal{M}|_V) = \lambda(\mathcal{M})$;*
(C_3) *$\mathcal{M}$ is nondefective and for every proper invariant subspace V of $\mathbb{R}^n$ with respect to $\mathcal{M}$ such that $\mathcal{M}|_V$ is irreducible, $\lambda(\mathcal{M}|_V) = \lambda(\mathcal{M})$.*

Proof Without loss of generality we can assume that $\lambda(\mathcal{M}) = 0$.

Assume that (C_1) holds true and let us prove (C_2). Denote by v a Barabanov norm for $\mathcal{M}$. The extremality of v implies that $(\Sigma_{\mathcal{M}})$ is stable, hence $\mathcal{M}$ is nondefective. Assume by contradiction that $\lambda(\mathcal{M}|_V) < 0$ for some proper invariant subspace V with respect to $\mathcal{M}$. Then for every nonzero initial condition x_0 in V and every trajectory $x(\cdot)$ of $(\Sigma_{\mathcal{M}})$ starting from x_0, $\lim_{t\to+\infty} x(t) = 0$. This contradicts the existence of an extremal trajectory starting from x_0.

Notice that (C_2) trivially implies (C_3) and let us prove now that (C_3) implies (C_1).

Let v be defined as in (4.4). Then v is finite by nondefectiveness of $\mathcal{M}$ and (4.5)-(4.6) still hold true. In order to show that v is positive-definite (and, hence, a norm), assume by contradiction that there exists a nonzero vector $x \in \mathbb{R}^n$ such that $v(x) = 0$. The space V spanned by all the vectors $\Phi_A(t,0)x$ for $A \in \mathcal{S}_{\mathrm{arb}}(\mathcal{M})$ and $t \geq 0$ is invariant with respect to all matrices in $\mathscr{S}$. Moreover, $v|_V$ vanishes identically by (4.5). Note that there always exists a nontrivial invariant subspace W of V such that $\mathcal{M}|_W$ is irreducible and, by assumption (C_3), we have that $\lambda(\mathcal{M}|_W) = 0$. Then, by definition and Theorem 4.8, $v|_W$ is a Barabanov norm for $\mathcal{M}|_W$ contradicting the fact that $v|_W = 0$. Hence v is a norm on $\mathbb{R}^n$. Finally, given $x \in \mathbb{R}^n$, the construction of $A \in \mathcal{S}_{\mathrm{arb}}(\mathcal{M})$ such that $v(\Phi_A(t,0)x) = v(x)$ for every $t > 0$ is done exactly as in the proof of Theorem 4.8. □

4.2 Extremal Trajectories and Properties of Barabanov Norms

In this section, we characterize extremal trajectories for a Barabanov norm v by means of the Pontryagin maximum principle. We first recall basic facts and definitions associated with v.

In the next proposition, we recall the notion of dual norm and gather its basic properties. We use here the symbol ∂ to denote the subgradient of a function, as given in (3.2).

Proposition 4.14 *Let w be a norm on $\mathbb{R}^n$ and S its corresponding unit sphere. Consider the function w^* defined on $\mathbb{R}^n$ by*

$$w^*(\ell) = \max_{y \in S} \ell^\top y. \tag{4.8}$$

Then, the following properties hold true:

(1) *The function w^* is a norm on $\mathbb{R}^n$;*
(2) *The unit sphere S^* of w^* is the polar of S, that is, $S^* = \bigcup_{x \in S} \partial w(x)$;*
(3) *For $x \in S$ and $\ell \in S^*$, we have that $\ell \in \partial w(x)$ if and only if $x \in \partial w^*(\ell)$.*

Proof We first prove Item (1). The only nontrivial fact to establish is that w^* is positive-definite. For any nonzero $\ell \in \mathbb{R}^n$, the vector $\ell / w(\ell)$ belongs to S and then $w^*(\ell) \geq \|\ell\|^2 / w(\ell) > 0$.

For Item (2), we have by definition of the subdifferential of w that the polar of S is included in S^*. Conversely, S^* is the set of vectors $\ell \in \mathbb{R}^n$ such that $\max_{y \in S} \ell^\top y = 1$. For such an ℓ, pick $y \in S$ such that $\ell^\top y = 1$. Then $\ell \in \partial w(y)$. Hence S^* belongs to the polar of S.

As regards Item (3), it simply follows from the definitions. □

Remark 4.15 It immediately follows from the previous definitions that, if w (respectively, w^*) is not differentiable at some $x \in S$ (respectively, $\ell \in S^*$) then S^* (respectively, S) is not strictly convex, since it contains a nontrivial segment included in $\partial w(x)$ (respectively, $\partial w^*(\ell)$).

We next provide the definition of the adjoint system.

Definition 4.16 The *adjoint system associated with* $(\Sigma_{\mathcal{M}})$ is the system $(\Sigma_{-\mathcal{M}^\top})$, where $-\mathcal{M}^\top = \{-M^\top \mid M \in \mathcal{M}\}$, that is, the switched linear system

$$\dot{\ell}(t) = -A^\top(t)\ell(t), \qquad A \in \mathcal{S}_{\mathrm{arb}}(\mathcal{M}). \tag{4.9}$$

Lemma 4.17 *Let v be a Barabanov norm for $\mathcal{M}$. For every solution $\ell(\cdot)$ of* (4.9) *and $t \geq 0$, one has*

$$v^*(\ell(0)) \leq e^{\lambda(\mathcal{M})t} v^*(\ell(t)).$$

Proof Let $\ell(\cdot)$ be a solution of (4.9) associated with a signal $A(\cdot)$. By definition of $v^*(\ell(0))$, there exists $z_0 \in S$ such that $v^*(\ell(0)) = \ell(0)^\top z_0$. Consider the solution $z(\cdot) = \Phi_A(\cdot, 0)z_0$. Then $\ell(t)^\top z(t) = \ell(0)^\top z_0$ for every $t \geq 0$, which implies that $v^*(\ell(0)) = \ell(t)^\top z(t) = \ell(t)^\top \big(\frac{z(t)}{v(z(t))}\big) v(z(t))$ for every $t \geq 0$. Since $v(z(t)) \leq e^{\lambda(\mathcal{M})t} v(z_0) = e^{\lambda(\mathcal{M})t}$ and $\ell(t)^\top \big(\frac{z(t)}{v(z(t))}\big) \leq v^*(\ell(t))$, we get the required inequality. □

The next results provide a characterization of extremal trajectories relying on the use of the Pontryagin maximum principle, see Theorem A.22.

Lemma 4.18 *Let v be a Barabanov norm for $\mathcal{M}$. Consider an extremal trajectory $x(\cdot)$ for v associated with $A(\cdot)$ and fix $T \geq 0$ and $\hat{\ell} \in \partial v(x(T))$. Then there exists a nonzero solution $\ell(\cdot)$ of* (4.9) *associated with $A(\cdot)$ such that*

$$e^{\lambda(\mathcal{M})t}\ell(t) \in \partial v(x(t)), \qquad \forall t \in [0, T],$$

$$\ell^\top(t)A(t)x(t) = \max_{M \in \mathcal{M}} \ell^\top(t)Mx(t) = \lambda(\mathcal{M})v(x(0)), \qquad \textit{for a.e. } t \in [0, T],$$

$$e^{\lambda(\mathcal{M})T}\ell(T) = \hat{\ell}.$$

Proof Up to replacing $\mathcal{M}$ by $\mathcal{M} - \lambda(\mathcal{M})\mathrm{Id}_n$, we assume with no loss of generality that $\lambda(\mathcal{M}) = 0$. Let $x(\cdot)$, $A(\cdot)$, T, and $\hat{\ell}$ be as in the statement of the lemma. Let $x_0 = x(0)$ and consider the optimal control problem in Mayer form

$$\max \hat{\ell}^\top z(\tau), \tag{4.10}$$

among all trajectories $z(\cdot)$ of $(\Sigma_{\mathcal{M}})$ satisfying $z(0) = x_0$ and with free final time $\tau \geq 0$. Then, the pair $(x(\cdot), A(\cdot))$ is an optimal solution of Problem (4.10) with $\tau = T$. Indeed, let $z(\cdot)$ be a solution of $(\Sigma_{\mathcal{M}})$ defined on $[0, \tau]$ such that $z(0) = x_0$. Since $\hat{\ell} \in \partial v(x(T))$, then $v(z(\tau)) - v(x(T)) \geq \hat{\ell}^\top(z(\tau) - x(T))$. Since $v(z(\tau)) \leq v(z(0)) = v(x_0) = v(x(T))$, one gets $\hat{\ell}^\top z(\tau) \leq \hat{\ell}^\top x(T)$.

Then the Pontryagin maximum principle (Theorem A.22) ensures the existence of a nonzero Lipschitz curve $\ell\colon [0, T] \to \mathbb{R}^n$ satisfying

(1) $\dot{\ell}(t) = -A^\top(t)\ell(t)$, for a.e. $t \in [0, T]$,
(2) $\ell^\top(t)A(t)x(t) = \max_{M \in \mathcal{M}} \ell^\top(t)Mx(t) = 0$, for a.e. $t \in [0, T]$,
(3) $\ell(T) = \hat{\ell}$.

To conclude the proof of Lemma 4.18, it is enough to show that $\ell(t) \in \partial v(x(t))$ for all $t \in [0, T]$. Fix $t \in [0, T]$, $\bar{x} \in \mathbb{R}^n$, and let $y(\cdot)$ be the unique solution of $\dot{y}(\tau) = A(\tau)y(\tau)$ on $[0, T]$ such that $y(t) = \bar{x}$. Notice that $\ell^\top(y - x)$ is constant on $[0, T]$. Then one has

$$\begin{aligned} v(\bar{x}) - v(x(t)) &\geq v(y(T)) - v(x(t)) = v(y(T)) - v(x(T)) \\ &\geq \ell^\top(T)(y(T) - x(T)) = \ell^\top(t)(y(t) - x(t)) = \ell^\top(t)(\bar{x} - x(t)), \end{aligned}$$

because $v(y(T)) \le v(y(t)) = v(\bar{x})$ and $\ell(T) \in \partial v(x(T))$. As $t \in [0, T]$ and $\bar{x} \in \mathbb{R}^n$ are arbitrary, one deduces that $\ell(t) \in \partial v(x(t))$ for all $t \in [0, T]$. □

A necessary condition for a trajectory to be extremal is the following.

Theorem 4.19 *Let v be a Barabanov norm for $\mathcal{M}$. Let $x(\cdot)$ be an extremal trajectory for v associated with a switching law $A(\cdot)$. Then there exists a nonzero solution $\ell(\cdot)$ of* (4.9) *associated with $A(\cdot)$ such that*

$$e^{\lambda(\mathcal{M})t}\ell(t) \in \partial v(x(t)), \qquad \forall t \ge 0, \tag{4.11}$$

$$\ell^\top(t)A(t)x(t) = \max_{M\in\mathcal{M}} \ell^\top(t)Mx(t) = \lambda(\mathcal{M})v(x(0)), \qquad \textit{for a.e. } t \ge 0. \tag{4.12}$$

Proof We assume with no loss of generality that $\lambda(\mathcal{M}) = 0$. For every $k \in \mathbb{N}$, let $\hat{\ell}_k \in \partial v(x(k))$ and $\ell_k(\cdot)$ be the solution of (4.9) associated with $A(\cdot)$ obtained by Lemma 4.18 applied with $\hat{\ell} = \hat{\ell}_k$ and $T = k$. Since $\ell_k(0) \in \partial v(x(0))$ for every $k \in \mathbb{N}$ and $\partial v(x(0))$ is compact, then up to subsequence $\ell_k(0) \to \ell_* \in \partial v(x(0))$. We then consider the solution $\ell(\cdot)$ of (4.9) associated with $A(\cdot)$ such that $\ell(0) = \ell_*$. The sequence of solutions $\ell_k(\cdot)$ converges uniformly to $\ell(\cdot)$ on every compact time interval of $[0, +\infty)$. Hence by virtue of the compactness of $\partial v(x(t))$ and the fact that $\ell_k(t) \in \partial v(x(t))$ for every $t \le k$, we get $\ell(t) \in \partial v(x(t))$.

Regarding (4.12), notice that for every $t \le k$ and every $M \in \mathcal{M}$, we have ${\ell_k}^\top(t)Mx(t) \le {\ell_k}^\top(t)A(t)x(t) = 0$. Then passing to the limit as $k \to +\infty$ we get that $\ell^\top(t)Mx(t) \le \ell^\top(t)A(t)x(t) = 0$, which concludes the proof of the theorem.

□

Definition 4.20 Given a Barabanov norm v for $\mathcal{M}$, we say that $(x(\cdot), \ell(\cdot))$ is an *extremal pair for* v if $x(\cdot)$ is an extremal trajectory for v and $\ell(\cdot)$ is as in the statement of Theorem 4.19.

Remark 4.21 Given an extremal pair $(x(\cdot), \ell(\cdot))$, the curve $\ell(\cdot)$ takes values in S^*, thanks to Item (2) of Proposition 4.14.

We next propose to describe the set of extremal trajectories as solutions of a suitable differential inclusion. Recall that an absolutely continuous function from $[0, +\infty)$ to $\mathbb{R}^n$ is a solution of $(\Sigma_{\mathcal{M}})$ if and only if it satisfies

$$\dot{x}(t) \in F_{\mathcal{M}}(x(t)) := \{Mx(t) \mid M \in \mathcal{M}\}, \qquad \text{for a.e. } t \ge 0. \tag{4.13}$$

Notice that, for every $x \in \mathbb{R}^n$, $F_{\mathcal{M}}(x)$ is nonempty, and moreover compact (respectively, convex) if $\mathcal{M}$ is.

We now turn to a similar description for extremal trajectories. Let $x \in \mathbb{R}^n$ and $M \in \mathcal{M}$. We say that M satisfies the property $\mathcal{P}(x)$ if there exists $\ell \in \partial v(x)$ such that $\ell^\top Mx = \max_{N\in\mathcal{M}} \ell^\top Nx = \lambda(\mathcal{M})v(x)$. Consider now the linear differential inclusion

$$\dot{x} \in \hat{F}_{\mathcal{M}}(x) := \{Mx \mid M \in \mathcal{M},\ M \text{ satisfies } \mathcal{P}(x)\}, \tag{4.14}$$

and we say that a solution $x(\cdot)$ of $(\Sigma_{\mathcal{M}})$ is a solution of System (4.14) if it is associated with a switching law $A(\cdot)$ such that $A(t)$ satisfies $\mathcal{P}(x(t))$ for almost every time $t \geq 0$.

Based on Theorem 4.19, the following result provides a necessary and sufficient condition for a trajectory of $(\Sigma_{\mathcal{M}})$ to be extremal.

Proposition 4.22 *The trajectories of System* (4.14) *coincide with the extremal trajectories of* $(\Sigma_{\mathcal{M}})$.

Proof By virtue of Theorem 4.19, if $x(\cdot)$ is extremal then it is a trajectory of System (4.14).

Conversely, let $x(\cdot)$ be a solution of (4.14) associated with some switching law $A(\cdot)$. We next prove that $x(\cdot)$ is extremal and we assume without loss of generality that $\lambda(\mathcal{M}) = 0$. Notice that the non-increasing map $\varphi \colon t \mapsto v(x(t))$ is absolutely continuous since v is Lipschitz continuous and $x(\cdot)$ is absolutely continuous. Therefore, in order to prove that $\varphi(\cdot)$ is constant, it is enough to show that $\varphi(\cdot)$ has zero derivative almost everywhere. Let $t_0 \in [0, +\infty)$ be a point of differentiability of both $\varphi(\cdot)$ and $x(\cdot)$ and such that $A(t_0)$ satisfies $\mathcal{P}(x(t_0))$. Then there exists $\ell_0 \in \partial v(x(t_0))$ such that $\ell_0^\top A(t_0)x(t_0) = 0$. Since $\varphi(t) - \varphi(t_0) \geq \ell_0^\top(x(t) - x(t_0))$ for every $t \in [0, +\infty)$, one deduces that $\frac{\varphi(t)-\varphi(t_0)}{t-t_0} \geq \ell_0^\top\left(\frac{x(t)-x(t_0)}{t-t_0}\right)$ for $t > t_0$. Passing to the limit as $t \to t_0^+$, we get $\dot{\varphi}(t_0) \geq \ell_0^\top \dot{x}(t_0) = \ell_0^\top A(t_0)x(t_0) = 0$ and, since φ is non-increasing, then $\dot{\varphi}(t_0) \leq 0$. Hence $\dot{\varphi}(t) = 0$ for almost every $t \in [0, +\infty)$. This proves that $x(\cdot)$ is extremal. □

4.2.1 Uniqueness of Barabanov Norms

In this section we investigate the issue of uniqueness of the Barabanov norm. As immediate remark, notice that if v is a Barabanov norm, then λv is a Barabanov norm as well for every positive λ. Therefore, uniqueness of Barabanov norms can only hold up to multiplication by a positive scalar.

Moreover, consider the case where $\mathcal{M}$ admits a Barabanov norm v and has more than one invariant component, that is, $\mathbb{R}^n$ splits as the direct sum of two proper subspaces V_1, V_2 that are both invariant with respect to $\mathcal{M}$. Then the restriction v_j of v to V_j is a Barabanov norm for $\mathcal{M}|_{V_j}$, $j = 1, 2$, and any norm of the form $\mu_1 v_1(x_1) + \mu_2 v_2(x_2)$, with $\mu_1, \mu_2 > 0$, is also a Barabanov norm for $\mathcal{M}$. Even when $\mathcal{M}$ is irreducible, the Barabanov norm is not necessarily unique, as illustrated by Example 4.9.

The characterization of the sets $\mathcal{M}$ for which the Barabanov norm is unique remains an open and hard question. The main difficulty lies in the fact that Barabanov norms are usually difficult to compute, especially for systems of dimension larger than two. In the remaining of the section, we provide some sufficient conditions for uniqueness. For that purpose, we first recall the notion of ω-limit set.

Definition 4.23 The *ω-limit set* of a continuous curve $x\colon [0,+\infty) \to \mathbb{R}^n$ is the set made of the limit points of $x(\cdot)$, i.e.,

$$\omega(x(\cdot)) := \{z \in \mathbb{R}^n \mid \exists (t_k)_{k\in\mathbb{N}} \text{ such that } t_k \to +\infty \text{ and } x(t_k) \to z \text{ as } k \to +\infty\}.$$

We next list the basic properties of ω-limit sets.

Lemma 4.24 *Let $x\colon [0,+\infty) \to \mathbb{R}^n$ be a bounded and continuous curve. Then its ω-limit set $\omega(x(\cdot))$ is a nonempty, compact, and connected subset of $\mathbb{R}^n$.*

Proof Since $x(\cdot)$ is bounded, then $\omega(x(\cdot))$ is nonempty and bounded. It is closed since

$$\omega(x(\cdot)) = \cap_{\tau\geq 0}\overline{\{x(t) \mid t \geq \tau\}}.$$

To prove that $\omega(x(\cdot))$ is connected, assume that $\omega(x(\cdot))$ is contained in the union of two nonempty open disjoint subsets U_1, U_2 of $\mathbb{R}^n$. Let F be the complement of this union inside a closed ball containing the support of $x(\cdot)$ in its interior. By definition F is compact and for $t \geq 0$ large enough, $x(t) \notin F$ (otherwise $\omega(x(\cdot))$ would intersect F). Hence $\{x(t) \mid t \geq T\} \subset U_1 \cup U_2$ for some $T > 0$, and since it is connected, it must be contained in one of the open sets U_1 or U_2. Then $\omega(x(\cdot))$ is contained either in $\bar{U}_1$ or $\bar{U}_2$. But since $\omega(x(\cdot)) \subset U_1 \cup U_2$ and $\bar{U}_j \cap (U_1 \cup U_2) = U_j$ for $j = 1, 2$, the conclusion follows. □

Lemma 4.25 *Assume that $\lambda(\mathcal{M}) = 0$ and let $x(\cdot)$ be a solution of $(\Sigma_{\mathcal{M}})$. Then, given any Barabanov norm v for $\mathcal{M}$, the ω-limit set $\omega(x(\cdot))$ of $x(\cdot)$ is contained in a level set of v. Moreover, if $\mathcal{M}$ is compact and convex, then $\omega(x(\cdot))$ is either $\{0\}$ or the union of the supports of extremal trajectories for v.*

Proof The fact that v is constant on $\omega(x(\cdot))$ simply follows from the fact that $t \mapsto v(x(t))$ is nonincreasing. Assume now that $\mathcal{M}$ is compact and convex, so that $\mathcal{S}_{\rm arb}(\mathcal{M})$ is sequentially weak-$\star$ compact (Theorem A.11). Fix now $z \in \omega(x(\cdot))\setminus\{0\}$ and a sequence of times $t_i \to +\infty$ such that $\lim_{i\to\infty} x(t_i) = z$. Let $A(\cdot)$ be the signal in $\mathcal{S}_{\rm arb}(\mathcal{M})$ associated with $x(\cdot)$. Up to a subsequence, $A(t_i + \cdot)$ weak-$\star$ converges to some $\bar{A}(\cdot)$ in $\mathcal{S}_{\rm arb}(\mathcal{M})$. Therefore, $x(t_i + \cdot)$ converges uniformly on compact time intervals to a solution $\bar{x}(\cdot)$ associated with $\bar{A}(\cdot)$ such that $\bar{x}(0) = z$ and the support of $\bar{x}(\cdot)$ is contained in $\omega(x(\cdot))$. In particular, $t \mapsto v(\bar{x}(t))$ is constant, which means that $\bar{x}$ is extremal, hence the conclusion. □

In order to study whether a Barabanov norm v for $\mathscr{M}$ is unique, we can assume that $\lambda(\mathscr{M}) = 0$, so that every extremal trajectory for v starting in the unit Barabanov sphere S for v remains in it. We can then consider the union $\Omega(v)$ of all possible ω-limit sets of extremal trajectories for v on S, that is,

$$\Omega(v) = \bigcup_{\{x(\cdot) \text{ extremal on } S\}} \omega(x(\cdot)). \tag{4.15}$$

Theorem 4.26 *Let v be a Barabanov norm for $\mathscr{M}$ and assume that $\lambda(\mathscr{M}) = 0$. Let $\hat{\Omega}$ be a dense subset of $\Omega(v)$ such that, for every $z_1, z_2 \in \hat{\Omega}$, there exist an integer $N > 0$ and N extremal trajectories $x_1(\cdot), \ldots, x_N(\cdot)$ for v with $z_1 \in \omega(x_1(\cdot))$, $z_2 \in \omega(x_N(\cdot))$ and $\omega(x_i(\cdot)) \cap \omega(x_{i+1}(\cdot)) \neq \emptyset$ for $i = 1, \ldots, N-1$. Then v is the unique Barabanov norm for $\mathscr{M}$ up to multiplication by a positive scalar.*

Proof Let w be a Barabanov norm for $\mathscr{M}$. We next prove that $w = \bar{\lambda} v$, where

$$\bar{\lambda} := \min\{\lambda > 0 \mid v^{-1}(1) \subset w^{-1}([0, \lambda])\}. \tag{4.16}$$

Standard compactness arguments show that $\bar{\lambda}$ is well defined and $v^{-1}(1) \cap w^{-1}(\bar{\lambda})$ is nonempty. Moreover, by definition of $\bar{\lambda}$ and absolute homogeneity of v, one has that

$$v(y) \geq 1 \qquad \forall y \in w^{-1}(\bar{\lambda}). \tag{4.17}$$

Consider $x_0 \in v^{-1}(1) \cap w^{-1}(\bar{\lambda})$ and an extremal trajectory for w starting at x_0, denoted by $\hat{x}(\cdot)$. By (4.17) and monotonicity of $v(\hat{x}(\cdot))$ the support of $\hat{x}(\cdot)$ must be included in the set $v^{-1}(1) \cap w^{-1}(\bar{\lambda})$. One therefore gets that $\omega(\hat{x}(\cdot)) \subset v^{-1}(1) \cap w^{-1}(\bar{\lambda})$. Moreover, by monotonicity of $w(x(\cdot))$ along any trajectory $x(\cdot)$ of the system one has that w is constant on $\omega(x(\cdot))$. Thus, under the hypotheses of the theorem, w is constant on $\Omega(v)$ and, since $\omega(\hat{x}(\cdot)) \subset \Omega(v)$, its value is equal to $\bar{\lambda}$. Now, given a point $x \in v^{-1}(1)$, one can consider an extremal trajectory for v starting from x and, since the value of w on the corresponding ω-limit set is equal to $\bar{\lambda}$, it turns out that $w(x) \geq \bar{\lambda}$. Equation (4.16) then implies that $w(x) = \bar{\lambda}$. Hence we have proved that $v^{-1}(1) = w^{-1}(\bar{\lambda})$, which concludes the proof. □

From the previous result one gets the following consequence.

Corollary 4.27 *Let v be a Barabanov norm for $\mathscr{M}$ with unit sphere S. Assume that $\lambda(\mathscr{M}) = 0$ and that there exists a finite set of extremal trajectories $x_1(\cdot), \ldots, x_N(\cdot)$ on S such that $\Omega(v)$ is equal to $\cup_{i=1}^{N} \omega(x_i(\cdot))$ and is connected. Then v is the unique Barabanov norm for $\mathscr{M}$ up to multiplication by a positive scalar.*

Proof Let us observe that for every proper subset I of $\{1, \ldots, N\}$ one has $\big(\cup_{j\in I} \omega(x_j(\cdot))\big) \cap \big(\cup_{j\notin I} \omega(x_j(\cdot))\big) \neq \emptyset$. This is a consequence of the connectedness of $\Omega(v)$ and the fact that each $\omega(x_j(\cdot))$ is closed. In particular there exists $i \notin I$ such that $\big(\cup_{j\in I} \omega(x_j(\cdot))\big) \cap \omega(x_i(\cdot)) \neq \emptyset$. Let z_1, z_2 be arbitrary points of $\Omega(v)$. By what precedes we can inductively construct a reordering $i_1, \ldots, i_N$ of $1, \ldots, N$ such that $z_1 \in \omega(x_{i_1}(\cdot))$ and $\left(\cup_{j=1,\ldots,k-1} \omega(x_{i_j}(\cdot))\right) \cap \omega(x_{i_k}(\cdot)) \neq \emptyset$ for $k = 2, \ldots, N$. In particular, either $z_2 \in \omega(x_{i_1}(\cdot))$ or there exists $k_1 > 1$ such that $z_2 \in \omega(x_{i_{k_1}}(\cdot))$. In the latter case, there exists $1 \leq k_2 < k_1$ such that $\omega(x_{i_{k_2}}(\cdot)) \cap \omega(x_{i_{k_1}}(\cdot)) \neq \emptyset$. Thus we can inductively construct a finite sequence $k_1, \ldots, k_L$ such that $k_{j+1} < k_j$, $\omega(x_{i_{k_{j+1}}}(\cdot)) \cap \omega(x_{i_{k_j}}(\cdot)) \neq \emptyset$ and $k_L = 1$, and the conclusion follows from Theorem 4.26. □

Note that in Example 4.9 the lack of uniqueness of a Barabanov norm is consistent with the fact that for every $\beta > 0$ the set $\Omega(v_\beta)$ defined as in (4.15) is disconnected, since it reduces to the union of the four points $(-1, 0)$, $(0, 1)$, $(1, 0)$, and $(0, -1)$.

Remark 4.28 The assumptions of Corollary 4.27 are verified, for instance, when the set $\Omega(v)$ is formed by a unique limit cycle (but this is not the only case, as shown in the example below).

Example 4.29 Let $\mathcal{M} = [\![M_1, M_2]\!]$ with

$$M_1 = \begin{pmatrix} 0 & 1 & 0 \\ -1 & 0 & 0 \\ 0 & 0 & -1 \end{pmatrix}, \qquad M_2 = \begin{pmatrix} 0 & 0 & 1 \\ 0 & -1 & 0 \\ -1 & 0 & 0 \end{pmatrix}.$$

The flow $t \mapsto e^{tM_1}$ corresponds to a rotation around the x_3-axis and a contraction along that axis. The same properties hold for the flow $t \mapsto e^{tM_2}$ with respect to the x_2-axis. The set of modes $\mathcal{M}$ is irreducible and $(\Sigma_{\mathcal{M}})$ is stable but not asymptotically stable (it admits $\|x\|^2$ as a nonstrict common Lyapunov function and there are periodic trajectories). Thus the Barabanov norm in (4.4) is well defined. Note that the unit Barabanov sphere S must contain the two circles

$$\{(x_1, x_2, x_3) \in \mathbb{R}^3 \mid x_1^2 + x_2^2 = 1,\ x_3 = 0\}$$
$$\text{and } \{(x_1, x_2, x_3) \in \mathbb{R}^3 \mid x_1^2 + x_3^2 = 1,\ x_2 = 0\},$$

since they are supports of periodic trajectories of $(\Sigma_{\mathcal{M}})$. Let us check that the Barabanov norm is unique up to multiplication by a positive scalar. We claim that the ω-limit set of any extremal trajectory is contained in the union of the circles defined above. Since these circles coincide with the intersection of the sphere S with the planes defined by $x_3 = 0$ and $x_2 = 0$ it is enough to show that $\min\{|x_2(t)|, |x_3(t)|\}$ converges to zero as t tends to infinity. Indeed, let $V(x) := \|x\|^2$. Then a simple computation leads to

$$\frac{d}{dt} V(x(t)) \leq -2 \min\{x_2(t)^2, x_3(t)^2\},$$

and, since V is positive-definite, then $F(t) := \int_0^t \min\{x_2(\tau)^2, x_3(\tau)^2\} d\tau$ is a (monotone) bounded function. Since $\|x(t)\| \leq 1$ for all positive times and $\mathcal{M}$ is bounded, then $x(\cdot)$ is Lipschitz continuous as well as $\dot{F}(t) = \min\{x_2(t)^2, x_3(t)^2\}$. Hence by Barbalat's lemma (Theorem A.8) we have that $\lim_{t\to\infty} \min\{x_2(t)^2, x_3(t)^2\} = 0$, which proves the claim. The hypotheses of Corollary 4.27 are then satisfied.

4.2.2 Strict Convexity of Barabanov Norms

In this section, we focus on the strict convexity of Barabanov norms, which can be equivalently formulated as the strict convexity of their corresponding balls. In full generality, such a geometrical property does not hold as illustrated in Examples 4.9 and 4.10.

We next provide a technical result needed to establish the main result of this section, Proposition 4.31 below.

Lemma 4.30 *Let v be a Barabanov norm for $\mathcal{M}$ and S its unit sphere. Assume that $\lambda(\mathcal{M}) = 0$ and consider $A \in \mathcal{S}_{\mathrm{arb}}(\mathcal{M})$ and $x_0, x_1 \in S$ such that $x_0 \neq x_1$. Set $x_\beta = \beta x_1 + (1-\beta)x_0$ for $\beta \in (0, 1)$. Suppose that there exists $t \geq 0$ such that the open segment connecting $\Phi_A(t, 0)x_0$ to $\Phi_A(t, 0)x_1$ has nonempty intersection with S, that is, there exists $\alpha \in (0, 1)$ such that $\Phi_A(t, 0)x_\alpha \in S$. Then, $\Phi_A(t, 0)x_\beta \in S$ for every $\beta \in [0, 1]$. Moreover, if $\Phi_A(\cdot, 0)x_\alpha$ is an extremal trajectory of $(\Sigma_{\mathcal{M}})$ for some $\alpha \in (0, 1)$ then $\Phi_A(\cdot, 0)x_\beta$ is also an extremal trajectory for every $\beta \in [0, 1]$.*

Proof For the first part of the lemma, it is enough to show that $\Phi_A(t, 0)x_\beta \in S$ for $\beta \in (0, 1)$, as S is closed. Since v is an extremal norm, we have that

$$v(\Phi_A(t, 0)x_\beta) \leq v(x_\beta) \leq \beta v(x_1) + (1-\beta)v(x_0) = 1$$

for every $\beta \in (0, 1)$. We prove that $v(\Phi_A(t, 0)x_\beta) \geq 1$ under the assumption that $v(\Phi_A(t, 0)x_\alpha) = 1$ for some $\alpha \in (0, 1)$. The point $\Phi_A(t, 0)x_\alpha$ is a convex combination of $\Phi_A(t, 0)x_\beta$ with either $\Phi_A(t, 0)x_0$ or $\Phi_A(t, 0)x_1$. In the first case, there exists $\delta \in (0, 1]$ such that

$$\begin{aligned} 1 = v(\Phi_A(t, 0)x_\alpha) &\leq \delta v(\Phi_A(t, 0)x_\beta) + (1-\delta)v(\Phi_A(t, 0)x_0) \\ &\leq \delta v(\Phi_A(t, 0)x_\beta) + 1 - \delta, \end{aligned}$$

which implies that $v(\Phi_A(t, 0)x_\beta) \geq 1$. The second case can be treated with the same arguments. This proves the first part of the lemma. The second part is then a trivial consequence of the first one. □

We can now state our main result on the geometry of the balls of a Barabanov norm.

Proposition 4.31 *Assume that $\mathcal{M} - \lambda(\mathcal{M})\mathrm{Id}_n$ contains no singular matrix. Let v be a Barabanov norm for $\mathcal{M}$ and S its unit sphere. Then, the intersection of S with any hyperplane P has empty relative interior in P.*

Proof The conclusion is clearly true if $0 \in P$. So assume that the conclusion is false for some hyperplane P not containing 0. Assume, without loss of generality, that $\lambda(\mathcal{M}) = 0$. Let x_0 be a point in the interior of $S \cap P$ admitting an extremal trajectory $\Phi_A(\cdot, 0)x_0$ for some switching law $A \in \mathcal{S}_{\text{arb}}(\mathcal{M})$. Without loss of generality we assume that $t \mapsto \Phi_A(t, 0)$ is differentiable at $t = 0$ with $\frac{d}{dt}\Phi_A(t, 0)|_{t=0} = A(0) \in \mathcal{M}$. By assumption and by Lemma 4.30, for any $x \in S \cap P$ there exists a segment in $S \cap P$ connecting x and x_0 and containing x_0 in its interior. Then, again by Lemma 4.30, $\Phi_A(t, 0)x \in S \cap P$ for small t and for all x in the interior of $S \cap P$, which implies that $A(0)x$ is tangent to P, that is $\ell^\top A(0)x = 0$, where ℓ is nonzero and orthogonal to P. It follows that $\ell^\top A(0)y = 0$ for every y in a cone of $\mathbb{R}^n$ with nonempty interior, and thus for every $y \in \mathbb{R}^n$. We have reached a contradiction since $A(0)$ is nonsingular. □

Remark 4.32 As a consequence of the previous proposition, a set $\mathcal{M}$ does not admit a polyhedral Barabanov norm as soon as $\mathcal{M} - \lambda(\mathcal{M})\mathrm{Id}_n$ contains no singular matrix.

We have the following corollary of Proposition 4.31.

Corollary 4.33 *If $n = 2$ and $\mathcal{M} - \lambda(\mathcal{M})\mathrm{Id}_2$ contains no singular matrix, then every Barabanov norm for $\mathcal{M}$ has strictly convex balls.*

Proof Consider the line going through two distinct points x_0, x_1 belonging to the unit sphere S of a Barabanov norm v for $\mathcal{M}$. By Lemma 4.30 if there exists a point belonging to S and to the interior of the segment between x_0 and x_1, then the whole segment is contained in S. This contradicts Proposition 4.31 and we deduce that the interior of the segment between x_0 and x_1 is contained in the interior of the unit ball of v, which yields the conclusion. □

Addressing the issue of strict convexity of Barabanov norms appears to be a complicated task in higher dimensions. The following result shows that the Barabanov norms are strictly convex under some regularity condition on $\mathcal{M}$.

Theorem 4.34 *Let $\mathcal{M}$ be a C^1 convex compact domain of $M_n(\mathbb{R})$. Then every Barabanov norm for $\mathcal{M}$ has strictly convex balls.*

Proof Let v be a Barabanov norm for $\mathcal{M}$ with unit sphere S. Assume, without loss of generality, that $\lambda(\mathcal{M}) = 0$.

By contradiction, assume that there exist two distinct points $x_0, x_1 \in S$ and $\alpha \in (0, 1)$ such that $x_\alpha = \alpha x_1 + (1 - \alpha)x_0 \in S$. Let $x^\alpha(\cdot) = \Phi_A(\cdot, 0)x_\alpha$ be an extremal trajectory for v, with $A \in \mathcal{S}_{\text{arb}}(\mathcal{M})$. Then, by Lemma 4.30, for every $\beta \in [0, 1]$ there exists an extremal pair $(x^\beta(\cdot), \ell^\beta(\cdot))$ with $x^\beta(\cdot) = \Phi_A(\cdot, 0)x_\beta$. By using (4.12), for almost every $t \geq 0$ one gets

$$\min_{M \in \mathcal{M}} \operatorname{tr}\big((A(t) - M)x^\beta(t)\ell^\beta(t)^\top\big) = 0, \qquad \forall \beta \in [0, 1]. \tag{4.18}$$

It follows from (4.18) that $A(t)$ belongs to the boundary of $\mathcal{M}$. The convexity and the C^1 regularity of $\mathcal{M}$ imply that the outward normal to $\mathcal{M}$ at $A(t)$ is a nonzero matrix X, unique up to multiplication by a positive scalar, such that $\mathrm{tr}((A(t) - M)X) \geq 0$ for every $M \in \mathcal{M}$. Hence, by (4.18), X is positively proportional to $x^{\beta}(t)\ell^{\beta}(t)^{\top}$ for every $\beta \in [0, 1]$. This yields $x^0(t) = \pm x^1(t)$ for almost every $t \geq 0$. By continuity, and since $x_0 \neq x_1$, $x^0(t) = -x^1(t)$ for every $t \geq 0$. Hence one has $x_0 = -x_1$ and then $v(x_\alpha) = |1 - 2\alpha| < 1$, in contradiction with the fact that x_α is in S. □

4.2.3 *Regularity of Barabanov Norms*

Note that every Barabanov norm v is differentiable almost everywhere in $\mathbb{R}^n$ by Item 2 of Proposition A.15, and since v is also homogeneous, it is differentiable on a dense subset of its unit sphere.

We next provide the following basic proposition.

Proposition 4.35 *Let v be a Barabanov norm for $\mathcal{M}$. Let $x(\cdot)$ be an extremal trajectory for v starting at some point of differentiability x_0 of v. Then the following results hold true:*

1. *The norm v is differentiable at $x(t)$ for every $t \geq 0$;*
2. *There exists a unique $\ell(\cdot)$ such that $(x(\cdot), \ell(\cdot))$ is an extremal pair and $\ell(t) = \nabla v(x(t))$ for $t \geq 0$.*

Proof Let $x(\cdot)$ be associated with a switching law $A(\cdot)$. Consider $T > 0$ and $\hat{\ell}_1, \hat{\ell}_2 \in \partial v(x(T))$. By virtue of Lemma 4.18, there exist two nonzero Lipschitz continuous solutions $\ell_1(\cdot)$ and $\ell_2(\cdot)$ of System (4.9) such that

$$\begin{aligned} \dot{\ell}_k(t) &= -A^{\top}(t)\ell_k(t), \quad \text{for a.e. } t \in [0, T], \\ \ell_k(t) &\in \partial v(x(t)), \quad \forall t \in [0, T], \\ \ell_k(T) &= \hat{\ell}_k, \end{aligned}$$

for $k = 1, 2$. Since the norm v is differentiable at x_0, then $\ell_k(0) = \nabla v(x_0)$ for $k = 1, 2$ by Item 1 of Proposition A.15. Therefore by uniqueness of the Cauchy problem $\dot{\ell}(t) = -A^{\top}(t)\ell(t)$ with $\ell(0) = \nabla v(x_0)$ we have $\ell_1(T) = \ell_2(T)$, i.e., $\hat{\ell}_1 = \hat{\ell}_2$. Once again by Item 1 of Proposition A.15 we obtain that v is differentiable at $x(T)$ and, if $\ell(\cdot)$ is such that $(x(\cdot), \ell(\cdot))$ is an extremal pair, then $\ell(T) = \nabla v(x(T))$. The result follows since $T > 0$ is arbitrary. □

Let us fix a Barabanov norm v for $\mathcal{M}$ and denote by S its unit sphere. For $x_0 \in S$ we define the subset $\mathcal{V}_{x_0}$ of $\mathbb{R}^n$ by

$$\mathcal{V}_{x_0} := \left\{ y \in \mathbb{R}^n \;\middle|\; \begin{array}{l} \exists\, t_0 > 0, \exists\, x(\cdot) \text{ extremal for } v \text{ with } x(t_0) = x_0, \\ (t_j)_{j \in \mathbb{N}} \text{ s.t. } \lim\limits_{j \to +\infty} t_j = t_0^- \text{ and } y = \lim\limits_{j \to +\infty} \dfrac{e^{-\lambda(\mathcal{M})(t_j - t_0)} x(t_j) - x_0}{t_j - t_0} \end{array} \right\}. \tag{4.19}$$

Remark 4.36 The set $\mathcal{V}_{x_0}$ is invariant by shift of $\mathcal{M}$ by an element $\mu \mathrm{Id}_n$. Notice also that $\mathcal{V}_{x_0}$ can be empty since Theorem 4.19 does not guarantee the existence of extremal trajectories reaching x_0.

The next proposition provides a sufficient condition for the differentiability of a Barabanov norm v at a point x_0.

Proposition 4.37 *Let $x_0 \in S$. Assume that $\mathcal{V}_{x_0}$ contains $n-1$ linearly independent elements. Then v is differentiable at x_0.*

Proof Let us assume, without loss of generality, that $\lambda(\mathcal{M}) = 0$. We first show that $\mathcal{V}_{x_0} \subset \overline{\mathrm{co}}(\mathcal{M}) x_0 := \{M x_0 \mid M \in \overline{\mathrm{co}}(\mathcal{M})\}$. Indeed, assume by contradiction that there exists $y \in \mathcal{V}_{x_0}$ and $y \notin \overline{\mathrm{co}}(\mathcal{M}) x_0$. Thus, by virtue of the hyperplane separation theorem (Theorem A.14), there exist $w \in \mathbb{R}^n$ and $\eta \in \mathbb{R}$ such that $w^\top y < \eta < w^\top M x_0$ for all $M \in \overline{\mathrm{co}}(\mathcal{M})$. Thanks to the definition of $\mathcal{V}_{x_0}$ there exist $t_0 > 0$ and an extremal trajectory $x(\cdot)$ of $(\Sigma_{\mathcal{M}})$ such that $x(t_0) = x_0$ and a sequence $(t_j)_{j \in \mathbb{N}}$ such that $t_j \to t_0^-$ as j goes to infinity satisfying $y = \lim_{j \to +\infty} \frac{x(t_j) - x_0}{t_j - t_0}$. Letting $A(\cdot)$ be a switching law associated with $x(\cdot)$ we also have

$$\begin{aligned} \frac{x(t_j) - x_0}{t_j - t_0} &= \frac{1}{t_j - t_0} \int_{t_0}^{t_j} A(\tau) x(\tau) d\tau \\ &= \frac{1}{t_j - t_0} \int_{t_0}^{t_j} A(\tau)(x(\tau) - x_0) d\tau + \frac{1}{t_j - t_0} \int_{t_0}^{t_j} A(\tau) x_0 d\tau . \end{aligned}$$

Since

$$\lim_{j \to +\infty} \frac{1}{t_j - t_0} \int_{t_0}^{t_j} A(\tau)(x(\tau) - x_0) d\tau = 0,$$

it follows that $y = \lim_{j \to +\infty} \frac{1}{t_j - t_0} \int_{t_0}^{t_j} A(\tau) x_0 d\tau$. Notice that for each $\tau \in (t_j, t_0)$, $w^\top A(\tau) x_0 > \eta$. Hence, by integrating and passing to the limit as j goes to infinity, we get $w^\top y \geq \eta$, leading to a contradiction. Thus $\mathcal{V}_{x_0} \subset \overline{\mathrm{co}}(\mathcal{M}) x_0$.

Let us now check that if $\ell_0 \in \partial v(x_0)$ and $y \in \mathcal{V}_{x_0}$ then $\ell_0^\top y = 0$. By Lemma 4.18 one has that $\ell_0^\top A x_0 \leq 0$ for each $A \in \mathcal{M}$, and in particular $\ell_0^\top y \leq 0$. For the opposite inequality it is enough to observe that $\ell_0^\top (x(t) - x_0) \leq v(x(t)) - v(x_0) = 0$

for every extremal trajectory $x(\cdot)$ and thus $\ell_0^\top\left(\frac{x(t)-x_0}{t-t_0}\right) \geq 0$ for $t < t_0$. The desired inequality is obtained passing to the limit along the sequence $(t_j)_{j\in\mathbb{N}}$.

Finally, under the assumptions of Proposition 4.37, there is a unique vector ℓ_0 satisfying $\ell_0^\top y = 0$ for every $y \in \mathcal{V}_{x_0}$ and such that $\ell_0^\top x_0 = v(x_0) = 1$. Hence $\partial v(x_0) = \{\ell_0\}$ and the differentiability of v at x_0 is proved. □

Notice that, in the case $n = 2$, the previous result states that v is differentiable at any point x_0 reached by an extremal trajectory different from $t \mapsto e^{(t-t_0)\lambda(\mathcal{M})}x_0$. For $n = 3$, differentiability at x_0 is instead guaranteed as soon as two extremal trajectories reach x_0 from two different directions. If the two incoming extremal trajectories reach x_0 with the same direction, we can still prove differentiability of v at x_0 under the additional assumption that $\mathcal{M} - \lambda(\mathcal{M})\mathrm{Id}_3$ is made of nonsingular matrices, as shown below.

Definition 4.38 We say that two extremal trajectories $x_1(\cdot)$ and $x_2(\cdot)$ of $(\Sigma_\mathcal{M})$ intersect each other if there exist $t_1, t_2 > 0$ and $\varepsilon > 0$ such that $x_1(t_1) = x_2(t_2)$ and $x_1(s_1) \neq x_2(s_2)$ for every $s_1 \in [t_1 - \varepsilon, t_1)$ and $s_2 \in [t_2 - \varepsilon, t_2)$.

Proposition 4.39 *Assume that $n = 3$ and let $\mathcal{M}$ be convex and compact. Assume, moreover, that every matrix of $\mathcal{M} - \lambda(\mathcal{M})\mathrm{Id}_3$ is nonsingular. If two extremal trajectories $x_1(\cdot)$ and $x_2(\cdot)$ of $(\Sigma_\mathcal{M})$ intersect each other at some $z \in \mathbb{R}^3$, then v is differentiable at z.*

Proof With no loss of generality, we assume that $\lambda(\mathcal{M}) = 0$. Let $x_1(\cdot)$ and $x_2(\cdot)$ be two extremal trajectories of $(\Sigma_\mathcal{M})$ associated with $A_1(\cdot)$ and $A_2(\cdot)$, respectively, and let $t_1, t_2 > 0$ be such that $z = x_1(t_1) = x_2(t_2)$. Since $\lambda(\mathcal{M}) = 0$ and $\mathcal{M}$ is made of nonsingular matrices then, by Remark 1.19, we have $\mathcal{M}z \cap \mathbb{R}_+ z = \emptyset$.

Notice that the set $\mathcal{M}z$ is compact and convex while $\mathbb{R}_+ z$ is closed and convex, then, by Theorem A.14, there exist $w \in \mathbb{R}^n$ and $\beta \in \mathbb{R}$ such that $\lambda w^\top z < \beta < w^\top Mz$ for all $M \in \mathcal{M}$ and $\lambda \in \mathbb{R}_+$. In particular $\beta > 0$, $w^\top z \leq 0$, and $w^\top Mz > \beta > 0$ for all $M \in \mathcal{M}$.

Since $\mathcal{M}$ is compact, one deduces by a standard continuity argument that there exist a neighborhood O of z such that $w^\top Mx > \beta$ for every $x \in O$ and $M \in \mathcal{M}$. Fix $\eta_1, \eta_2 \geq 0$ such that, for $i = 1, 2$, $\eta_i < t_i$ and the map $x_i(t) \in O$ for every $t \in (t_i - \eta_i, t_i)$. Let $\xi > 0$ be sufficiently small such that $w^\top z - \xi > w^\top x_i(t_i - \eta_i)$ for $i = 1, 2$. Then, there exists a unique pair $(\tau_1, \tau_2) \in (t_1 - \eta_1, t_1) \times (t_2 - \eta_2, t_2)$ such that $w^\top z - \xi = w^\top x_i(\tau_i)$ for $i = 1, 2$. Denote by I the curve obtained by intersecting S and the affine plane $\{x \mid w^\top x = w^\top z - \xi\}$. Notice that I is the union of two curves joining $x_1(\tau_1)$ and $x_2(\tau_2)$. Up to taking ξ small enough, we assume without loss of generality that only one of the two curves, denoted by C, is entirely contained in O.

Let $\mathcal{T}$ be the open region of S contained in O and whose boundary is the union of C, $\mathcal{P}_1$, and $\mathcal{P}_2$, where $\mathcal{P}_i := \{x_i(t) \mid t \in [\tau_i, t_i]\}$ for $i = 1, 2$. Consider now a point of differentiability $\bar{x}$ of v in $\mathcal{T}$. We next claim that if $\bar{y}(\cdot)$ is an extremal trajectory of $(\Sigma_\mathcal{M})$ such that $\bar{y}(0) = \bar{x}$, then $\bar{y}$ intersects $\mathcal{P}_1 \cup \mathcal{P}_2$ (see Fig. 4.2). To see that, first notice that there exists some $\bar{t} > 0$ such that $\bar{y}(\bar{t}) \in C \cup \mathcal{P}_1 \cup \mathcal{P}_2$. Indeed, $\bar{y}(t)$

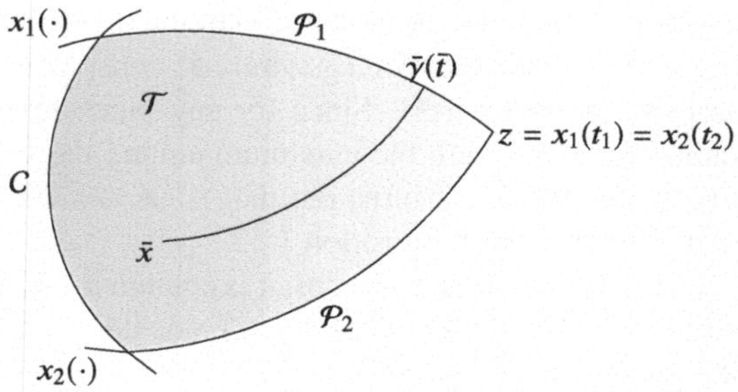

Fig. 4.2 Proof of Proposition 4.39

cannot be in $\mathcal{T}$ for all $t \geq 0$, since the derivative of the Lipschitz continuous map $t \mapsto w^\top \bar{\gamma}(t)$ admits a positive lower bound as long as $\bar{\gamma}(t)$ is in $\mathcal{T}$. Finally $\bar{\gamma}(\bar{t}) \notin C$ since $w^\top \bar{\gamma}(\bar{t}) > w^\top \bar{\gamma}(0) \geq w^\top z - \xi$. This proves that $\bar{\gamma}(\bar{t}) \in \mathcal{P}_1 \cup \mathcal{P}_2$.

We conclude the proof of Proposition 4.39 by assuming, without loss of generality, that $\bar{\gamma}(\bar{t}) \in \mathcal{P}_1$. Then $\bar{\gamma}(\bar{t}) = x_1(\bar{t}_1)$ for some $\bar{t}_1 \in (\tau_1, t_1]$. Since v is differentiable at $\bar{x}$ then by Proposition 4.35 v is differentiable at $\bar{\gamma}(\bar{t})$. In other words v is differentiable at $x_1(\bar{t}_1)$ implying again that v is differentiable at $x_1(t_1) = z$. □

4.3 Barabanov Norms as a Tool

In this section, we present some results whose proofs use the Barabanov norm as a central technical tool.

4.3.1 Berger–Wang Formula and Continuity of the Growth Rate

Recall that the joint spectral exponent associated with a class of signals $\mathcal{S}$ is defined as

$$\mu(\mathcal{S}) = \limsup_{t \to \infty} \sup_{A \in \mathcal{S}} \frac{\log(\rho(\Phi_A(t, 0)))}{t}$$

and that $\mu(\mathcal{S}) \leq \lambda(\mathcal{S})$ (see Sect. 1.3.3). We use $\mu(\mathcal{M})$ to denote $\mu(\mathcal{S}_{\text{arb}}(\mathcal{M}))$ and we prove next the equality between $\mu(\mathcal{M})$ and $\lambda(\mathcal{M})$.

Theorem 4.40 (Berger–Wang formula) *The equality $\mu(\mathcal{M}) = \lambda(\mathcal{M})$ holds for every bounded $\mathcal{M} \subset M_n(\mathbb{R})$.*

Proof According to Remark 1.21, one has $\lambda(\mathcal{M}) = \lambda(\overline{\text{co}}(\mathcal{M}))$. A similar reasoning also yields $\mu(\mathcal{M}) = \mu(\overline{\text{co}}(\mathcal{M}))$. Therefore, we can assume with no loss of generality that $\mathcal{M}$ is convex and compact.

Assume first that the conclusion of the theorem holds under the extra assumption that $\mathcal{M}$ is irreducible. For a general $\mathcal{M}$, consider its irreducible blocks $\mathcal{M}_1, \dots, \mathcal{M}_r$ as in Propositions 1.31. Since for any block upper-triangular matrix the spectral radius coincides with the maximum among the spectral radii of the corresponding irreducible blocks, it turns out that $\mu(\mathcal{M}) = \max_{i=1,\dots,r} \mu(\mathcal{M}_i)$. The conclusion then follows from Proposition 1.32.

According to what precedes, it is enough to show that for every convex, compact, and irreducible subset $\mathcal{M}$ of $M_n(\mathbb{R})$, one has $\lambda(\mathcal{M}) \leq \mu(\mathcal{M})$. Relying on (1.8) and the analogous equation for $\mu(\mathcal{M})$, we can assume without loss of generality that $\lambda(\mathcal{M}) = 0$ and we are left to prove that $\mu(\mathcal{M}) \geq 0$.

According to Theorem 4.8, there exists a Barabanov norm v for $\mathcal{M}$ and use $\|\cdot\|_v$ to denote its induced matrix norm. Pick $x_0 \in \mathbb{R}^n$ with $v(x_0) = 1$. By definition of Barabanov norm, there exists $A \in \mathcal{S}_{\text{arb}}(\mathcal{M})$ such that $v(\Phi_A(t, 0)x_0) = 1$ for every $t \geq 0$. Let $(t_j)_{j\in\mathbb{N}}$ be an increasing sequence of times going to infinity such that $t_{j+1} - t_j \geq 1$ for every $j \in \mathbb{N}$ and, if $x_j = \Phi_A(t_j, 0)x_0$, then the sequence $(x_j)_{j\in\mathbb{N}}$ converges to some y_0 with $v(y_0) = 1$.

Notice that the sequence $B_j = \Phi_A(t_{j+1}, t_j)$ is contained in the compact subset of $M_n(\mathbb{R})$ of matrices with unit $\|\cdot\|_v$-norm. Then, up to extracting a subsequence, B_j converges to some $B \in M_n(\mathbb{R})$ with $\|B\|_v = 1$. In particular, $\rho(B) \leq \|B\|_v = 1$. Moreover, one has $By_0 = y_0$ by taking limits in $B_j x_j = x_{j+1}$, so that $\rho(B)$ is actually equal to one. Hence, $\rho(B_j)$ converges to one by continuity of the spectral radius. Pick $\varepsilon > 0$ and j such that B_j has an eigenvalue λ_j with modulus larger than $1 - \varepsilon$. Consider the periodic signal $\hat{A} \in \mathcal{S}_{\text{arb}}(\mathcal{M})$ obtained by concatenating $A|_{[0,t_{j+1}-t_j]}(t_j + \cdot)$ infinitely many times. Then

$$\begin{aligned}\mu(\mathcal{M}) &\geq \limsup_{t\to\infty} \frac{\log(\rho(\Phi_{\hat{A}}(t,0)))}{t} \geq \lim_{k\to\infty} \frac{\log(\rho(B_j^k))}{k(t_{j+1}-t_j)} \geq \frac{\log(1-\varepsilon)}{t_{j+1}-t_j} \\ &\geq \log(1-\varepsilon).\end{aligned}$$

We conclude by letting ε tend to zero. □

Let us now deduce from the Berger–Wang formula that the growth rate $\lambda(\mathcal{M})$ is continuous with respect to the set of matrices $\mathcal{M}$.

Theorem 4.41 *The function $\mathcal{M} \mapsto \lambda(\mathcal{M})$ defined on the set of compact subsets of $M_n(\mathbb{R})$ endowed with the Hausdorff distance is continuous.*

Proof We first notice that the upper semi-continuity of $\lambda(\cdot)$ is a direct consequence of Remark 1.23 (see also Corollary 3.19).

We are then left to prove that $\lambda(\cdot)$ is lower semi-continuous on the set of compact subsets of $M_n(\mathbb{R})$ endowed with the Hausdorff distance, i.e., given $\mathcal{M}_1 \subset M_n(\mathbb{R})$ compact and $\varepsilon > 0$, there exists a neighborhood W_ε of $\mathcal{M}_1$ such that if $\mathcal{M}_2 \in W_\varepsilon$ then

$$\lambda(\mathcal{M}_2) \geq \lambda(\mathcal{M}_1) - \varepsilon. \tag{4.20}$$

According to Theorem 4.40, we can equivalently replace $\lambda(\cdot)$ by $\mu(\cdot)$ in (4.20). By definition of $\mu(\mathcal{M}_1)$, there exist $t > 0$ and $A \in \mathcal{S}_{\mathrm{arb}}(\mathcal{M}_1)$ such that

$$\frac{\log(\rho(\Phi_A(t,0)))}{t} > \mu(\mathcal{M}_1) - \frac{\varepsilon}{2}. \tag{4.21}$$

By Corollary 1.8, we can assume without loss of generality that A is in $\mathcal{S}_{\mathrm{pc}}(\mathcal{M}_1)$.

For every $\eta > 0$, let W^η be the set of compact subsets $\mathcal{M}_2$ of $M_n(\mathbb{R})$ such that $d_H(\mathcal{M}_1, \mathcal{M}_2) < \eta$. Notice that for every $\mathcal{M}_2 \in W^\eta$ there exists $A_{\mathcal{M}_2} \in \mathcal{S}_{\mathrm{pc}}(\mathcal{M}_2)$ which is t-periodic and such that $\|A(s) - A_{\mathcal{M}_2}(s)\| < \eta$ for every $s \in [0,t)$. Choosing $\eta(\varepsilon) > 0$ small enough, if $\mathcal{M}_2 \in W_\varepsilon := W^{\eta(\varepsilon)}$ then

$$\frac{\log(\rho(\Phi_{A_{\mathcal{M}_2}}(t,0)))}{t} > \frac{\log(\rho(\Phi_A(t,0)))}{t} - \frac{\varepsilon}{2}. \tag{4.22}$$

Since

$$\begin{aligned}\mu(\mathcal{M}_2) &\geq \limsup_{s\to\infty} \frac{\log(\rho(\Phi_{A_{\mathcal{M}_2}}(s,0)))}{s} \geq \limsup_{k\to\infty} \frac{\log(\rho(\Phi_{A_{\mathcal{M}_2}}(kt,0)))}{kt} \\ &= \frac{\log(\rho(\Phi_{A_{\mathcal{M}_2}}(t,0)))}{t},\end{aligned}$$

one concludes the proof of (4.20) by (4.21) and (4.22). □

4.3.2 Resonance and Defectiveness

We have seen in Lemma 4.7 that if $\mathcal{M}$ is irreducible then it is nondefective, that is, $(\Sigma_{\mathcal{M}-\lambda(\mathcal{M})\mathrm{Id}_n})$ is bounded. In the general case such boundedness does not hold, as illustrated by the case of nontrivial nilpotent matrices (see Remark 1.22).

We therefore focus in this section to the study of the qualitative behaviour of trajectories of $(\Sigma_{\mathcal{M}})$ in the case where $\mathcal{M}$ is reducible.

Proposition 4.42 *Let $\mathcal{M}_1, \ldots, \mathcal{M}_r$ be the irreducible blocks of $\mathcal{M}$. Let $\hat{r} \leq r$ be the number of $\mathcal{M}_i$ such that $\lambda(\mathcal{M}_i) = \lambda(\mathcal{M})$. Then, there exists $C > 0$ such that, for every $A \in \mathcal{S}_{\mathrm{arb}}(\mathcal{M})$ one has*

$$\|\Phi_A(t,0)\| \leq C(1 + t^{\hat{r}-1})e^{t\lambda(\mathcal{M})}, \quad t \geq 0. \tag{4.23}$$

Proof Let $\{0\} = E_0 \subset E_1 \subset \cdots \subset E_r = \mathbb{R}^n$ be the maximal invariant flag associated with the irreducible blocks $\mathcal{M}_1, \ldots, \mathcal{M}_r$. Let j be the largest index in $\{1, \ldots, r\}$ such that $\lambda(\mathcal{M}_j) = \lambda(\mathcal{M})$.

Consider three subspaces V_1, V_2, V_3 of $\mathbb{R}^n$ with $V_1 = E_{j-1}$, $V_1 \oplus V_2 = E_j$ and $V_1 \oplus V_2 \oplus V_3 = \mathbb{R}^n$.

Let $m_i = \dim V_i$, $i = 1, 2, 3$, and $\{v_1, \ldots, v_n\}$ be a vector basis of $\mathbb{R}^n$ such that $V_1 = \text{span}\{v_1, \ldots, v_{m_1}\}$, $V_2 = \text{span}\{v_{m_1+1}, \ldots, v_{m_1+m_2}\}$, and $V_3 = \text{span}\{v_{m_1+m_2+1}, \ldots, v_n\}$. Notice that m_1 and m_3 could be equal to zero, while $m_2 \geq 1$. Considering the basis $\{v_1, \ldots, v_n\}$, we write every matrix $M \in \mathcal{M}$ and every signal $A \in \mathcal{S}_{\text{arb}}(\mathcal{M})$ in the block form

$$M = \begin{pmatrix} \tilde{M}_{11} & \tilde{M}_{12} & \tilde{M}_{13} \\ 0 & \tilde{M}_{22} & \tilde{M}_{23} \\ 0 & 0 & \tilde{M}_{33} \end{pmatrix}, \qquad A = \begin{pmatrix} \tilde{A}_{11} & \tilde{A}_{12} & \tilde{A}_{13} \\ 0 & \tilde{A}_{22} & \tilde{A}_{23} \\ 0 & 0 & \tilde{A}_{33} \end{pmatrix},$$

with the convention that some of the blocks are empty if $m_1 m_3 = 0$. The corresponding fundamental matrix may also be decomposed accordingly

$$\Phi_A(t, 0) = \begin{pmatrix} \tilde{R}_{11}(t, 0) & \tilde{R}_{12}(t, 0) & \tilde{R}_{13}(t, 0) \\ 0 & \tilde{R}_{22}(t, 0) & \tilde{R}_{23}(t, 0) \\ 0 & 0 & \tilde{R}_{33}(t, 0) \end{pmatrix},$$

where each block is either empty or may be represented as follows

$$\tilde{R}_{ii}(t, 0) = \Phi_{\tilde{A}_{ii}}(t, 0) \qquad i = 1, 2, 3,$$

$$\tilde{R}_{i,i+1}(t, 0) = \int_0^t \Phi_{\tilde{A}_{ii}}(t, \tau)\tilde{A}_{i,i+1}(\tau)\Phi_{\tilde{A}_{i+1,i+1}}(\tau, 0)d\tau \quad \text{for } i = 1, 2,$$

$$\tilde{R}_{13}(t, 0) = \int_0^t \int_0^{\tau_1} \Phi_{\tilde{A}_{11}}(t, \tau_1)\tilde{A}_{12}(\tau_1)\Phi_{\tilde{A}_{22}}(\tau_1, \tau_2)\tilde{A}_{23}(\tau_1)\Phi_{\tilde{A}_{33}}(\tau_2, 0)d\tau_2\, d\tau_1,$$

using, for the non-diagonal terms, the variation of constants formula (Lemma A.6). We then deduce that there exists a positive constant C such that, for every $t \geq 0$,

$$\|\tilde{R}_{i,i+1}(t, 0)\| \leq C \int_0^t \|\Phi_{\tilde{A}_{ii}}(t, \tau)\| \|\Phi_{\tilde{A}_{i+1,i+1}}(\tau, 0)\| d\tau \qquad i = 1, 2, \tag{4.24}$$

$$\|\tilde{R}_{13}(t, 0)\| \leq C \int_0^t \int_0^{\tau_1} \|\Phi_{\tilde{A}_{11}}(t, \tau_1)\| \|\Phi_{\tilde{A}_{22}}(\tau_1, \tau_2)\| \|\Phi_{\tilde{A}_{33}}(\tau_2, 0)\| d\tau_2\, d\tau_1. \tag{4.25}$$

To conclude the proof of the proposition it is thus enough to show that there exists $C > 0$ so that for every $1 \leq i \leq j \leq 3$,

$$\begin{cases} \tilde{R}_{ij} \text{ is empty,} & \text{if } m_1 m_3 = 0, \\ \|\tilde{R}_{ij}(t, 0)\| \leq C(1 + t^{\hat{r}-1})e^{t\lambda(\mathcal{M})}, \ \forall t \geq 0, & \text{otherwise.} \end{cases} \tag{4.26}$$

Define, for $i = 1, 2, 3$, $\tilde{\mathcal{M}}_i := \{\tilde{M}_{ii} \mid M \in \mathcal{M}\}$ and notice that $\tilde{\mathcal{M}}_2$ is conjugate to $\mathcal{M}_j$ thanks to Proposition 1.31. In particular, $\tilde{\mathcal{M}}_2$ is irreducible and $\lambda(\tilde{\mathcal{M}}_2) =$

$\lambda(\mathcal{M})$. We deduce by Lemma 4.7 that there exists $C > 0$ such that for every $\tilde{A}_{22} \in \mathcal{S}_{\text{arb}}(\tilde{\mathcal{M}_2})$ one has

$$\|\Phi_{\tilde{A}_{22}}(t+s,s)\| \leq Ce^{t\lambda(\mathcal{M})}, \qquad s,t \geq 0. \tag{4.27}$$

The proof goes now by induction on $\hat{r}$. If $\hat{r} = 1$, by construction and Proposition 1.32 one has that for $i = 1, 3$, either $m_i = 0$ or $\lambda(\tilde{\mathcal{M}_i}) < \lambda(\mathcal{M})$, which implies, in that case, that there exist $C > 0$ and $\lambda < \lambda(\mathcal{M})$ such that for every $\tilde{A}_{ii} \in \mathcal{S}_{\text{arb}}(\tilde{\mathcal{M}_i})$ and $t \geq 0$,

$$\|\Phi_{\tilde{A}_{ii}}(t+s,s)\| \leq Ce^{t\lambda}, \qquad s,t \geq 0. \tag{4.28}$$

In particular, Eqs. (4.27) and (4.28) yield Eq. (4.26) for $1 \leq i = j \leq 3$. Moreover, by plugging estimates (4.27) and (4.28) in (4.24) and (4.25), the conclusion in the case $\hat{r} = 1$ follows by direct integration.

Assume now that $\hat{r} \geq 2$. By the induction hypothesis applied to $\hat{r} - 1$, there exists $C > 0$ such that, for every $\tilde{A}_{11} \in \mathcal{S}_{\text{arb}}(\tilde{\mathcal{M}_1})$,

$$\|\Phi_{\tilde{A}_{11}}(t+s,s)\| \leq C(1+t^{\hat{r}-2})e^{t\lambda(\mathcal{M})}, \quad s,t \geq 0,$$

while estimate (4.28) holds for $i = 3$ if $m_3 > 0$. Plugging now the above estimates together with (4.27) in (4.24) and (4.25), one derives (4.26) for $\hat{r}$, which concludes the induction argument. □

To sharpen the estimate in Proposition 4.42, we introduce the notion of resonance between irreducible blocks of $\mathcal{M}$. In the definition, we use the notation $\pi_{11}, \ldots, \pi_{rr}$ for the linear matrix applications introduced in (1.14).

Definition 4.43 Denote by $\mathcal{M}_1, \ldots, \mathcal{M}_r$ the irreducible blocks of $\mathcal{M}$. We say that *$\mathcal{M}_i$ and $\mathcal{M}_j$ are in resonance* if they satisfy the two following conditions:

(*a*) $i \neq j$ and $\lambda(\mathcal{M}_i) = \lambda(\mathcal{M}_j) = \lambda(\mathcal{M})$;
(*b*) there exists a signal $A \in \mathcal{S}_{\text{arb}}(\overline{\text{co}}(\mathcal{M}))$ such that $A_{ii} = \pi_{ii}(A) \in \mathcal{S}_{\text{arb}}(\overline{\text{co}}(\mathcal{M}_i))$ and $A_{jj} = \pi_{jj}(A) \in \mathcal{S}_{\text{arb}}(\overline{\text{co}}(\mathcal{M}_j))$ give rise to extremal trajectories for some Barabanov norms v_i and v_j associated with $\overline{\text{co}}(\mathcal{M}_i)$ and $\overline{\text{co}}(\mathcal{M}_j)$, respectively.

The concept of resonance does not actually depend on the specific Barabanov norms appearing in condition (*b*) in the above definition, as it follows from the next lemma.

Lemma 4.44 *Assume that two irreducible blocks $\mathcal{M}_i$ and $\mathcal{M}_j$ of $\mathcal{M}$ are in resonance (cf. Definition 4.43) and let w_i and w_j be Barabanov norms for $\overline{\text{co}}(\mathcal{M}_i)$ and $\overline{\text{co}}(\mathcal{M}_j)$. Then there exists a signal $\tilde{A} \in \mathcal{S}_{\text{arb}}(\overline{\text{co}}(\mathcal{M}))$ such that $\tilde{A}_{ii} = \pi_{ii}(\tilde{A})$ and $\tilde{A}_{jj} = \pi_{jj}(\tilde{A})$ give rise to extremal trajectories for w_i and w_j, respectively.*

Proof Without loss of generality, we assume that $\mathcal{M}$ is convex, compact, and such that $\lambda(\mathcal{M}) = 0$. Let $A(\cdot)$ be as in condition (*b*) of Definition 4.43. Let m_i and m_j

be such that $\mathcal{M}_i \subset M_{m_i}(\mathbb{R})$ and $\mathcal{M}_j \subset M_{m_j}(\mathbb{R})$. Consider a sequence $(t_k)_{k\in\mathbb{N}}$ of positive times going to infinity such that $\Phi_{A_{ii}}(t_k, 0)$ and $\Phi_{A_{jj}}(t_k, 0)$ converge to some matrices $\bar{\Phi}_i$ and $\bar{\Phi}_j$, respectively.

Since w_i and w_j are extremal, for every $M_i \in M_{m_i}(\mathbb{R})$ and $M_j \in M_{m_j}(\mathbb{R})$, one has that $t \mapsto \|\Phi_{A_{ii}}(t, 0)M_i\|_{w_i}$ and $t \mapsto \|\Phi_{A_{jj}}(t, 0)M_j\|_{w_j}$ are nonincreasing. This implies that the limits as t tends to infinity of these two nonnegative functions exist and are equal to $\|\bar{\Phi}_i M_i\|_{w_i}$ and $\|\bar{\Phi}_j M_j\|_{w_j}$, respectively.

By Corollary 1.4 and by sequential weak-$\star$ compactness of $\mathcal{S}_{\text{arb}}(\mathcal{M})$ Theorem A.11, there exists $\tilde{A} \in \mathcal{S}_{\text{arb}}(\mathcal{M})$ such that $\Phi_{A_{ii}}(t_k + \cdot, t_k)$ and $\Phi_{A_{jj}}(t_k + \cdot, t_k)$ converge (up to subsequences) uniformly on compact time intervals to $\Phi_{\tilde{A}_{ii}}(\cdot, 0)$ and $\Phi_{\tilde{A}_{jj}}(\cdot, 0)$, respectively, where $\tilde{A}_{ii} = \pi_{ii}(\tilde{A})$ and $\tilde{A}_{jj} = \pi_{jj}(\tilde{A})$. Hence, for every $t \geq 0$,

$$\begin{aligned}
\|\Phi_{\tilde{A}_{ii}}(t, 0)\|_{w_i} &= \|\lim_{k\to\infty} \Phi_{A_{ii}}(t_k + t, t_k)\|_{w_i} \\
&= \|\lim_{k\to\infty} \Phi_{A_{ii}}(t_k + t, 0)\Phi_{A_{ii}}(t_k, 0)^{-1}\|_{w_i} \\
&= \lim_{k\to\infty} \|\Phi_{A_{ii}}(t_k + t, 0)\bar{\Phi}_i^{-1}\|_{w_i} = 1,
\end{aligned}$$

and, similarly, $\|\Phi_{\tilde{A}_{jj}}(t, 0)\|_{w_j} = 1$.

As noticed in Remark 4.4, this implies that $\tilde{A}_{ii}$ and $\tilde{A}_{jj}$ give rise to extremal trajectories for w_i and w_j respectively. □

Let us give an example of systems with irreducible blocks in resonance. Since extremal trajectories for one-dimensional irreducible subsystems are trivial, we focus on two-dimensional irreducible blocks.

Example 4.45 Consider $n = 4$ and $\mathcal{M} = \{M^0, M^1\}$ with

$$M^0 = \begin{pmatrix} M_{11}^0 & M_{12}^0 \\ 0 & M_{22}^0 \end{pmatrix}, \quad M^1 = \begin{pmatrix} M_{11}^1 & M_{12}^1 \\ 0 & M_{22}^1 \end{pmatrix}, \tag{4.29}$$

where all the blocks are 2×2. We choose M_{11}^0, M_{11}^1 as the matrices M_1, M_2 in (2.3) with $a = \bar{a}_4 \approx 1.367$ (see p. 29). We also set $M_{22}^0 = M_{11}^0$ and $M_{22}^1 = M_{11}^1$. Then the two irreducible blocks of $\mathcal{M}$ are $\mathcal{M}_1 = \mathcal{M}_2 = \{M_{11}^0, M_{11}^1\}$ and $\lambda(\mathcal{M}_1) = \lambda(\mathcal{M}_2) = 0$. Moreover, $\mathcal{M}_1$ and $\mathcal{M}_2$ are in resonance, and the resonance is given by a periodic piecewise constant signal $A(\cdot)$.

Let us now show that, for almost every choice of the matrices M_{12}^0 and M_{12}^1, $\mathcal{M}$ is defective. In order to do so, denote by $t_0 > 0$ (respectively, $t_1 > 0$) the length of the intervals where $A(t) = M^0$ (respectively, $A(t) = M^1$) and set $A_{ij} = \pi_{ij}(A)$, $i, j = 1, 2$, where the maps π_{ij} are defined as in (1.14). Write $T = t_0 + t_1$ and let

x_1^k, x_2^k be the eigenvectors of $\Phi_{A_{kk}}(T,0)$ for $k=1,2$. We claim that $\mathcal{M}$ is defective if

$$\int_0^T \Pi_{x_1^1}\Big(\Phi_{A_{11}}(T,\tau)A_{12}(\tau)\Phi_{A_{22}}(\tau,0)\,x_1^2\Big)\,d\tau \neq 0, \tag{4.30}$$

where $\Pi_{x_1^1}(x)$ is the component of the vector $x\in\mathbb{R}^2$ along the direction x_1^1 with respect to the basis $\{x_1^1, x_2^1\}$.

Consider the trajectory $(x_1(\cdot), x_2(\cdot))$ of $(\Sigma_{\mathcal{M}})$ starting at $(0, x_1^2)^\top$ and corresponding to $A(\cdot)$,

$$x_1(t)=\int_0^t \Phi_{A_{11}}(t,\tau)A_{12}(\tau)\Phi_{A_{22}}(\tau,0)x_1^2 d\tau, \quad x_2(t)=\Phi_{A_{11}}(t,0)x_1^2. \tag{4.31}$$

Notice that, by definition of $\Pi_{x_1^1}$, if $v=v_1x_1^1+v_2x_2^1\in\mathbb{R}^2$, we have

$$\begin{aligned}\Pi_{x_1^1}\big(\Phi_{A_{11}}(T,0)\,v\big) &= v_1\,\Pi_{x_1^1}\big(\Phi_{A_{11}}(T,0)\,x_1^1\big)+v_2\,\Pi_{x_1^1}\big(\Phi_{A_{11}}(T,0)\,x_2^1\big)=-v_1\\ &= -\Pi_{x_1^1}(v).\end{aligned}$$

By T-periodicity of $A_{11}(\cdot)$ and $A_{12}(\cdot)$, one deduces that, for every $m\in\mathbb{N}$,

$$\begin{aligned}&\Pi_{x_1^1}\Big(\int_0^{mT}\Phi_{A_{11}}(mT,\tau)\,A_{12}(\tau)\,\Phi_{A_{22}}(\tau,0)\,x_1^2\,d\tau\Big)=\\ &\qquad (-1)^{m+1}m\int_0^T \Pi_{x_1^1}\Big(\Phi_{A_{11}}(T,\tau)\,A_{12}(\tau)\,\Phi_{A_{22}}(\tau,0)\,x_1^2\Big)\,d\tau\,.\end{aligned}$$

Then, under Condition (4.30), $|\Pi_{x_1^1}(x_1(mT))|\geq C_1 m$, so that $\|x(mT)\|\geq C_2 m$ and therefore $\|x(t)\|\geq C_3 t$, for suitable constants $C_1, C_2, C_3>0$ independent of $m\in\mathbb{N}$ and $t\geq 0$.

Let us now prove that (4.30) is satisfied for all (M_{12}^0, M_{12}^1) in the complement of a hyperplane of $M_2(\mathbb{R})\times M_2(\mathbb{R})$. This is a straightforward consequence of the following fact: for $y\in\mathbb{R}^2\setminus\{0\}$ orthogonal to x_2^1, the linear map

$$\begin{aligned}\Psi: M_2(\mathbb{R})\times M_2(\mathbb{R}) &\to \mathbb{R}\\ (M_{12}^0, M_{12}^1) &\mapsto \int_0^T y^\top\Phi_{A_{11}}(T,\tau)\,A_{12}(\tau)\,\Phi_{A_{22}}(\tau,0)\,x_1^2\,d\tau\end{aligned}$$

is onto. In order to see that, notice that the trajectory $\tau\mapsto\Phi_{A_{22}}(\tau,0)\,x_1^2$ lies, as τ varies in $[0,t_0]$, in one of the four positive cones defined by the lines $\mathbb{R}x_1^2$ and $\mathbb{R}x_2^2$. Each of these cones is strictly contained in a half-plane. In particular, there exists $w\in\mathbb{R}^2$ such that $w^\top\Phi_{A_{22}}(\tau,0)\,x_1^2$ is strictly positive as τ varies in $[0,t_0]$. Similarly,

$\tau \mapsto \Phi_{A_{11}}(T,\tau)^\top y$ lies in a half-plane for τ in $[0, t_0]$, hence there exists $v \in \mathbb{R}^2$ such that $y^\top \Phi_{A_{11}}(T,\tau)\, v$ is strictly positive as τ varies in $[0, t_0]$. The argument is concluded by noticing that the pair $(M_{12}^0, M_{12}^1) = (vw^\top, 0)$ has a nonzero image for the linear map Ψ.

We can now state, in terms of the resonance notion introduced above, the following necessary condition for defectiveness.

Theorem 4.46 *Assume that $\mathcal{M}$ is defective. Then $\mathcal{M}$ is reducible and, for any maximal invariant flag, it admits two irreducible blocks in resonance.*

The argument for the above theorem relies on the following two lemmas.

Lemma 4.47 *Assume that $\lambda(\mathcal{M}) = 0$. Let $\mathcal{M}_1, \dots, \mathcal{M}_r$ be the irreducible blocks associated with $\mathcal{M}$. If none of such blocks are in resonance, then there exist $C \in (0, 1)$ and $T > 0$ such that, for every $A \in \mathcal{S}_{\mathrm{arb}}(\mathcal{M})$ and every $i, j \in \{1, \dots, r\}$ with $i \neq j$,*

$$\|\Phi_{\pi_{ii}(A)}(T, 0)\|_i \,\|\Phi_{\pi_{jj}(A)}(T, 0)\|_j \le C, \tag{4.32}$$

where, for $l = i, j$, $\|\cdot\|_l$ and π_{ll} denote the matrix norm induced by any fixed Barabanov norm associated with $\mathcal{M}_l$ and the map introduced in (1.14), *respectively.*

Proof Equation (4.32) is true for every $T > 0$ for some $C \in (0, 1)$ as soon as $\lambda(\mathcal{M}_i) < 0$ or $\lambda(\mathcal{M}_j) < 0$, as it follows from (4.3). Suppose now that $\lambda(\mathcal{M}_i) = \lambda(\mathcal{M}_j) = 0$. Reasoning by contradiction, there exist a sequence $(A^{(k)})_{k\in\mathbb{N}}$ in $\mathcal{S}_{\mathrm{arb}}(\mathcal{M})$ and a sequence $(T^{(k)})_{k\in\mathbb{N}}$ in $(0, \infty)$, with $\lim_{k\to\infty} T^{(k)} = \infty$ such that $\|\Phi_{A_{ll}^{(k)}}(T^{(k)}, 0)\|_l > 1 - \frac{1}{k}$ for $l = i, j$, where $A_{ll}^{(k)} = \pi_{ll}(A^{(k)})$.

By Corollary 1.4 and by the sequential weak-$\star$ compactness of $\mathcal{S}_{\mathrm{arb}}(\mathcal{M})$ (Theorem A.11), there exists $\tilde{A} \in \mathcal{S}_{\mathrm{arb}}(\mathcal{M})$ such that $\|\Phi_{\pi_{ll}(\tilde{A})}(t, 0)\|_l = 1$ for every $t \ge 0$ and $l = i, j$. By Remark 4.4, $\tilde{A}$ corresponds to a resonance between $\mathcal{M}_i$ and $\mathcal{M}_j$, contradicting the hypotheses of the lemma. This completes the proof of the lemma. □

Lemma 4.48 *For every positive integers s, q, set*

$$\Xi_q^s = \{k \in \mathbb{N}^s \mid k_1 \le k_2 \le \cdots \le k_s \le q\}.$$

Given a finite sequence $\alpha = (\alpha^i)_{i=1}^q$ in $(0, 1]^{s+1}$ denote the components of each α^i by $\alpha_1^i, \dots, \alpha_{s+1}^i$. For every $k = (k_1, \dots, k_s) \in \Xi_q^s$, define

$$P(\alpha, k) = \prod_{i=1}^{k_1} \alpha_1^i \prod_{i=k_1+1}^{k_2} \alpha_2^i \cdots \prod_{i=k_s+1}^{q} \alpha_{s+1}^i,$$

and

$$S_q^s(\alpha) = \sum_{k \in \Xi_q^s} P(\alpha, k).$$

Then, for any fixed $C \in (0, 1)$, there exists a constant L depending on C such that $S_q^s(\alpha) \leq L$ for every positive integer q and for every sequence α as above satisfying in addition $\alpha_j^i \alpha_l^i \leq C$ for every $j \neq l$ and every $i = 1, \ldots, q$.

Proof Let $k^{(1)}, k^{(2)} \in \Xi_q^s$ and observe that

$$P(\alpha, k^{(1)}) P(\alpha, k^{(2)}) \leq C^{\max_{l=1,\ldots,s} |k_l^{(1)} - k_l^{(2)}|}. \tag{4.33}$$

Indeed, assume without loss of generality that $k_{l_*}^{(1)} < k_{l_*}^{(2)}$, where

$$|k_{l_*}^{(1)} - k_{l_*}^{(2)}| = \max_{l=1,\ldots,s} |k_l^{(1)} - k_l^{(2)}|.$$

Let i be an integer satisfying $k_{l_*}^{(1)} < i \leq k_{l_*}^{(2)}$. If $\alpha_{j_1}^i$ and $\alpha_{j_2}^i$ are terms corresponding to the subscript i in the factorization of $P(\alpha, k^{(1)})$ and $P(\alpha, k^{(2)})$, respectively, then $k_{j_1-1}^{(1)} < i \leq k_{j_1}^{(1)}$ and $k_{j_2-1}^{(2)} < i \leq k_{j_2}^{(2)}$. It follows that $j_2 \leq l_* < j_1$, implying that $\alpha_{j_1}^i \alpha_{j_2}^i \leq C$ from the hypothesis of the lemma. Thus (4.33) follows.

For $h \in \mathbb{N}$, let us define the set

$$\mathcal{I}_h = \{k \in \Xi_q^s \mid P(\alpha, k) > C^h\},$$

and observe that, if $k^{(1)}, k^{(2)} \in \mathcal{I}_h$, then $P(\alpha, k^{(1)}) P(\alpha, k^{(2)}) > C^{2h}$ so that, from (4.33), we deduce that

$$\max_{i=1,\ldots,s} |k_i^{(1)} - k_i^{(2)}| < 2h, \qquad \forall k^{(1)}, k^{(2)} \in \mathcal{I}_h.$$

In particular, the set $\mathcal{I}_h$ contains at most $(2h)^s$ elements. Since $\Xi_q^s = \cup_{h=1}^{\infty} (\mathcal{I}_h \setminus \mathcal{I}_{h-1})$, we get

$$S_q(\alpha) = \sum_{h=1}^{\infty} \sum_{\hat{k} \in \mathcal{I}_h \setminus \mathcal{I}_{h-1}} P(\alpha, \hat{k}) \leq \sum_{h=1}^{\infty} (2h)^s C^{h-1} < +\infty.$$

The lemma is proved by setting $L = \sum_{h=1}^{\infty} (2h)^s C^{h-1}$. □

We can now complete the proof of Theorem 4.46.

Proof of Theorem 4.46 Let us prove that, if $\lambda(\mathcal{M}) = 0$ and there are no irreducible blocks of $\mathcal{M}$ in resonance, then $(\Sigma_{\mathcal{M}})$ is uniformly bounded. Without loss of generality we assume that $\mathcal{M}$ is convex and compact. Fix a maximal invariant flag and a system of coordinates such that all the matrices of $\mathcal{M}$ are of the form (1.13) and let $x = (x_1, \dots, x_r)$ where $x_i \in \mathbb{R}^{n_i - n_{i-1}}$ for $i = 1, \dots, r$. Consider a signal $A \in \mathcal{S}_{\text{arb}}(\mathcal{M})$ and write $A_{ij} = \pi_{ij}(A)$ and $\Phi_i(\cdot, \cdot)$ for $\Phi_{A_{ii}}(\cdot, \cdot)$, where π_{ij} is as in (1.14), $1 \le i \le j \le r$.

By the variation of constants formula (Lemma A.6), we get

$$x_i(t) = \Phi_i(t, 0)x_i(0) + \sum_{s=1}^{r-i} \sum_{i<i_s<\cdots<i_1 \le r} I(t, i, i_s, \dots, i_1)x_{i_1}(0), \qquad i = 1, \dots, r,$$

where the integral $I(t, i, i_s, \dots, i_1)$ is defined as

$$\int_{0 \le \tau_1 \le \cdots \le \tau_s \le t} \Phi_i(t, \tau_s)A_{i,i_s}(\tau_s)\, \Phi_{i_s}(\tau_s, \tau_{s-1}) \cdots A_{i_2,i_1}(\tau_1)\Phi_{i_1}(\tau_1, 0)\, d\tau_s \cdots d\tau_1. \tag{4.34}$$

We deduce at once that

$$\|\Phi_A(t, 0)\| \le \sum_{i=1}^{r} \|\Phi_i(t, 0)\| + \sum_{i=1}^{r} \sum_{s=1}^{r-i} \sum_{i<i_1<\cdots<i_s \le r} \|I(t, i, i_s, \dots, i_1)\|. \tag{4.35}$$

Denote by $\|\cdot\|_i$, $i = 1, \dots, r$, the matrix norm $\|\cdot\|_{v_i}$ induced by a Barabanov norm v_i for $\mathcal{M}_i$. Since two norms defined on a finite dimensional vector space are equivalent, there exists $K_i > 0$ such that $\|M\| \le K_i \|M\|_i$ for $i = 1, \dots, k$. Therefore the term $\sum_{i=1}^{r} \|\Phi_i(t, 0)\|$ is upper-bounded by $\sum_{i=1}^{r} K_i$.

We are left to estimate each term $\|I(t, i, i_s, \dots, i_1)\|$. Since $\mathcal{M}$ is bounded, there exists $K > 0$ independent of A such that the following holds true,

$$\begin{aligned}
&\|I(t, i, i_s, \dots, i_1)\| \\
&\quad \le \int_{0 \le \tau_1 \le \cdots \le \tau_s \le t} \|\Phi_i(t, \tau_s)A_{i,i_s}(\tau_s)\, \Phi_{i_s}(\tau_s, \tau_{s-1}) \\
&\qquad \cdots A_{i_2,i_1}(\tau_1)\Phi_{i_1}(\tau_1, 0)\|\, d\tau_1 \cdots d\tau_s \\
&\quad \le K \int_{0 \le \tau_1 \le \cdots \le \tau_s \le t} \|\Phi_i(t, \tau_s)\|_i\, \|\Phi_{i_s}(\tau_s, \tau_{s-1})\|_{i_s} \\
&\qquad \cdots \|\Phi_{i_1}(\tau_1, 0)\|_{i_1}\, d\tau_1 \cdots d\tau_s.
\end{aligned} \tag{4.36}$$

Notice that $\|\Phi_j(\tau_2, \tau_1)\|_j \le 1$ for every $j = 1, \dots, s$ and every $\tau_1 \le \tau_2$.

Given $T > 0$ there exists $K' \geq 1$ depending on T but not on t nor on A, such that

$$\int_{0\leq\tau_1\leq\cdots\leq\tau_s\leq t} \|\Phi_i(t,\tau_s)\|_i \, \|\Phi_{i_s}(\tau_s,\tau_{s-1})\|_{i_s} \cdots \|\Phi_{i_1}(\tau_1,0)\|_{i_1} \, d\tau_1\cdots d\tau_s$$

$$\leq K' \int_{0\leq\tau_1\leq\cdots\leq\tau_s\leq\lceil\frac{t}{T}\rceil} \left\|\Phi_i\left(\left\lceil\frac{t}{T}\right\rceil T, \left\lceil\frac{\tau_s}{T}\right\rceil T\right)\right\|_i \left\|\Phi_{i_s}\left(\left\lceil\frac{\tau_s}{T}\right\rceil T, \left\lceil\frac{\tau_{s-1}}{T}\right\rceil T\right)\right\|_{i_s}$$

$$\cdots \left\|\Phi_{i_1}\left(\left\lceil\frac{\tau_1}{T}\right\rceil T, 0\right)\right\|_{i_1} d\tau_1\cdots d\tau_s$$

$$\leq K'T^s \sum_{(m_1,\dots,m_s)\in\Xi^s_{\lceil\frac{t}{T}\rceil}} \left\|\Phi_i\left(\left\lceil\frac{t}{T}\right\rceil T, m_s T\right)\right\|_i \|\Phi_{i_s}(m_s T, m_{s-1}T)\|_{i_s}$$

$$\cdots \|\Phi_{i_1}(m_1 T, 0)\|_{i_1}$$

$$\leq K'T^s \sum_{(m_1,\dots,m_s)\in\Xi^s_{\lceil\frac{t}{T}\rceil}} \prod_{j=1}^{m_1}\alpha^j_{i_1}\cdots\prod_{j=m_{s-1}+1}^{m_s}\alpha^j_{i_s}\prod_{j=m_s+1}^{\lceil\frac{t}{T}\rceil}\alpha^j_i =: S_{A,i}(t), \tag{4.37}$$

where Ξ^s_μ is the set of s-tuples $m = (m_1,\dots,m_s) \in \mathbb{N}^s$ with $m_1 \leq \cdots \leq m_s \leq \mu$ and $\alpha^l_j = \|\Phi_j(lT,(l-1)T)\|_j$ for $j = 1,\dots,r$ and $l \in \mathbb{N}$.

Assume now that $\mathcal{M}$ does not have irreducible blocks in resonance. According to Lemma 4.47, there exist $C \in (0,1)$ and $T > 0$ independent of A such that the sequence $(\alpha^l)_{l=1}^{\lceil\frac{t}{T}\rceil}$, where $\alpha^l = (\alpha^l_j)_{1\leq j\leq r} \in (0,1]^r$, satisfy the assumptions of Lemma 4.48. Applying the latter, one concludes $S_{A,i}(t)$ is upper bounded by a constant independent of A, $i \in \{1,\dots,r\}$, and $t \geq 0$, hence concluding the proof of the theorem. □

Remark 4.49 The converse of Theorem 4.46 is clearly not true, namely, two blocks may be in resonance even if $\mathcal{M}$ is nondefective. This can be easily seen considering the case of $\mathcal{M}$ made of a single diagonal matrix.

On the other hand, considering $\mathcal{M}$ as in Example 4.45, we have seen that, for almost every choice of the pair (M^0_{12}, M^1_{12}), $\mathcal{M}$ is defective.

However, it is not true in general that almost every interconnection between blocks in resonance lead to a defective set of modes. In order to see this, let $\mathcal{M} = \{M^0, M^1\}$ with M^0, M^1 as in (4.29) and

$$M^0_{11} = 0_{2\times 2}, \quad M^0_{22} = \begin{pmatrix} 0 & -1 \\ 1 & 0 \end{pmatrix}, \quad M^1_{11} = M^1_{22} = \begin{pmatrix} -s & 1 \\ 1 & -s \end{pmatrix}, \quad s > 1.$$

The blocks $\mathcal{M}_1 = \{M^0_{11}, M^1_{11}\}$ and $\mathcal{M}_2 = \{M^0_{22}, M^1_{22}\}$ are in resonance for the signal $A \equiv M^0$. However, for s large enough $(\Sigma_{\mathcal{M}})$ is uniformly bounded for every choice of M^0_{12} and M^1_{12} in $\{N \in M_2(\mathbb{R}) \mid \|N\| < 1\}$ (Lemma 3.24).

We close this section by providing a sharpened version of the estimate in Proposition 4.42.

Proposition 4.50 *Consider a maximal invariant flag and the associated irreducible blocks $\mathcal{M}_1, \dots, \mathcal{M}_r$. Let I be the set of all indices $i \in \{1, \dots, r\}$ for which there exists $j \in \{1, \dots, r\}$ with $\mathcal{M}_i$ and $\mathcal{M}_j$ in resonance. Let $L = 1$ if I is empty and $L = \#I$ otherwise. Then, there exists $C > 0$ such that, for every $A \in \mathcal{S}_{\text{arb}}(\mathcal{M})$,*

$$\|\Phi_A(t, 0)\| \leq C(1 + t^{L-1})e^{t\lambda(\mathcal{M})}, \quad t \geq 0. \tag{4.38}$$

Proof of Theorem 4.46 Let $E_1 \subset \cdots \subset E_r = \mathbb{R}^n$ be the maximal invariant flag associated with the irreducible blocks $\mathcal{M}_1, \dots, \mathcal{M}_r$. The proof goes by induction on L. For $L = 1$, the conclusion follows by applying Theorem 4.46. Assume now that $L \geq 2$ and that the thesis is true up to $L - 1$.

Let j be the largest index in I. Consider two subspaces V_1, V_2 of $\mathbb{R}^n$ with $V_1 = E_{j-1}$ and $V_1 \oplus V_2 = \mathbb{R}^n$.

Let $m = \dim V_1$, and $\{v_1, \dots, v_n\}$ be a vector basis such that $V_1 = \text{span}\{v_1, \dots, v_m\}$ and $V_2 = \text{span}\{v_{m+1}, \dots, v_n\}$. Write every matrix $M \in \mathcal{M}$ and every signal $A \in \mathcal{S}_{\text{arb}}(\mathcal{M})$ in the corresponding block form

$$M = \begin{pmatrix} \tilde{M}_{11} & \tilde{M}_{12} \\ 0 & \tilde{M}_{22} \end{pmatrix}, \qquad A = \begin{pmatrix} \tilde{A}_{11} & \tilde{A}_{12} \\ 0 & \tilde{A}_{22} \end{pmatrix}.$$

Define, for $i = 1, 2$, $\tilde{\mathcal{M}}_i := \{\tilde{M}_{ii} \mid M \in \mathcal{M}\}$ and notice that, by construction, the blocks of $\tilde{\mathcal{M}}_1$ and $\tilde{\mathcal{M}}_2$ are $\mathcal{M}_1, \dots, \mathcal{M}_{j-1}$ and $\mathcal{M}_j, \dots, \mathcal{M}_r$ respectively. Let J be the subset of $\{1, \dots, j-1\}$ such that for every $i \in J$ there exists $j \in J$ with $\mathcal{M}_i$ and $\mathcal{M}_j$ in resonance. Then, $\#J \leq L - 1$ since there exist at most $L - 1$ diagonal blocks of $\tilde{\mathcal{M}}_1$ resonant with other blocks, and there is no resonance between blocks of $\tilde{\mathcal{M}}_2$. By the induction hypothesis, there exists $C > 0$ such that, for every $t, s \geq 0$, one has

$$\|\Phi_{\tilde{A}_{11}}(t + s, s)\| \leq C(1 + t^{L-2})e^{t\lambda(\mathcal{M})}, \qquad \|\Phi_{\tilde{A}_{22}}(t, 0)\| \leq Ce^{t\lambda(\mathcal{M})}. \tag{4.39}$$

The nondiagonal term in $\Phi_A(t, 0)$ satisfies an estimate analogous to Eq. (4.24) and, by using Eq. (4.39), a simple computation concludes the induction proof. □

4.4 Notes and References

The main subject of this chapter, the notion of Barabanov norm, takes its name from N.E. Barabanov, who first introduced it for continuous-time systems in the paper [Bar88a] (after having developed the analogous concept for time-discrete systems in [Bar88b]). The original norm construction by Barabanov in the case where $\mathcal{M}$ is convex, irreducible, and compact is obtained through the formula

$$v_B(x) := \sup_{A \in \mathcal{S}_{\text{arb}}(\mathcal{M})} \left(\limsup_{t \to +\infty} \|\Phi_A(t,0)x\| \right), \qquad x \in \mathbb{R}^n. \tag{4.40}$$

Based on this definition, Barabanov gives in [Bar88a] an alternative proof of Theorem 4.8.

A further alternative proof, simplifying some steps of the original one, has been proposed by F. Wirth in [Wir02], where a Barabanov norm is identified using the relation obtained in Eq. (4.7) as a definition. Let us mention that the notion of nondefectiveness used in [Wir02] is slightly different from the one used here. The word nondefective is actually well-established and unambiguous for single matrices, but generalizations to the case of sets of modes differ from one author to another. For instance, in [GZ01] nondefectiveness is defined in terms of the existence of an extremal norm. The characterization of compact and convex sets of modes admitting a Barabanov norm (Proposition 4.13) has first appeared in [CMS22].

The notion of adjoint system and its connection with extremal trajectories had already been studied in [Bar88a]. In particular, Theorem 4.19 has been established in [Bar88a, Theorem 4] with a completely different argument than ours. Our proof rather follows that in [CGM15]. The latter reference also contains the first appearance of the results on uniqueness, strict convexity, and regularity of Barabanov norms contained in this chapter, as well as several open problems. Example 4.29 was first discussed in [Bar08, Example 1]. More results on the uniqueness of the Barabanov norm in the discrete-time setting can be found, for instance, in [Mor12, Pro22].

The proof of the Berger–Wang formula first appeared in the discrete-time setting in [BW92]. The fact that the equality in the Berger–Wang formula yields the continuity of the maximal Lyapunov exponent has been first noticed in [HS95], again in the discrete-time setting. Finer continuity results can be found in [Wir02], where the author proves that the maximal Lyapunov exponent is locally Lipschitz continuous when restricted to irreducible and compact sets of modes.

The notion of resonance for irreducible subsystems of a switched linear system has been introduced in the paper [CMS12], which contains finer results on the maximal degree of polynomial divergence of a system with zero maximal Lyapunov exponent, generalizing Proposition 4.50. Such results are based on the notion of *chord*, which consists in a family of resonances involving several subsystems. Further extensions of Proposition 4.50, dealing with non-integer orders of divergence of trajectories of a switched linear system, can be found in [Mor22, Pro16]. The last part of Remark 4.49 has been observed in [PJ15], in the framework of the disproof of a conjecture given in [CMS12] on the genericity (in a suitable class of systems) of the resonance phenomenon.

Chapter 5
Switched Systems in Dimension Three

In this chapter, we mainly focus on systems in $\mathbb{R}^3$ for the class of arbitrarily switching signals. For that purpose, we assume that $\mathscr{M}$ is irreducible, since otherwise, relying on Proposition 1.31, the stability analysis can essentially be reduced to switched systems in dimension two (and addressed using the techniques introduced in Chap. 2). By Remark 1.21, we may further suppose that $\mathscr{M}$ is convex and compact, hence, according to Theorem 4.8, $\mathscr{M}$ admits a Barabanov norm v and one can study its extremal trajectories, whose role in stability analysis is similar to that of worst-case trajectories for two-dimensional systems (see Chap. 2). Up to replacing the set $\mathscr{M}$ with $\mathscr{M} - \lambda(\mathscr{M})\mathrm{Id}_3$, the unit sphere S associated with v turns out to be a compact Lipschitz manifold of dimension two, invariant for the flow of extremal trajectories. Exploiting this fact, we prove in the first part of the chapter a Poincaré–Bendixson-type result, showing in particular the existence of a periodic extremal trajectory on S. This is achieved under an assumption on $\mathscr{M}$, referred to in the sequel as the forward uniqueness property. We also prove that the sets of modes for which such a property holds true are dense among the convex and compact subsets of $M_3(\mathbb{R})$. The second part of the chapter deals with a particular class of three-dimensional switched systems associated with a segment of matrices for which the structure of periodic extremal trajectories is investigated in detail.

5.1 A Poincaré–Bendixson Theorem on Two-Dimensional Barabanov Spheres

In this section we first consider an assumption on the set $\mathscr{M}$ stating that, for every Barabanov norm v, forward extremal trajectories associated with v are unique, and then we show that such an assumption is generic in a sense specified below.

Y. Chitour et al., *Dynamics and Stability of Continuous-Time Switched Linear Systems*, PNLDE Subseries in Control 105,
https://doi.org/10.1007/978-3-031-82982-6_5

Finally, we prove a Poincaré–Bendixson theorem for three-dimensional switched linear systems.

5.1.1 The Forward Uniqueness Property

Let us give a precise definition for the forward uniqueness property. The definition actually makes sense in any dimension $n \geq 1$.

Definition 5.1 (Forward Uniqueness Property) We say that $(\mathcal{M}, v)$, with $\mathcal{M} \subset M_n(\mathbb{R})$ convex, compact, and irreducible, and v a Barabanov norm for $\mathcal{M}$, satisfies the *forward uniqueness property* if for every $(x_0, \ell_0) \in \mathbb{R}^n \times \mathbb{R}^n$ with $x_0 \neq 0$ and $\ell_0 \in \partial v(x_0)$ there exists at most one extremal pair for v starting at (x_0, ℓ_0).

Recall that, according to Definition 4.20, an extremal pair for v is a solution $\big(x(\cdot), \ell(\cdot)\big)$ of $\dot{x}(t) = A(t)x(t)$ and $\dot{\ell}(t) = -A^\top(t)\ell(t)$ for $t \in [0, +\infty)$, with $\big(x(0), \ell(0)\big) = (x_0, \ell_0)$, $x_0 \neq 0$, and $\ell_0 \in \partial v(x_0)$, for some $A \in \mathcal{S}_{\mathrm{arb}}(\mathcal{M})$ satisfying $\max_{M\in\mathcal{M}} \ell^\top(t)Mx(t) = \ell^\top(t)A(t)x(t) = \lambda(\mathcal{M})v(x_0)$ for almost every $t \geq 0$.

In the rest of this section we focus on the construction of a relevant family of convex subsets of $M_n(\mathbb{R})$ satisfying the forward uniqueness property. We start by introducing the notion of strongly convex functions and sets.

Definition 5.2 A twice continuously differentiable function $\varphi \colon \mathbb{R}^k \to [0, +\infty)$ is said to be *strongly convex* if there exists $\alpha > 0$ such that the Hessian of φ satisfies $\mathrm{Hess}_x \varphi \geq \alpha \mathrm{Id}_k$ for every $x \in \mathbb{R}^k$. We say that a subset X of $\mathbb{R}^k$ is *strongly convex* if there exists a C^2 differentiable strongly convex function $\varphi \colon \mathbb{R}^k \to [0, +\infty)$ such that $X = \{x \in \mathbb{R}^k \mid \varphi(x) \leq 1\}$.

It can be easily shown that a C^2 function φ is strongly convex if and only if $x \mapsto \varphi(x) - \alpha\|x\|^2$ is convex for some $\alpha > 0$. Note that the minimum of a strongly convex function always exists and consists of a single point, hence a strongly convex set containing at least two points has nonempty interior. Furthermore, it is easy to see that any strongly convex set is bounded, hence compact.

We will next consider strongly convex subsets of $M_n(\mathbb{R})$. Note that, identifying $M_n(\mathbb{R})$ with $\mathbb{R}^{n^2}$, the corresponding Euclidean norm of a matrix M coincides with its Frobenius norm $\|M\|_F = \mathrm{tr}(M^\top M)^{1/2}$. Also, every strongly convex subset of $M_n(\mathbb{R})$ containing at least two matrices, having nonempty interior, is irreducible. We have the following density result for strongly convex subsets of $M_n(\mathbb{R})$.

Proposition 5.3 *The subset of $M_n(\mathbb{R})$ made of strongly convex sets of matrices is dense in the set of compact convex subsets of $M_n(\mathbb{R})$.*

Proof Pick a compact convex subset $\mathcal{M}$ of $M_n(\mathbb{R})$ and $\varepsilon > 0$. Let $\mathcal{M}_\varepsilon^1$ be the closed $\frac{\varepsilon}{3}$-neighborhood of $\mathcal{M}$ in $M_n(\mathbb{R})$, i.e.,

$$\mathcal{M}_\varepsilon^1 = \left\{N \in M_n(\mathbb{R}) \mid \exists M \in \mathcal{M} \text{ s.t. } \|N - M\| \leq \frac{\varepsilon}{3}\right\}.$$

It is immediate to see that $\mathscr{M}_\varepsilon^1$ is also compact and convex and, in addition, it has nonempty interior. Given M_0 in the interior of $\mathscr{M}_\varepsilon^1$ we define $\varphi_0 \colon M_n(\mathbb{R}) \to [0, +\infty)$ by setting $\varphi_0(M) = \inf\{\lambda > 0 \mid M_0 + \frac{1}{\lambda}(M - M_0) \in \mathscr{M}_\varepsilon^1\}$. It is clear that φ_0 admits a strict minimum equal to zero at $M = M_0$. Moreover, $\varphi_0(M) \leq 1$ for every $M \in \mathscr{M}_\varepsilon^1$. Conversely, $\varphi_0(M) \leq 1$ implies that $M \in \mathscr{M}_\varepsilon^1$, since for every $\lambda \in (0, 1]$, we can write

$$M = (1 - \lambda)M_0 + \lambda \left(M_0 + \frac{1}{\lambda}(M - M_0) \right),$$

that is, M is a convex combination of M_0 and $M_0 + \frac{1}{\lambda}(M - M_0)$. Furthermore, for every $M_1, M_2 \in M_n(\mathbb{R}) \setminus \{M_0\}$ and $\alpha \in [0, 1]$, setting $\xi = \frac{\alpha\varphi_0(M_1)}{\alpha\varphi_0(M_1)+(1-\alpha)\varphi_0(M_2)} \in [0, 1]$, one has

$$\begin{aligned}
M_0 + &\frac{1}{\alpha\varphi_0(M_1) + (1-\alpha)\varphi_0(M_2)} (\alpha M_1 + (1-\alpha)M_2 - M_0) \\
&= \xi \left(M_0 + \frac{1}{\varphi_0(M_1)}(M_1 - M_0) \right) + (1 - \xi) \left(M_0 + \frac{1}{\varphi_0(M_2)}(M_2 - M_0) \right),
\end{aligned}$$

which is an element of the convex set $\mathscr{M}_\varepsilon^1$. A similar fact also holds in the case $M_1 = M_0$. We then deduce that φ_0 is convex. The next step consists in regularizing the map φ_0, and is achieved by a classical procedure (of which we only provide a sketch here). Let $\rho \colon M_n(\mathbb{R}) \to [0, \infty)$ be a smooth function with compact support such that $\int_{M_n(\mathbb{R})} \rho(M)dM = 1$. For $\eta > 0$, denote by $\varphi_\eta \colon M_n(\mathbb{R}) \to [0, \infty)$ the convolution

$$\varphi_\eta(N) = \eta^{-n^2} \int_{M_n(\mathbb{R})} \rho\left(\frac{M}{\eta}\right) \varphi_0(N - M)dM, \quad N \in M_n(\mathbb{R}).$$

It turns out that φ_η is smooth and that $(\varphi_\eta)_{\eta>0}$ converges to φ_0 uniformly on any compact subset of $M_n(\mathbb{R})$ as η goes to zero. We next show that φ_η is also convex. Indeed, one has for every N_1, N_2 in $M_n(\mathbb{R})$ and $\lambda \in [0, 1]$

$$\begin{aligned}
\varphi_\eta(\lambda N_1 + (1-\lambda)N_2) &= \eta^{-n^2} \int_{M_n(\mathbb{R})} \rho\left(\frac{M}{\eta}\right) \varphi_0(\lambda N_1 + (1-\lambda)N_2 - M)dM \\
&= \eta^{-n^2} \int_{M_n(\mathbb{R})} \rho\left(\frac{M}{\eta}\right) \varphi_0\big(\lambda(N_1 - M) \\
&\quad + (1-\lambda)(N_2 - M)\big)dM \\
&\leq \lambda\varphi_\eta(N_1) + (1-\lambda)\varphi_\eta(N_2).
\end{aligned}$$

Notice that $M \mapsto \varphi_0(M_0 + M)$ is absolutely homogeneous, implying that $\{M \in M_n(\mathbb{R}) \mid \varphi_0(M) < \alpha\}$ coincides with the interior of $\{M \in M_n(\mathbb{R}) \mid \varphi_0(M) \leq \alpha\}$ for every $\alpha \geq 0$. This fact, together with the uniform convergence of $(\varphi_\eta)_{\eta>0}$ to φ_0 on compact sets, implies that there exists $\eta_* > 0$ small enough such that the Hausdorff distance between $\mathcal{M}_\varepsilon^1$ and $\mathcal{M}_\varepsilon^2$ is smaller than $\frac{\varepsilon}{3}$, where $\mathcal{M}_\varepsilon^2$ denotes the set of matrices M for which $\varphi_{\eta_*}(M) \leq 1$. Finally, there exists $a > 0$ such that, letting $\mathcal{M}_\varepsilon^3$ be the set of matrices M for which $\varphi_{\eta_*}(M) + a\|M\|_F^2 \leq 1$, the Hausdorff distance between $\mathcal{M}_\varepsilon^2$ and $\mathcal{M}_\varepsilon^3$ is smaller than $\frac{\varepsilon}{3}$. This is again a consequence of the fact that $\{M \in M_n(\mathbb{R}) \mid \varphi_0(M) < \alpha\}$ coincides with the interior of $\{M \in M_n(\mathbb{R}) \mid \varphi_0(M) \leq \alpha\}$ for $\alpha \geq 0$. The set $\mathcal{M}_\varepsilon^3$ is strongly convex and the Hausdorff distance between $\mathcal{M}$ and $\mathcal{M}_\varepsilon^3$ is smaller than ε, concluding the proof. □

We next investigate the relationship between strong convexity and the forward uniqueness property. We will use the following technical result.

Lemma 5.4 *Let $X \subset \mathbb{R}^k$ be strongly convex and φ as in Definition 5.2. Denote by Y the boundary of X, that is, $Y = \{x \in \mathbb{R}^k \mid \varphi(x) = 1\}$. Then, the map $x \mapsto \frac{\nabla\varphi(x)}{\|\nabla\varphi(x)\|}$ from Y to S^{k-1}, that is, the map that associates with a point $x \in Y$ the corresponding outward unit normal vector, is a diffeomorphism.*

Proof Note that $\nabla\varphi(x)$ is a nonzero vector for every x in Y since the gradient of a differentiable convex function vanishes only at global minima, hence the map $x \mapsto \frac{\nabla\varphi(x)}{\|\nabla\varphi(x)\|}$ is well-defined from Y to S^{k-1}. It is also a bijection. Indeed, for every unit vector $v \in S^{k-1}$, there exists a unique point $z \in Y$ such that, for some $\mu \in \mathbb{R}$, $\{y \in \mathbb{R}^k \mid v^\top y = \mu\}$ is the tangent space to X at z and v is the outward unit normal vector at z to X, as it follows from the strong convexity of X. To conclude the proof it remains to check that, for every $x \in \mathbb{R}^k$, the Jacobian of the map $x \mapsto \frac{\nabla\varphi(x)}{\|\nabla\varphi(x)\|}$, restricted to T_xY, the tangent space to Y at x, is nonsingular. A direct computation shows that

$$\frac{d}{dx}\frac{\nabla\varphi(x)}{\|\nabla\varphi(x)\|} = \frac{1}{\|\nabla\varphi(x)\|}\mathrm{Hess}_x\varphi - \frac{\nabla\varphi(x)\nabla\varphi^\top(x)}{\|\nabla\varphi(x)\|^3}\mathrm{Hess}_x\varphi.$$

Every nonzero vector $v \in T_xY$ is orthogonal to $\nabla\varphi(x)$, hence

$$v^\top \frac{d}{dx}\frac{\nabla\varphi(x)}{\|\nabla\varphi(x)\|} v = \frac{1}{\|\nabla\varphi(x)\|}\mathrm{Hess}_x\varphi(v, v) \geq \frac{\alpha\|v\|^2}{\|\nabla\varphi(x)\|} > 0,$$

that is T_xY has trivial intersection with the kernel of the Jacobian matrix $\frac{d}{dx}\frac{\nabla\varphi(x)}{\|\nabla\varphi(x)\|}$. This concludes the proof of the lemma. □

Proposition 5.5 *Assume that $\mathcal{M} \subset M_n(\mathbb{R})$ is strongly convex and contains at least two matrices, and let v be a Barabanov norm for $\mathcal{M}$. Then $(\mathcal{M}, v)$ satisfies the forward uniqueness property.*

Proof According to Theorem 4.19, every extremal pair $(x(\cdot), \ell(\cdot))$ for v associated with a switching law $A \in \mathcal{S}_{\text{arb}}(\mathcal{M})$ satisfies

$$\max_{M \in \mathcal{M}} \ell^\top(t) M x(t) = \ell^\top(t) A(t) x(t) = \lambda(\mathcal{M}) v(x_0), \qquad \text{for a.e. } t \geq 0. \tag{5.1}$$

Writing $\ell^\top(t) M x(t)$ as $\operatorname{tr}(M P(t))$ where $P(t) = x(t)\ell^\top(t)$ it follows that (5.1) is equivalent to the property that, for almost every $t \geq 0$, the hyperplane

$$\{Q \in M_n(\mathbb{R}) \mid \operatorname{tr}(Q P(t)) = \lambda(\mathcal{M}) v(x_0)\}$$

is tangent to the boundary of $\mathcal{M}$ at $A(t) \in \mathcal{M}$ and $P^\top(t)$ coincides, up to rescaling by a time-dependent positive factor, with the outward unit normal vector to $\mathcal{M}$ at $A(t)$ (which is well-defined since the boundary of $\mathcal{M}$ is C^2). By Lemma 5.4 there exists a C^1 diffeomorphism ψ between the boundary of $\mathcal{M}$ and the set of matrices of unit Frobenius norm, so that the matrix $A(t)$ may be expressed as

$$\psi^{-1}\left(\frac{P^\top(t)}{\|P^\top(t)\|_F}\right) = \psi^{-1}\left(\frac{\ell(t) x^\top(t)}{\|\ell(t)\| \, \|x(t)\|}\right).$$

Hence, any extremal pair is an integral curve of the vector field

$$(x, l) \mapsto F(x, \ell) = \begin{pmatrix} A(x, \ell) x \\ -A(x, \ell)^\top \ell \end{pmatrix}, \qquad \text{where} \qquad A(x, \ell) = \psi^{-1}\left(\frac{\ell x^\top}{\|\ell\| \, \|x\|}\right),$$

which is well-defined and C^1 on $(\mathbb{R}^n \setminus \{0\})^2$. We deduce that, for every $x_0 \neq 0$ and $\ell_0 \in \partial v(x_0)$, there exists a unique extremal pair $(x(\cdot), \ell(\cdot))$ with $(x(0), \ell(0)) = (x_0, \ell_0)$, proving the forward uniqueness property for $(\mathcal{M}, v)$. □

5.1.2 A Poincaré–Bendixson Theorem in Dimension Three

The aim of this section consists in proving the following Poincaré–Bendixson theorem for three-dimensional switched linear systems.

Theorem 5.6 *Let $\mathcal{M}$ be a convex, compact, and irreducible subset of $M_3(\mathbb{R})$ and denote by v a Barabanov norm for $\mathcal{M}$. Assume, moreover, that $(\mathcal{M}, v)$ satisfies the forward uniqueness property, that $\lambda(\mathcal{M}) = 0$, and that $\mathcal{M}$ is made of nonsingular matrices. Then, the ω-limit set of any extremal pair for v is the support of a periodic extremal pair for v.*

Fig. 5.1 Transverse section.

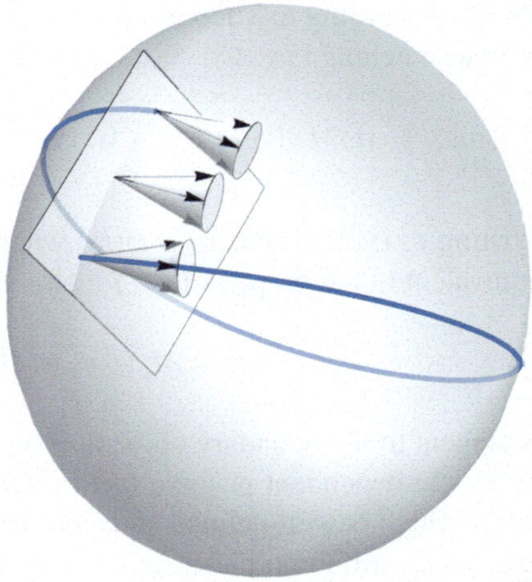

For that purpose, we introduce some notation as well as several technical lemmas. We fix in this section a Barabanov norm v for $\mathcal{M} \subset M_3(\mathbb{R})$, we denote by S the corresponding unit sphere, and we assume that $\lambda(\mathcal{M}) = 0$. We use the notation $F_{\mathcal{M}}$ and $\hat{F}_{\mathcal{M}}$ introduced in (4.13) and (4.14). Recall that

$$\dot{x} \in \hat{F}_{\mathcal{M}}(x) \tag{5.2}$$

is a linear differential inclusion whose solutions are the extremal trajectories for v. Then Theorem 5.6 ensures, in particular, the existence of periodic trajectories for the differential inclusion (5.2) on the possibly nonsmooth manifold S.

Definition 5.7 A point $z \in S$ is said to be a *nonstationary point of* $(\Sigma_{\mathcal{M}})$ if $0 \notin F_{\mathcal{M}}(z)$. We say that an affine plane P in $\mathbb{R}^3$ *crosses* S *transversally* if $S \setminus P$ is not connected. Given an affine plane P crossing S transversally and a connected nonempty open subset Ξ of $P \cap S$, we say that Ξ is a *transverse section of* S *for* $(\Sigma_{\mathcal{M}})$ if for every $x \in \Xi$ and $y \in F_{\mathcal{M}}(x)$ it holds $w^\top y > 0$, where w is a fixed orthogonal nonzero vector to P (see Fig. 5.1).

Notice that, by the connectedness and Lipschitz regularity of S and the definition of transverse crossing, any transversal section Ξ of S for $(\Sigma_{\mathcal{M}})$ is a connected nontrivial Lipschitz curve in $\mathbb{R}^3$ lying on S.

The following lemma allows one to follow a strategy which is similar to the classical one in order to prove a Poincaré–Bendixson result.

Lemma 5.8 *Assume that the hypotheses of Theorem 5.6 hold true. For every nonstationary point $z \in S$ of $(\Sigma_{\mathcal{M}})$, there exist a plane P_z through the origin and a neighborhood W of z in S such that either (a) for every $z' \in W$, $(z' + P_z) \cap W$ is a transverse section of S for $(\Sigma_{\mathcal{M}})$ or (b) no extremal trajectory starting in $S \setminus W$ enters in W.*

Proof Let z be a nonstationary point of $(\Sigma_{\mathcal{M}})$. Since the set $F_{\mathcal{M}}(z)$ is convex and compact, one gets by applying the hyperplane separation theorem (Theorem A.14), there exists a plane P which strictly separates 0 and $F_{\mathcal{M}}(z)$ and let $w \in \mathbb{R}^3$ and $\nu > 0$ such that $P = \{x \in \mathbb{R}^3 \mid w^\top x = \nu\}$. Notice that for $\tilde{w}$ sufficiently close to w, the plane $\{x \in \mathbb{R}^3 \mid \tilde{w}^\top x = \nu\}$ still strictly separates 0 and $F_{\mathcal{M}}(z)$. Hence, setting $P_z = \{x \mid w^\top x = 0\}$, and up to slightly modifying w, $(z + P_z) \cap S$ has nonempty interior in S, and hence, we can assume that either $z + P_z$ crosses S transversally or $(z + P_z) \cap S = \{z\}$. The point 0 lies then in the region $\{x \mid w^\top x < \nu\}$, while the set $F_{\mathcal{M}}(z)$ lies in the region $\{x \mid w^\top x > \nu\}$. Since P is closed, $P \cap F_{\mathcal{M}}(z) = \emptyset$, and $F_{\mathcal{M}}$ is continuous as a function from $\mathbb{R}^3$ to the space of compact subsets of $\mathbb{R}^3$ endowed with the Hausdorff distance, there exists $\varepsilon_0 > 0$ so that if x satisfies $\|x - z\| < \varepsilon_0$ then $F_{\mathcal{M}}(x)$ is a subset of the region $\{y \in \mathbb{R}^3 \mid w^\top y > \nu\}$. Defining $W = \{x \in S \mid \|x - z\| < \varepsilon_0\}$, one has $w^\top y > \nu$ for every $x \in W$ and $y \in F_{\mathcal{M}}(x)$.

We distinguish two cases, depending on whether $z + P_z$ crosses S transversally or $(z + P_z) \cap S = \{z\}$. In the first case, up to possibly reducing ε_0, we can conclude that property (a) of the statement holds true. In the second case, $w^\top z < w^\top x$ for every $x \in S \setminus \{z\}$. Hence, there exists $\delta > 0$ such that $Z = \{x \in S \mid w^\top (x - z) < \delta\}$ is a neighborhood of z in S contained in W. Notice that for every extremal trajectory $x(\cdot)$ on S, $\frac{d}{dt}(w^\top x(t)) > \nu$ for almost every time t such that $x(t) \in W$. Then no extremal trajectory starting in $S \setminus Z$ can enter in Z, hence (b) holds. □

The following lemmas are also needed in the argument of Theorem 5.6. The first one is an immediate consequence of Proposition 4.39.

Lemma 5.9 *Assume that* $(\mathcal{M}, v)$ *satisfies the forward uniqueness property and that every matrix of* $\mathcal{M} - \lambda(\mathcal{M})\mathrm{Id}_3$ *is nonsingular. If* $x_1(\cdot)$ *and* $x_2(\cdot)$ *are two extremal trajectories for* v *intersecting each other at some* $z \in \mathbb{R}^3$ *(in the sense of Definition 4.38), then* v *is differentiable at* z *and there exists a unique extremal trajectory starting from* z.

Proof The fact that v is differentiable at z follows by Proposition 4.39. By using Theorem 4.19, each extremal trajectory starting from z can be lifted to an extremal pair starting from $(z, \nabla v(z))$. By the forward uniqueness property, such an extremal pair is necessarily unique, hence there exists a unique extremal trajectory starting from z. □

The second lemma goes as follows.

Lemma 5.10 *Assume that the hypotheses of Theorem 5.6 hold true. Let* $x(\cdot)$ *be an extremal trajectory for* v. *Assume that* $x(t_0 + \cdot)$ *is non-injective for every* $t_0 \geq 0$. *Then* $x(\cdot)$ *is periodic on* $[T, +\infty)$ *for some* $T \geq 0$.

Proof Let $t_1 \geq 0$ and $\tau > 0$ be such that $x(t_1) = x(t_1 + \tau)$. By assumption, t_1 can be taken arbitrarily large.

Let $t_2 \in [0, t_1]$ be maximal such that $x(t_1 - s) = x(t_1 + \tau - s)$ for every $s \in [0, t_2]$. Assume for now that $t_2 < t_1$. Then $t \mapsto x(t)$ and $t \mapsto x(\tau + t)$ intersect in the sense of Definition 4.38 at $t = t_1 - t_2$. Applying Lemma 5.9, it follows that $x(t) = x(\tau + t)$ for every $t \geq t_1 - t_2$, i.e., $x(\cdot)$ is τ-periodic on $[t_1 - t_2, +\infty)$.

We are left to consider the case where for every $t_1 \geq 0$ and $\tau > 0$ such that $x(t_1) = x(t_1 + \tau)$ one has $x(\cdot)|_{[0,t_1]} = x(\tau + \cdot)|_{[0,t_1]}$. In particular, for every $t_1 \geq 0$ and $\tau > 0$ one has that

$$x(t_1) = x(t_1 + \tau)$$
$$\Longrightarrow \Big(\forall k \in \mathbb{N} \text{ such that } k\tau \leq t_1, \quad x(\cdot)|_{[0,\tau]} = x((k-1)\tau + \cdot)|_{[0,\tau]}\Big). \tag{5.3}$$

Let $\bar{t} > 0$ be such that $x(\bar{t}) = x(0)$. We claim that, for every $k \in \mathbb{N}$, there exist $0 \leq \tau_1 < \tau_2$ such that $x(\tau_1) = x(\tau_2)$ and $\tau_2 - \tau_1 \geq k\bar{t}$. Indeed take $t_2 > t_1 > 2k\bar{t}$ such that $x(t_1) = x(t_2)$. Then either $t_2 - t_1 \geq k\bar{t}$ as required, or by applying (5.3), there exists an integer $h \geq 2$ such that $x(0) = x(h(t_2 - t_1))$ and $h(t_2 - t_1) > k\bar{t}$. The claim follows in the latter case by taking $\tau_1 = 0$ and $\tau_2 = h(t_2 - t_1)$.

Using the claim, we prove that $x(\cdot)$ is $\bar{t}$-periodic. We first apply (5.3) with $t_1 = \tau_1$, $\tau = \tau_2 - \tau_1$ and k arbitrary. One deduces that $x|_{[0,t_1]}(\cdot) = x(\tau_2 - \tau_1 + \cdot)|_{[0,t_1]}$, implying in particular that

$$x(\tau_2 - \tau_1) = x(0) = x(\bar{t}) = x(\tau_2 - \tau_1 + \bar{t}).$$

Applying again (5.3) with $t_1 = \tau_2 - \tau_1$, $\tau = \bar{t}$, and any $k \in \mathbb{N}$, one obtains that $x(t) = x(t + \bar{t})$ for $t \in [0, (k-1)\bar{t}]$, and the conclusion follows since k can be taken arbitrarily large. □

Thus, without loss of generality, we can assume in the proof of Theorem 5.6 that the trajectory $x(\cdot)$ is injective on $[0, +\infty)$. The following lemma provides a key monotonicity property for the intersection of an extremal trajectory on S with a transverse section Ξ of S for $(\Sigma_{\mathcal{M}})$.

Lemma 5.11 *Given an injective extremal trajectory $x(\cdot)$ on S and a transverse section Ξ of S for $(\Sigma_{\mathcal{M}})$, the following results hold true.*

- *If $x(\cdot)$ intersects Ξ several times, the intersection points are placed monotonically on Ξ, i.e., for any increasing sequence of positive times $(t_i)_{i\geq 0}$ such that $x(t_i) \in \Xi$ for every $i \geq 0$, $x(t_i)$ belongs to the arc in Ξ between $x(t_{i-1})$ and $x(t_{i+1})$ for every $i \geq 1$.*
- *The ω-limit set $\omega(x(\cdot))$ can intersect Ξ at no more than one point. Moreover, if z is the point of intersection of $\omega(x(\cdot))$ and Ξ then $x(\cdot)$ intersects Ξ only along a sequence of points $x(t_i)$, for $i \in \mathbb{N}$, such that the sequence $(t_i)_{i\in\mathbb{N}}$ is increasing with $t_i \to +\infty$ and the sequence $(x(t_i))_{i\in\mathbb{N}}$ tends to z monotonically on Ξ.*

Proof By definition of transverse section and by continuity of $F_{\mathcal{M}}$, there exists a nonzero vector w so that $y \mapsto w^\top y$ is constant on Ξ and, for every $x \in \Xi$, there exists a positive scalar δ such that $\delta \leq w^\top My$ for every $M \in \mathcal{M}$ and every y in a neighborhood O of x. Hence, for every pair of times t, t' such that $x(s) \in O$ for s between t and t', one has $|w^\top(x(t) - x(t'))| \geq \delta|t - t'|$. As a consequence, the set of times at which $x(\cdot)$ intersects Ξ is discrete. Let us assume, without loss of generality, that $(x(t_i))_{i\geq 0}$ is the set of all intersections of $x(\cdot)$ and Ξ. For every $i \geq 1$, consider the arc Γ_i made of the union of the arc on Ξ with extremities $x(t_{i-1})$ and $x(t_i)$ and $\{x(t) \mid t \in [t_{i-1}, t_i]\}$.

By assumption, Γ_i is a non-self-intersecting continuous loop so that $S \setminus \Gamma_i$ has two connected components C_i and D_i, open in S. We use C_i to denote the connected component of $S \setminus \Gamma_i$ such that the directions in $\hat{F}_{\mathcal{M}}(x)$ point inside C_i for every x in the arc of Ξ connecting $x(t_{i-1})$ and $x(t_i)$. We choose an order on Ξ (induced by an injective continuous parameterization of Ξ) such that $x(t_{i-1}) < x(t_i)$. We are left to prove that $x(t_i) < x(t_{i+1})$.

Let us first prove that $x(t) \in C_i$ for $t > t_i$. Indeed, reasoning by contradiction yields a time $t_* > t_i$ so that $x(t_*) \in \Gamma_i$ and $x(t) \in C_i$ for $t \in (t_i, t_*)$. By injectivity of $x(\cdot)$, $x(t_*)$ belong to the open arc of Ξ connecting $x(t_{i-1})$ and $x(t_i)$. However, by transversality of Ξ, $x(t) \notin C_i$ for t in a left neighborhood of t_*, contradicting the definition of t_*. The argument above shows, in particular, that $x(t_{i+1})$ cannot lie in the arc of Ξ connecting $x(t_{i-1})$ and $x(t_i)$.

To prove now the required monotonicity property, i.e., that $x(t_i) < x(t_{i+1})$, it remains to show that $x(t_{i+1})$ does not belong to the arc Ξ^- of Ξ made of the points y such that $y < x(t_{i-1})$.

The latter fact finally follows from the claim that $\Xi^- \subset D_i$. To see that, notice first that the claim follows by exhibiting one point of Ξ^- in D_i since Ξ^- does not intersect Γ_i. On the other hand, there exists a sufficiently small neighborhood U of $x(t_{i-1})$ so that the curve $x(\cdot)$ separates $S \cap U$ into two connected components, one containing $\Xi^- \cap U$ and the other one $C_i \cap U$. Therefore $\Xi^- \cap D_i$ is nonempty, which concludes the proof of the claim and that of the first item of the lemma.

The second item follows from the fact that if a monotone sequence has a limit point, then the latter is unique. □

The next two lemmas are instrumental for the rest of the argument.

Lemma 5.12 *Assume that the hypotheses of Theorem 5.6 hold true and let $x(\cdot)$ be a solution of $(\Sigma_{\mathcal{M}})$. For any $z \in \omega(x(\cdot)) \setminus \{0\}$ there exists a periodic extremal trajectory $x_z(\cdot)$ contained in $\omega(x(\cdot))$ with $x_z(0) = z$.*

Proof According to Lemma 4.25, there exists an extremal trajectory $x_z(\cdot)$ starting at z whose support is contained in $\omega(x(\cdot))$. Fix $\bar{z}$ in $\omega(x_z(\cdot)) \subset \omega(x(\cdot))$. Let Ξ be a transverse section of S for $(\Sigma_{\mathcal{M}})$ passing through $\bar{z}$, whose existence is guaranteed by Lemma 5.8 (since a point for which (b) as in the statement of Lemma 5.8 holds true cannot be in the ω-limit set of an extremal trajectory). Note that all the points of intersection of $x_z(\cdot)$ with Ξ are in $\omega(x(\cdot))$. Therefore, by Lemma 5.11, all these points coincide with $\bar{z}$. It follows then by Lemma 5.10 that $x_z(\cdot)$ is periodic on $[t^*, +\infty)$ for some $t^* \geq 0$. Let us assume, without loss of generality, that t^* is minimal. We are left to prove that $t^* = 0$. By contradiction, assume that $t^* > 0$, so that $x_z(\cdot)$ and $x_z(t^* + \cdot)$ intersect at $x_z(t^*) = x_z(t^* + T)$, where T is a period of x_z. Consider a neighborhood W of $x_z(t^*)$ and a plane $\bar{P}$ through the origin as in Lemma 5.8 (with $x_z(t^*)$, which belongs to $\omega(x_z(\cdot))$, playing the role of z). Then, since $F_{\mathcal{M}}$ is continuous and its values are convex and compact, we deduce that there exists $C > 0$ such that for $\varepsilon > 0$ small enough, every trajectory $y: [0, C\varepsilon] \to \mathbb{R}^n$ of $\dot{y} \in \hat{F}_{\mathcal{M}}(y)$ ($\subset F_{\mathcal{M}}(y)$) such that $y(C\varepsilon) = x_z(t^*)$ crosses the transverse section $(x_z(t^* + T - \varepsilon) + \bar{P}) \cap W$ at some time s belonging to the interval $(0, C\varepsilon)$. Since $x_z(\cdot)$ cannot cross a transverse section at two distinct points (Lemma 5.11), then $x_z(t^* - \varepsilon) = x_z(t^* + T - \varepsilon)$ for $\varepsilon > 0$ small enough. As a consequence, $x_z(\cdot)$ is periodic on $[t^* - \varepsilon, +\infty)$, contradicting the minimality of t^*. □

Lemma 5.13 *Assume that the hypotheses of Theorem 5.6 hold true. Let $\Gamma \subset S$ be a cycle (i.e., the support of a periodic trajectory of $(\Sigma_{\mathcal{M}})$) and let S_1 and S_2 denote the two open connected components of $S \setminus \Gamma$. If $z(\cdot)$ is an extremal trajectory on S such that $z(0) \in S_1$ then $z(t) \in S_1 \cup \Gamma$ for all $t \geq 0$. Moreover if $z(t_*) \in \Gamma$ for some t_* then $z(t) \in \Gamma$ for $t \geq t_*$.*

Proof We only treat the case where $\{z(t) \mid t \geq 0\} \cap \Gamma \neq \emptyset$. Then $z(t_*) \in \Gamma$ for some t_*. Denote by $x(\cdot)$ a periodic extremal trajectory with support Γ and let $t_0 = \min\{t \in [0, t_*] \mid z(t) \in \Gamma\}$. Since $z(t_0) = x(t_1)$ for some $t_1 > 0$ then, by Lemma 5.9, we have $z(t_0 + t) = x(t_1 + t)$ for all $t \geq 0$. Therefore $z(t) \in S_1$ for $t \in [0, t_0)$ and $z(t) \in \Gamma$ for all $t \geq t_0$, which concludes the proof. □

We can now conclude the section with a proof of Theorem 5.6.

Proof of Theorem 5.6 Let $x(\cdot)$ be an extremal trajectory for v. Let $x_z(\cdot)$ be as in Lemma 5.12 and T_z be its period. We define

$$\Gamma := \{x(t) \mid t \geq 0\}, \qquad \Gamma_z = \{x_z(t) \mid t \in [0, T_z]\}.$$

If $\Gamma \cap \Gamma_z \neq \emptyset$ then $\omega(x(\cdot)) = \Gamma_z$ by Lemma 5.13 and Theorem 5.6 is proved. We can then assume without loss of generality that $\Gamma \cap \Gamma_z = \emptyset$ for every $z \in \omega(x(\cdot))$.

Assume now that $\omega(x(\cdot))$ contains infinitely many disjoint cycles. We select among them a countable sequence $(\Gamma_k)_{k \in \mathbb{N}}$ and, for each k, we pick a point $x_k \in \Gamma_k$. Since Γ_k is contained in the compact set $\omega(x(\cdot))$, then up to a subsequence one has that $x_k \to \bar{x} \in \omega(x(\cdot))$ as $k \to +\infty$. Let $\Gamma_{\bar{x}}$ be a cycle passing through $\bar{x}$ and Ξ a transverse section passing through $\bar{x}$. Then for k large enough Γ_k intersects Ξ at some $y_k \in \omega(x(\cdot))$. By virtue of Lemma 5.11, $\omega(x(\cdot))$ intersects Ξ at only one point, which implies that $y_k = \bar{x}$ for k sufficiently large. Hence $\Gamma_k = \Gamma_{\bar{x}}$, contradicting the fact that the Γ_k are two by two disjoint.

We thus get that $\omega(x(\cdot))$ is made of finitely many cycles. Since $\omega(x(\cdot))$ is connected and the cycles are pairwise disjoint and closed, we conclude the proof of Theorem 5.6. □

5.2 Extremal Trajectories in the Corank One Case

In this section we analyze special three-dimensional switched systems associated with irreducible segments $\mathcal{M}$ of matrices. We prove that, up to the normalization $\lambda(\mathcal{M}) = 0$, the periodic extremal trajectories characterized in Theorem 5.6 correspond to piecewise constant signals made of two or four pieces.

We assume throughout this section that $\mathcal{M}$ is a segment connecting two matrices in $M_3(\mathbb{R})$ having corank one difference, that is,

$$\mathcal{M} := [\![M, M + bc^\top]\!] = \{M + ubc^\top \mid u \in [0, 1]\},$$

where $M \in M_3(\mathbb{R})$ and $b, c \in \mathbb{R}^3$. Furthermore, we suppose that M and $M + bc^\top$ are Hurwitz, the pair (M, b) is controllable and the pair (M, c) is observable, that is,

$$\text{span}\{b, Mb, M^2b\} = \text{span}\{c, M^\top c, (M^\top)^2c\} = \mathbb{R}^3,$$

and $\lambda(\mathcal{M}) = 0$. Trajectories of the corresponding switched linear system are solutions $x\colon [0, +\infty) \to \mathbb{R}^3$ of

$$\dot{x}(t) = Mx(t) + u(t)bc^\top x(t), \tag{5.4}$$

where $u(\cdot)$ is a measurable function taking values in $[0, 1]$. In the following we identify, when convenient, the switched linear system $(\Sigma_{\mathcal{M}})$ with System (5.4).

With the above assumptions, it is easy to see that the set $\mathcal{M}$ must be irreducible. Indeed from the controllability of the pairs (M, b) and $(M^\top, c)$ it follows that $\text{span}\{M^kbc^\top M^lx \mid k, l = 0, 1, 2\} = \text{span}\{M^kb \mid k = 0, 1, 2\} = \mathbb{R}^3$ for every nonzero $x \in \mathbb{R}^3$ so that $\{M, bc^\top\}$, and hence $\mathcal{M}$, is irreducible. The set $\mathcal{M}$ then admits by Theorem 4.8 a Barabanov norm v, and we are interested in characterizing the asymptotic behaviour of the extremal trajectories for v.

To describe our results, we need the following definitions.

Definition 5.14 Consider a trajectory $x\colon [0, +\infty) \to \mathbb{R}^3$ of (5.4) and its corresponding signal $u\colon [0, +\infty) \to [0, 1]$.

- A time $t \geq 0$ is said to be *regular* if there exists $\varepsilon > 0$ such that u is constant on $(t - \varepsilon, t + \varepsilon) \cap [0, +\infty)$ (up to a subset of measure zero). A time $t \geq 0$ that is not regular is called a *switching time*.
- A switching time $t > 0$ is said to be *isolated* if there exists $\varepsilon > 0$ such that u is constant both on $(t - \varepsilon, t)$ and on $(t, t + \varepsilon)$.
- A *bang arc* is a piece of trajectory defined on a time interval where u is constant. If all the switching times of a trajectory are isolated, then the trajectory is made of bang arcs and is said to be a *bang-bang trajectory*.

The aim of this section is to prove the following result.

Theorem 5.15 *Let v be a Barabanov norm for the switched linear system $(\Sigma_{\mathcal{M}})$. Then the ω-limit set of any extremal pair $(x(\cdot), \ell(\cdot))$ for v is the support of a periodic extremal pair $(\bar{x}(\cdot), \bar{\ell}(\cdot))$, where $\bar{x}(\cdot)$ and $\bar{\ell}(\cdot)$ have the same minimal period T. Moreover, for any switching time t_0, the restriction of $\bar{x}(\cdot)$ to $[t_0, t_0 + T]$ is made either by two bang arcs or by four bang arcs, the first case only happening if $b^\top\bar{\ell}(\cdot)$ and $c^\top\bar{x}(\cdot)$ have common zeroes.*

We will use in the sequel the following lemma.

Lemma 5.16 *The set* $\mathcal{M} = [\![M, M + bc^\top]\!]$ *is made of non-singular matrices and* $1 + c^\top M^{-1}b > 0$.

Proof For every $u \in [0, 1]$, we set $\delta_u = \det(M + ubc^\top)$. Then we have $\delta_u = \det(\mathrm{Id}_3 + uM^{-1}bc^\top)\det(M)$. Up to rewriting the matrix $\mathrm{Id}_3 + uM^{-1}bc^\top$ in a basis containing the vector c and a basis for the plane orthogonal to c, we get that

$$\delta_u = (1 + uc^\top M^{-1}b)\det(M).$$

This implies that $u \mapsto \delta_u$ is affine. Since both values at $u = 0$ and $u = 1$ are negative, δ_u must be negative for every $u \in [0, 1]$, hence the conclusion. □

We fix in what follows a Barabanov norm v associated with the switched linear system $(\Sigma_{\mathcal{M}})$. We denote by v^* the dual norm to v and by S and S^*, the unit spheres for v and v^*, respectively.

We associate with an extremal pair $(x(\cdot), \ell(\cdot))$ for v the real-valued functions $\phi_b, \phi_c, \phi \colon [0, +\infty) \to \mathbb{R}$ defined by

$$\phi_b(t) = b^\top \ell(t), \quad \phi_c(t) = c^\top x(t), \quad \phi(t) = \phi_b(t)\phi_c(t).$$

The function $\phi(\cdot)$ is called the *switching function* associated with the extremal pair $(x(\cdot), \ell(\cdot))$.

Remark 5.17 The maximality condition (4.12) is equivalent to

$$\max_{u\in[0,1]} u\phi(t) = u(t)\phi(t) = -\ell^\top(t)Mx(t) \quad a.e.\ t \geq 0. \tag{5.5}$$

Notice that if $\phi(\bar{t}) \neq 0$ for some $\bar{t} \geq 0$ then there exists $\varepsilon > 0$ such that

$$u(t) = \frac{1 + \mathrm{sign}(\phi(t))}{2} = \frac{1 + \mathrm{sign}(\phi(\bar{t}))}{2}, \quad \forall t \in (\bar{t} - \varepsilon, \bar{t} + \varepsilon) \cap [0, +\infty).$$

Hence $\bar{t}$ is a regular time.

As a consequence of the remark, an extremal pair $(x(\cdot), \ell(\cdot))$ for v is solution of the differential inclusion

$$\begin{aligned} \dot{x}(t) &= Mx(t) + u(t)bc^\top x(t), \\ \dot{\ell}(t) &= -M^\top \ell(t) - u(t)cb^\top \ell(t), \\ u(t) &\in U(\phi(t)), \end{aligned} \tag{5.6}$$

where

$$U(\varphi) = \begin{cases} \{1\} & \text{if } \varphi > 0, \\ [0,1] & \text{if } \varphi = 0, \\ \{0\} & \text{if } \varphi < 0. \end{cases}$$

Notice that $\ell(\cdot)$ is solution of the switched system $(\Sigma_{-\mathcal{M}^\top})$ associated with $-\mathcal{M}^\top := [\![-M^\top, -M^\top - cb^\top]\!]$. A trajectory of $(\Sigma_{-\mathcal{M}^\top})$ is the time-reverse of a trajectory of $(\Sigma_{\mathcal{M}^\top})$, which is itself a switched linear system of the form (5.4) satisfying $\lambda(\mathcal{M}^\top) = \lambda(\mathcal{M}) = 0$ by Item 9 in Proposition 1.20. Moreover, if $(x, \ell) \colon [0, +\infty) \to \mathbb{R}^3 \times \mathbb{R}^3$ is an extremal pair for v then $t \mapsto (\ell(-t), x(-t))$ is an extremal pair for v^*.

In the following we study the structure of the zeros of the switching function $\phi(\cdot)$ associated with an extremal pair $(x(\cdot), \ell(\cdot))$.

Lemma 5.18 *Let $\bar{t} \geq 0$ be such that $\phi_b(\bar{t}) = 0$. Then ϕ_b is differentiable at $\bar{t}$ and $\dot{\phi}_b(\bar{t}) = -b^\top M^\top \ell(\bar{t})$. Moreover, the following statements hold true:*

(*a*) *If $\dot{\phi}_b(\bar{t}) \neq 0$ then there exists $\varepsilon > 0$ such that $\phi_b(t) \neq 0$ for $t \in (\bar{t} - \varepsilon, \bar{t} + \varepsilon) \cap [0, +\infty)$, $t \neq \bar{t}$, and* $\operatorname{sign}(\phi_b|_{(\bar{t}-\varepsilon,\bar{t}) \cap [0,+\infty)}) = -\operatorname{sign}(\phi_b|_{(\bar{t},\bar{t}+\varepsilon)})$.

(*b*) *If $\dot{\phi}_b(\bar{t}) = 0$ then there exists $\varepsilon > 0$ such that $\phi_b(t) \neq 0$ for $t \in (\bar{t} - \varepsilon, \bar{t} + \varepsilon) \cap [0, +\infty)$, $t \neq \bar{t}$, and* $\operatorname{sign}(\phi_b|_{(\bar{t}-\varepsilon,\bar{t}) \cap [0,+\infty)}) = \operatorname{sign}(\phi_b|_{(\bar{t},\bar{t}+\varepsilon)})$.

Proof Since $\ell(\cdot)$ is absolutely continuous and according to (5.6), $\dot{\phi}_b(t)$ is well-defined and equal to $-b^\top M^\top \ell(t) - u(t)(b^\top c)\phi_b(t)$ for almost every $t > 0$. Hence, $\dot{\phi}_b(t)$ has a limit, equal to $-b^\top M^\top \ell(\bar{t})$, as t tends to $\bar{t}$. As a consequence, ϕ_b is differentiable at $\bar{t}$ and $\dot{\phi}_b(\bar{t}) = -b^\top M^\top \ell(\bar{t})$.

Item (*a*) is just a consequence of the definition of differentiability. Let us prove Item (*b*). Since $\dot{\phi}_b(\bar{t}) = 0$, one has that $\phi_b(t) = o(t - \bar{t})$ in a neighborhood of $\bar{t}$. Therefore, $\ell(t) = \ell(\bar{t}) - M^\top \ell(\bar{t})(t - \bar{t}) + o(t - \bar{t})$, which in turns imply that $\dot{\phi}_b(t) = b^\top (M^\top)^2 \ell(\bar{t})(t - \bar{t}) + o(t - \bar{t})$ almost everywhere in a neighborhood of $\bar{t}$. One has that $b^\top (M^\top)^2 \ell(\bar{t}) \neq 0$, since otherwise the vectors b, Mb, and $M^2 b$ would all be perpendicular to the nonzero vector $\ell(\bar{t})$, contradicting the fact that the pair (M, b) is controllable. Therefore $\dot{\phi}_b(t)$ has the same sign as $b^\top (M^\top)^2 \ell(\bar{t})(t - \bar{t})$ almost everywhere in a neighborhood of $\bar{t}$. Setting $\mu = \operatorname{sign}(b^\top (M^\top)^2 \ell(\bar{t}))$, one deduces that there exists $\varepsilon > 0$ such that $\mu\phi_b(t) > \phi_b(\bar{t}) = 0$ for $t \in (\bar{t} - \varepsilon, \bar{t} + \varepsilon) \cap [0, +\infty)$, $t \neq \bar{t}$. □

The previous result tells us in particular that the zeros of the function ϕ_b are isolated. A result analogous to Lemma 5.18 holds when one replaces the function ϕ_b by ϕ_c, since the pair $(M^\top, c)$ is controllable, and consequently the zeros of the function ϕ_c are also isolated. We then have the following proposition.

Proposition 5.19 *Let $(x(\cdot), \ell(\cdot))$ be an extremal pair for the switched linear system $(\Sigma_{\mathcal{M}})$. The zeros of the switching function $\phi(\cdot)$ are isolated and, in a neighborhood of any such zero $\bar{t}$, the sign of $\phi(\cdot)$ solely depends on the pair*

$(x(\bar{t}), \ell(\bar{t}))$. *Moreover, the functions* $\phi_b(\cdot)$ *and* $\phi_c(\cdot)$ *change sign infinitely many times on* $[0, +\infty)$.

Proof The first part of the proposition is an immediate consequence of what precedes.

Regarding the second part, we only provide an argument for $\phi_b(\cdot)$ since the corresponding one for $\phi_c(\cdot)$ is identical. We assume by contradiction that the function $\phi_b(\cdot)$ has a constant sign on $[T, +\infty)$ for some $T \geq 0$. Without loss of generality, suppose that $\phi_b(\cdot)$ is positive.

Define on $[T, +\infty)$ the increasing C^1 function $F(t) = \int_T^t \phi_b(s)ds$. We now claim that $\lim_{t\to+\infty} F(t)$ exists and is finite. Notice that for almost every $s > 0$ we have

$$\begin{aligned} b^\top (M^\top)^{-1}\dot{\ell}(s) &= b^\top (M^\top)^{-1}(-M^\top \ell(s) - u(s)cb^\top \ell(s)) \\ &= -\phi_b(s)(1 + u(s)b^\top (M^\top)^{-1}c). \end{aligned} \tag{5.7}$$

By Eq. (5.7), we have $-b^\top (M^\top)^{-1}\dot{\ell}(s) \geq \alpha\phi_b(s)$ where $\alpha := \min\{1, 1 + b^\top (M^\top)^{-1}c\} > 0$, thanks to Lemma 5.16. This implies that

$$F(t) \leq -\frac{1}{\alpha}\int_T^t b^\top (M^\top)^{-1}\dot{\ell}(s)ds = \frac{1}{\alpha}b^\top (M^\top)^{-1}(\ell(T) - \ell(t)).$$

Hence $F(\cdot)$ is bounded since $\ell(\cdot)$ is uniformly bounded. This shows the claim.

Notice that $\dot{F} = \phi_b$ is absolutely continuous with bounded derivative. By Barbalat's lemma (Theorem A.8) we conclude that $\phi_b(t)$ tends to zero as t goes to infinity.

Consider the sequence of solutions $\ell_n(\cdot) := \ell(n + \cdot)$. Then $\ell_n(\cdot)$ converges uniformly on every compact interval to some solution $\ell_*(\cdot)$ of $(\Sigma_{-\mathcal{M}^\top})$. Notice that $\ell_*(\cdot)$ is contained in S^* (the unit sphere of the norm v^*) since the sequence $(\ell_n(\cdot))_{n\geq 0}$ is entirely contained in S^*. By the fact that $b^\top \ell(s) = \phi_b(s) \to 0$ as $s \to +\infty$, we deduce that $b^\top \ell_*(s) = 0$ for every $s \geq 0$. Therefore $\dot{\ell}_*(s) = -M^\top \ell_*(s)$ for every $s \geq 0$. Hence, since $M^\top$ is Hurwitz we conclude that $\|\ell_*(s)\|$ tends to infinity as s goes to infinity, which contradicts the boundedness of $\ell_*(\cdot)$. □

As a consequence, there is no chattering phenomenon for extremal trajectories. Another interesting consequence of Proposition 5.19 is the following generalization of Lemma 5.16.

Lemma 5.20 *The set* $\mathcal{M} = [\![M, M + bc^\top]\!]$ *is made of Hurwitz matrices.*

Proof Note first that, by using Remark 1.19, one deduces from the fact that $\lambda(\mathcal{M}) = 0$ that the real part of any eigenvalue of an arbitrary matrix in $\mathcal{M}$ is nonpositive. If the conclusion of the corollary does not hold true, then there exists $\bar{u} \in (0, 1)$ such that $M_{\bar{u}} := M + \bar{u}bc^\top$ admits a purely imaginary eigenvalue, which is nonzero since $M_{\bar{u}}$ is non-singular according to Lemma 5.16. Therefore, there exists $\bar{x} \in S$ such that the curve $x : t \mapsto e^{tM_{\bar{u}}}\bar{x}$ is periodic. As a consequence $x(\cdot)$

must be extremal since it is an admissible trajectory of System (5.4). Moreover, there exists $\ell(\cdot)$ such that $(x(\cdot), \ell(\cdot))$ is solution of System (5.6). Since the corresponding function $u(\cdot)$ remains constant and equal to $\bar{u} \in (0, 1)$, the function $\phi(\cdot) = (c^\top x(\cdot))(b^\top \ell(\cdot))$ must be identically equal to zero, which is not possible by Proposition 5.19. □

The following result allows us to apply Theorem 5.6 to the switched linear system $(\Sigma_{\mathcal{M}})$.

Proposition 5.21 *The pair $(\mathcal{M}, v)$ satisfies the forward uniqueness property introduced in Definition 5.1.*

Proof Let $x_0, \ell_0 \in \mathbb{R}^3$ be such that $x_0 \neq 0$ and $\ell_0 \in \partial v(x_0)$. Assume that there exist two extremal pairs $(x_1(\cdot), \ell_1(\cdot))$ and $(x_2(\cdot), \ell_2(\cdot))$ for v starting at (x_0, ℓ_0), associated with the signals $u_1, u_2 \colon [0, +\infty) \to [0, 1]$. We next prove that $(x_1(\cdot), \ell_1(\cdot)) = (x_2(\cdot), \ell_2(\cdot))$. If it were not the case, then

$$t_0 = \sup\{t \geq 0 \mid (x_1, \ell_1)|_{[0,t]} = (x_2, \ell_2)|_{[0,t]}\}$$

would be finite. Denote by ϕ_1 and ϕ_2 the switching functions associated with $(x_1(\cdot), \ell_1(\cdot))$ and $(x_2(\cdot), \ell_2(\cdot))$, respectively. Notice that $\phi_1(t_0) = (b^\top \ell_*)(c^\top x_*) = \phi_2(t_0)$. If $(b^\top \ell_*)(c^\top x_*) \neq 0$ then ϕ_1 and ϕ_2 are nonzero and have the same sign in a neighborhood of t_0. In particular, there exists $\varepsilon > 0$ such that $u_1 = u_2$ on $(t_0, t_0 + \varepsilon)$. If $(b^\top \ell_*)(c^\top x_*) = 0$ then by Proposition 5.19 there exists a small $\varepsilon > 0$ such that $u_1(t) = \big(1 + \mathrm{sign}(\phi_1(t))\big)/2 = \big(1 + \mathrm{sign}(\phi_2(t))\big)/2 = u_2(t)$ is constant on $(t_0, t_0 + \varepsilon)$. In both cases $(x_1(\cdot), \ell_1(\cdot)) = (x_2(\cdot), \ell_2(\cdot))$ on $(t_0, t_0 + \varepsilon)$, contradicting the definition of t_0. □

We give below a series of technical results which are instrumental for the proof of Theorem 5.15. We start with a preliminary lemma whose statement uses the following convention: given a vector $z \in \mathbb{R}^3$, we use P_z to denote the plane perpendicular to z.

Lemma 5.22 *Let $N = e^{t_p M_p} \cdots e^{t_1 M_1}$ with $M_1, \ldots, M_p$ in $\{M, M + bc^\top\}$, $M_1 \neq M_p$, and $t_1, \ldots, t_p > 0$. Assume that 1 is a double eigenvalue of N. Then either $\ker(N - \mathrm{Id}_3) = P_c$ and $e^{t_1 M_1} P_c \neq P_c$ or $\ker(N^\top - \mathrm{Id}_3) = P_b$ and $e^{t_p M_p^\top} P_b \neq P_b$.*

Proof Since $\sup_{k \geq 0} \|N^k\|$ is finite, the Jordan blocks corresponding to the eigenvalue 1 must be trivial. Moreover, N cannot be the identity, since M and $M + bc^\top$ are Hurwitz and have at least one real eigenvalue. Therefore, both $\ker(N - \mathrm{Id}_3)$ and $\ker(N^\top - \mathrm{Id}_3)$ are two-dimensional subspaces of $\mathbb{R}^3$.

Let $u_\# \colon [0, +\infty) \to \{0, 1\}$ be the $(t_1 + \cdots + t_p)$-periodic signal such that $u_\#(s) = 0$ (respectively, $u_\#(s) = 1$) for every $s \in [t_1 + \cdots + t_{j-1}, t_1 + \ldots, t_j)$ if $M_j = M$ (respectively, $M_j = M + bc^\top$), $j = 1, \ldots, p$. Notice that for every $\bar{x} \in \ker(N - \mathrm{Id}_3) \cap S$ the trajectory $x(\cdot)$ of (5.4) starting at $\bar{x}$ and corresponding to $u_\#$ is periodic and, hence, extremal. By Theorem 4.19 there exists $\ell(\cdot)$ with initial condition $\bar{\ell} \in \partial v(\bar{x})$ such that $(x(\cdot), \ell(\cdot))$ is an extremal pair. According to (5.6), the value of $\ell(\cdot)$

at time $t_1 + \cdots + t_p$ is equal to $(N^\top)^{-1}\bar{\ell}$. Since $\ell(\cdot)$ stays in S^* (in particular, it remains bounded), $\ell(k(t_1 + \cdots + t_p)) = (N^\top)^{-k}\bar{\ell}$ for $k \in \mathbb{N}$, and N is similar to the diagonal matrix with diagonal entries $(1, 1, \nu)$ with $\nu \in (0, 1)$, then necessarily $\bar{\ell} \in \ker((N^\top)^{-1} - \mathrm{Id}_3) = \ker(N^\top - \mathrm{Id}_3)$, i.e., the extremal pair $(x(\cdot), \ell(\cdot))$ is periodic. Notice that ϕ is periodic and vanishes at all switching times, implying that $\phi(0) = 0$ (since $M_1 \neq M_p$), that is, $b^\top\bar{\ell} = 0$ or $c^\top\bar{x} = 0$.

Assume by contradiction that $\ker(N - \mathrm{Id}_3) = P_c$ and $e^{t_1 M_1} P_c = P_c$. Then $e^{t_1 M_1}(P_c \cap S) \subset P_c \cap S$, since $u_\#$ corresponds to an extremal trajectory for every initial condition $x \in P_c \cap S$. As a consequence, $e^{kt_1 M_1} x$ belongs to $P_c \cap S$ for every $k \geq 0$ and $x \in P_c \cap S$, contradicting the fact that M_1 is Hurwitz.

Applying the same contradiction argument to the transpose system $(\Sigma_{\mathcal{M}^\top})$, with N replaced by $N^\top$, we prove that $e^{t_p M_p^\top} P_b \neq P_b$ whenever $\ker(N^\top - \mathrm{Id}_3) = P_b$.

Let us now assume that $\ker(N - \mathrm{Id}_3) \neq P_c$ and prove that $\ker(N^\top - \mathrm{Id}_3) = P_b$. By assumption, $\ker(N - \mathrm{Id}_3) \cap P_c = \mathrm{span}(x_0)$ for some $x_0 \neq 0$. Hence, for every $\bar{x} \in \ker(N - \mathrm{Id}_3)$ not collinear to x_0 one has $c^\top\bar{x} \neq 0$. Therefore, for every $\bar{x} \in \ker(N - \mathrm{Id}_3) \setminus \mathrm{span}(x_0)$ and every adjoint vector $\bar{\ell}$ such that $(\bar{x}, \bar{\ell})$ is the initial condition of an extremal pair, we have $b^\top\bar{\ell} = 0$. If there exists $\bar{x} \in \ker(N - \mathrm{Id}_3)$ not collinear to x_0 admitting two linearly independent adjoint vectors as above, one gets at once that $\ker(N^\top - \mathrm{Id}_3) = P_b$. Otherwise one can define a map $\bar{x} \mapsto \ell(\bar{x})$ such that $\ell(\bar{x}) \in \ker(N^\top - \mathrm{Id}_3)$ and $b^\top\ell(\bar{x}) = 0$.

In order to show that $\ker(N^\top - \mathrm{Id}_3) = P_b$, it remains to show that there exist $x_1, x_2 \in \ker(N - \mathrm{Id}_3)$ not collinear to x_0 such that $\ell(x_1), \ell(x_2)$ are linearly independent. If this were not the case, then for every $x \in \ker(N - \mathrm{Id}_3) \setminus \mathrm{span}(x_0)$ the vector $\ell(x)$ would be collinear to some fixed $\ell_* \in S^*$. Since S^* is the unit sphere of the norm v^* and every $\ell(x)$ belongs to S^*, one would get that $\ell(x) \in \{\pm\ell_*\}$ for every $x \in \ker(N - \mathrm{Id}_3) \setminus \mathrm{span}(x_0)$. One would deduce that for every $x \in \ker(N - \mathrm{Id}_3) \cap S$ not collinear to x_0, one has that $\ell_*^\top x \in \{\pm 1\}$ and, by continuity, also $\ell_*^\top x_0 \in \{\pm 1\}$. Since $\ker(N - \mathrm{Id}_3) \cap S$ is connected and the closed sets $\{x \mid \ell_*^\top x = 1\}$ and $\{x \mid \ell_*^\top x = -1\}$ are disjoint, $\ker(N - \mathrm{Id}_3) \cap S$ must be contained inside one of them, contradicting the central symmetry of $\ker(N - \mathrm{Id}_3) \cap S$. □

The following lemma describes the spectrum of the fundamental matrix corresponding to a periodic trajectory of $(\Sigma_{\mathcal{M}})$.

Lemma 5.23 *Let $x^0(\cdot)$ be a solution of $(\Sigma_{\mathcal{M}})$ with associated switching signal $A^0(\cdot)$. Assume that $x^0(\cdot)$ and $A^0(\cdot)$ are periodic of period $T > 0$. Then the eigenvalue 1 of $\Phi_{A^0}(T, 0)$ is simple and the two other eigenvalues have modulus strictly less than one.*

Proof Since both M and $M + bc^\top$ are Hurwitz, the trace of $M + ubc^\top$ is negative for $u \in [0, 1]$, which implies that $\det(\Phi_{A^0}(T, 0)) < 1$.

Since $x^0(\cdot)$ is periodic, it is necessarily extremal and we deduce from Proposition 5.19 that $A^0(\cdot)$ is piecewise constant with values in $\{M, M + bc^\top\}$. Arguing by contradiction, we are left with the following alternative: either 1 is a double eigenvalue of $\Phi_{A^0}(T, 0)$ or both 1 and -1 are eigenvalues of $\Phi_{A^0}(T, 0)$. In both

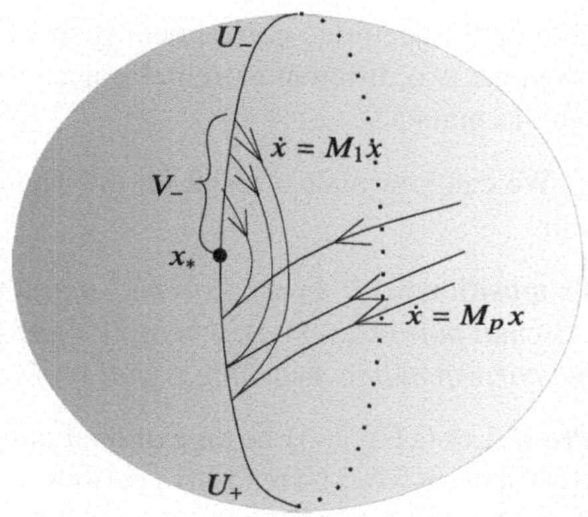

Fig. 5.2 Proof of Lemma 5.23

cases there exists a finite concatenation $N = e^{t_p M_p} \cdots e^{t_1 M_1}$ with $M_1, \ldots, M_p \in \{M, M + bc^\top\}$ and $t_1, \ldots, t_p > 0$ such that 1 is a double eigenvalue of N.

Reasoning as in Lemma 5.22, for every extremal pair $(x^0(\cdot), \ell^0(\cdot))$ the curve $\ell^0(\cdot)$ is periodic and $\ell^0(0) \in \ker((N^\top)^{-1} - \mathrm{Id}_3)$. Thanks to Lemma 5.22 we can assume that $\ker(N - \mathrm{Id}_3) = P_c$ and $e^{t_1 M_1} P_c \neq P_c$, up to replacing in what follows $\mathscr{M}$ by $\mathscr{M}^\top$ and N by $(N^\top)^{-1}$.

Notice that $P_c \cap P_{M_1^\top c} \cap S$ is made of two antipodal points since the pair $(M^\top, c)$ is controllable. Moreover, we can choose $x_* \in P_c \cap P_{M_1^\top c} \cap S$ such that $c^\top M_1^2 x_* > 0$. Observe that $P_c \cap S \setminus \{x_*, -x_*\}$ is the disjoint union of two open subsets U^+, U^- in $P_c \cap S$ such that $c^\top M_1 x > 0$ on U^+ and $c^\top M_1 x < 0$ on U^-. Consider a trajectory $t \mapsto e^{t M_1} x$ for $x \in U^-$ close to x_*, in such a way that $c^\top M_1^2 x > 0$.

We claim that there exists a small enough open neighborhood V of x_* such that for every $y \in V^- := V \cap U^-$ there exists $t(y) > 0$ and $x(y) \in U^+$ such that $x(y) = e^{t(y) M_1} y$ and $t(\cdot)$ is continuous with $\lim_{y \to x_*} t(y) = 0$. The claim is an immediate consequence of the implicit function theorem applied at the point $(0, x_*)$ to the function $F \colon \mathbb{R} \times P_c \to \mathbb{R}$ defined by

$$F(t, x) = \begin{cases} c^\top \frac{e^{t M_1} - \mathrm{Id}_3}{t} x & t \neq 0, \\ c^\top M_1 x & t = 0. \end{cases}$$

Indeed, $\frac{\partial}{\partial t} F(t, x)\big|_{(t,x) = (0, x_*)}$ is well defined and equal to $c^\top M_1^2 x_* / 2 > 0$, and $e^{t M_1} x$ stays on S for $t \in [0, t_1]$ and $x \in P_c \cap S$. By definition of N and since $\ker(N - \mathrm{Id}_3) = P_c$, we deduce that $e^{-t_p M_p}(P_c \cap S) \subset S$ and then every point $x(y)$ with $y \in V^-$ is reached at $t = t_p$ by an extremal trajectory $t \mapsto e^{t M_p} z$ for some $z \in S$. Since $M_p \neq M_1$ every point $x(y)$ with $y \in V^-$ is reached by two extremal trajectories corresponding to M and $M + bc^\top$ (see Fig. 5.2). Since $c^\top M_1 x(y) = c^\top M_p x(y) > 0$ for $y \in V^-$, we get that $c^\top e^{-t M_1} x(y) < 0$ and $c^\top e^{-t M_p} x(y) < 0$ for positive t small enough and $y \in V^-$. As a consequence, the points $e^{-t M_1} x(y)$ with positive t small enough and $y \in V^-$ are also reached by two extremal trajectories corresponding to M and $M + bc^\top$. Since from these points originate two different extremal trajectories, then we have reached a contradiction with Lemma 5.9. □

We can now state a result on the asymptotic behavior of any periodic extremal pair.

Proposition 5.24 *Every extremal pair* $(x(\cdot), \ell(\cdot))$ *for* v *converges to a periodic extremal pair* $(\bar{x}(\cdot), \bar{\ell}(\cdot))$*, where* $\bar{x}(\cdot)$ *and* $\bar{\ell}(\cdot)$ *have the same minimal period* T *and the corresponding switching signal is of minimal period* $T/2$ *or* T.

Proof Let $(x(\cdot), \ell(\cdot))$ be an extremal pair associated with a switching signal $A(\cdot)$. By Theorem 5.6, $x(\cdot)$ tends to a periodic solution $\bar{x}(\cdot)$ of $(\Sigma_{\mathcal{M}})$ with minimal period $T > 0$. Denote by $\bar{A}(\cdot)$ the signal associated with $\bar{x}(\cdot)$. One first deduces that $\bar{A}(\cdot)$ is periodic of period T since $\dot{\bar{x}}(\cdot)$ is periodic and the zeros of $c^\top \bar{x}(\cdot)$ are isolated. We next claim that the minimal period of $\bar{A}(\cdot)$ is $T/2$ or T. To see that, first recall that $\bar{A}(\cdot)$ is piecewise constant by Proposition 5.19. Hence, either $\bar{A}(\cdot)$ is constant or it has a minimal period $\bar{T} > 0$. In the first case, $\bar{A}(\cdot)$ must be constant and this contradicts the fact that M and $M + bc^\top$ are Hurwitz, since $\bar{x}(\cdot)$ is periodic. In the second case, $T = p\bar{T}$ for some positive integer p and $\Phi_{\bar{A}}(T, 0) = (\Phi_{\bar{A}}(\bar{T}, 0))^p$. Let ω be an eigenvalue of $\Phi_{\bar{A}}(\bar{T}, 0)$ of modulus equal to one. According to Lemma 5.23, 1 is a simple eigenvalue of $\Phi_{\bar{A}}(T, 0)$ and the other two eigenvalues have modulus strictly less than 1. Then ω must be real, equal to 1 or -1. Recalling now that T is the minimal period of $\bar{x}(\cdot)$, we get at once that $\bar{T} = T$ or $T/2$.

The trajectory $\ell(\cdot)$ belongs to S^* which is compact. Therefore its ω-limit set is made of solutions of $(\Sigma_{-\mathcal{M}^\top})$. Proving the proposition amounts to showing that this ω-limit set reduces to the support of a unique periodic solution $\bar{\ell}(\cdot)$ of $(\Sigma_{-\mathcal{M}^\top})$ with minimal period T.

Let ℓ_* be in the ω-limit set of $\ell(\cdot)$. Then $\ell_* = \lim_{k\to\infty} \ell(t_k)$ with $t_k \to \infty$ as $k \to \infty$. Up to extracting a subsequence, we can assume that $x(t_k)$ converges to some x_* and $A(t_k + \cdot)$ weak-$\star$ converges to some $\hat{A}$ as $k \to \infty$. By what precedes, up to reparameterizing $\bar{x}$, we can assume that $x_* = \bar{x}(0)$. Moreover, since $t \mapsto \Phi_{\hat{A}}(t, 0)x_*$ is an extremal trajectory contained in $\omega(x(\cdot))$, we have $\hat{A} = \bar{A}$. Hence, the trajectory $\bar{\ell}(\cdot)$ of $(\Sigma_{-\mathcal{M}^\top})$ with initial condition ℓ_* corresponding to the signal $-\bar{A}^\top(\cdot)$ is contained in $\omega(\ell(\cdot))$.

Let $N(t) = \Phi_{\bar{A}}(0, t) = \Phi_{\bar{A}}(t, 0)^{-1}$ for $t \geq 0$. By Lemma 5.23, 1 is a simple eigenvalue of $N(T)$, while the other eigenvalues have modulus strictly larger than one. Let ℓ_1 be an eigenvector of $N(T)$ associated with the eigenvalue 1. Then $\ell_* = \alpha\ell_1 + w$, where $w \in \ker(N(T) - \mathrm{Id}_3)^\perp$ and $N(kT)\ell_* = \alpha\ell_1 + N(T)^k w$ for every $k \in \mathbb{N}$. Since the sequence $(N(kT)\ell_*)_{k\in\mathbb{N}}$ is uniformly bounded we deduce that $w = 0$ and $\bar{\ell}(\cdot)$ is periodic of period T.

Since there exists at most one value $\alpha \in \mathbb{R}$ such that $\alpha\ell_1 \in \partial v(x_*)$, we have proved that $\omega(\ell(\cdot))$ is the support of a unique periodic trajectory of period T. We have to prove that the minimal period of $\bar{\ell}(\cdot)$ is T. If $\bar{T} = T$, we have finished. Otherwise we have that $N(\bar{T})\ell_*$ is either equal to ℓ_* or $-\ell_*$. In the first case, ℓ_*

belongs to both $\partial v(x_*)$ and $\partial v(-x_*)$, which is a contradiction. The proof of the proposition is complete. □

Remark 5.25 Assume that the switched linear system $(\Sigma_{\mathcal{M}})$ admits a T-periodic trajectory $x(\cdot)$ with support Γ such that $\Gamma \cap (-\Gamma) \neq \emptyset$. Then, necessarily $\Gamma = -\Gamma$ and $x(\cdot)$ is $\frac{T}{2}$-antiperiodic, that is, $x(\cdot + \frac{T}{2}) = -x(\cdot)$. Moreover, up to time-translations and dilations, $x(\cdot)$ is the unique $\frac{T}{2}$-antiperiodic trajectory. Indeed, up to dilations we can assume that Γ is contained in the unit sphere S of a Barabanov norm for $(\Sigma_{\mathcal{M}})$. Then Γ divides S in two regions S_-, S_+ with $S_+ = -S_-$, which implies that any other trajectory $x'(\cdot)$ with support $\Gamma' \subset S$ such that $\Gamma' = -\Gamma'$ must intersect Γ. Hence, by Lemma 5.9, there exists $\bar{t}$ such that $x'(\cdot) = x(\bar{t} + \cdot)$.

Lemma 5.26 *The trajectory $\bar{\ell}(\cdot)$ defined in Proposition 5.24 cannot intersect itself.*

Proof Consider the trajectory $\bar{m}(\cdot)$ of $(\Sigma_{\mathcal{M}^\top})$ defined as $\bar{m}(t) = \bar{\ell}(T - t)$ for every $t \in [0, T]$ and extended by periodicity. Then $v^*(\bar{m}(t)) = v^*(\bar{m}(0))$ for every $t \in [0, T]$, since $v^*(\cdot)$ is a Barabanov norm for $(\Sigma_{\mathcal{M}^\top})$ and $\bar{\ell}(\cdot)$ is T-periodic. Since $\lambda(\mathcal{M}^\top) = \lambda(\mathcal{M}) = 0$ and by using $v^*(\cdot)$ instead of $v(\cdot)$ in Lemma 5.9, we deduce that $\bar{\ell}(\cdot)$ cannot intersect itself. □

We conclude the proof of Theorem 5.15 by proving that for a periodic extremal pair $(\bar{x}(\cdot), \bar{\ell}(\cdot))$ as in Proposition 5.24 the corresponding switching function changes sign two or four times on every period T.

Proof of Theorem 5.15 We are left to prove that each of the functions $c^\top \bar{x}(\cdot)$ and $b^\top \bar{\ell}(\cdot)$ changes sign exactly twice on each minimal period. We only deal with the function $b^\top \bar{\ell}(\cdot)$, since by the same arguments the conclusion can also be obtained for the function $c^\top \bar{x}(\cdot)$.

Notice first that, thanks to Lemma 5.18, the function $b^\top \bar{\ell}(\cdot)$ changes sign at some time $\bar{t} > 0$ if and only if $b^\top \bar{\ell}(\bar{t}) = 0$ and $b^\top \dot{\bar{\ell}}(\bar{t}) \neq 0$. By Proposition 5.19 the function $b^\top \bar{\ell}(\cdot)$ changes sign infinitely many times. Then, with no loss of generality, we assume that $b^\top \bar{\ell}(0) = 0$ and $b^\top \dot{\bar{\ell}}(0) = -b^\top M^\top \bar{\ell}(0) > 0$. Define now the set $\Xi_- = P_b \cap \{\ell \in S^* \mid b^\top M^\top \ell < 0\}$, where we recall that P_b denotes the orthogonal plane to b. We claim that Ξ_- is a transverse section of S^* for the switched linear system $(\Sigma_{\mathcal{M}^\top})$ passing through $\bar{\ell}(0)$. Indeed, one has

$$b^\top (M^\top + ucb^\top)\bar{\ell}(0) = b^\top M^\top \bar{\ell}(0), \qquad \forall u \in [0, 1].$$

Since, moreover, the trajectory obtained by repeating $[0, T] \ni t \mapsto \bar{\ell}(T - t)$ by periodicity is extremal for v^* (for $(\Sigma_{\mathcal{M}^\top})$), we deduce from Lemma 5.11 that

$$\{\bar{\ell}(t) \mid t \in [0, T]\} \cap \Xi_- = \{\bar{\ell}(0)\}. \tag{5.8}$$

Again by Proposition 5.19, the function $b^\top \bar{\ell}(\cdot)$ should change sign on $(0, T)$, i.e., there exists $t^* \in (0, T)$ such that $b^\top \bar{\ell}(t^*) = 0$ and $b^\top \dot{\bar{\ell}}(t^*) = -b^\top M^\top \bar{\ell}(t^*) < 0$. By defining the set $\Xi_+ = P_b \cap \{\ell \in S^* \mid b^\top M^\top \ell > 0\}$, the same reasoning as above shows that $\{\bar{\ell}(t) \mid t \in [0, T]\} \cap \Xi_+ = \{\bar{\ell}(t^*)\}$. The next zero of $b^\top \bar{\ell}$ yields an intersection point $\ell(t_*)$, with $t_* \in (t^*, T]$, between $\{\bar{\ell}(t) \mid t \in [0, T]\}$ and Ξ_-, and one gets from (5.8) that $\bar{\ell}(0) = \bar{\ell}(t_*)$. Then, according to Lemma 5.26, $T = t_*$ and hence the function $b^\top \bar{\ell}(\cdot)$ changes sign exactly twice on each minimal period. □

5.3 Notes and References

This chapter mostly revisits some results presented in [CGM15]. The forward uniqueness property, introduced in [CGM15] as Condition G, is inspired from a slightly more restrictive condition in [Bar08], referred to as Condition C, where uniqueness of the trajectory is assumed also in negative time. The sufficient criterion for forward uniqueness given in Proposition 5.5 is presented in this monograph for the first time.

The map considered in Lemma 5.4, associating with points of an oriented hypersurface the corresponding outward normal unit vector, is usually called the Gauss map of the hypersurface and is introduced in relation with curvature notions, see e.g. [Spi99].

In the classical setting of planar systems, one can find under the name of Poincaré–Bendixson theorem several related statements describing the structure of ω-limit sets of bounded trajectories. One of these results goes as follows (cf. [Tes12, Lemma 7.13]): if the ω-limit set of an integral curve of a planar vector field $x \mapsto f(x)$ is compact and does not contain zeroes of f, then it coincides with a nontrivial periodic orbit of f. A version of the Poincaré–Bendixson theorem for planar differential inclusions can be found in [Fil88] (see Theorem 3, page 137, therein). The Poincaré–Bendixson result given in Theorem 5.6 first appeared in [CGM15]. This result does not directly follow from the corresponding results for planar differential inclusions since the differential inclusion (5.2) is defined on a non-smooth manifold instead of $\mathbb{R}^2$ and the set $\hat{F}_{\mathcal{M}}(x)$ is not necessarily convex. Existence of periodic trajectories for the systems considered in Sect. 5.2 has been proven in [PR91a], while a Poincaré–Bendixson result in the more general case in which $\mathcal{M}$ is finite has been stated in [PR96] without details. The existence of a periodic trajectory with four bang arcs for the systems considered in Sect. 5.2 has been shown in [Bar94, Rap93].

For $\mathcal{M}$ compact and convex, the problem of the existence of periodic extremal trajectories with a finite number of bang arcs or, more precisely, of a switching law $A \in \mathcal{S}_{\mathrm{pc}}(\mathcal{M})$ such that $\lambda(\mathcal{M}) = \frac{1}{T} \log\big(\rho(\Phi_A(T, 0))\big)$ for some $T > 0$, can be seen as the continuous-time analogue of the finiteness conjecture. In the discrete-time setting the latter conjecture has been posed in [LW95] and disproved in [BM02, BTV03, Koz05, HMST11]. In the continuous-time setting, partial results can be found in [Gur03].

Chapter 6
Restrained Switching Signals

The goal of this chapter is to study how the results of Chap. 4 can be extended when the class of arbitrary switching signals is replaced by a more general one, as, for instance, one of those introduced in Sect. 1.1.1. This raises nontrivial issues, since the construction of converse Lyapunov functions as in Sect. 3.3 and of Barabanov norms as in Chap. 4 heavily relies on the fact that the concatenation of two admissible signals is again an admissible signal. Said equivalently, $\{\Phi_A(t, 0) \mid A \in \mathcal{S},\ t \geq 0\}$ is a semigroup when $\mathcal{S} = \mathcal{S}_{\text{arb}}$, but not in general. Such a semigroup property is the key point in proving that the constructed Lyapunov functions and Barabanov norms are nonincreasing along admissible trajectories of the switched system.

In Sect. 6.1, some of the results of Chap. 4, and in particular the notion of Barabanov norms, are adapted to the non-concatenable case. The construction is based on the notion of *skeleton* of a class of switching signals. Skeletons identify subclasses of the class of admissible signals which have some suitable partial concatenability property. When skeletons are large enough to capture the asymptotic behavior of the system, a Barabanov-like construction using the skeletons behaves well enough on all trajectories of the system.

The notion of skeleton is used in Sect. 6.2 to prove that, for the classes of signals introduced in Sect. 1.1.1 and under a mild assumption on the set of modes $\mathcal{M}$, the maximal Lyapunov exponent corresponding to a subset of trajectories associated with periodic signals is still equal to maximal Lyapunov exponent associated with the class.

Finally, in Sect. 6.3 we study which properties of the class of signals are needed to conclude exponential stability in the case where the modes of the system share a nonstrict quadratic Lyapunov function, i.e., a quadratic form whose time-evaluation along trajectories of the switched system is nonincreasing.

Y. Chitour et al., *Dynamics and Stability of Continuous-Time Switched Linear Systems*, PNLDE Subseries in Control 105,
https://doi.org/10.1007/978-3-031-82982-6_6

6.1 Skeletons and Quasi-Extremal Trajectories

We consider in this section a switched linear system of the type $(\Sigma_{\mathcal{S}})$ introduced in Chap. 1.

A key object in what follows is the identification of a subclass $\mathcal{F}_\infty$ of $\mathcal{S}$, with which we are able to associate a semigroup of fundamental matrices that captures the asymptotic behaviour of $\mathcal{S}$.

6.1.1 Skeletons as Concatenable Subfamilies

Let us provide a definition of *skeleton of* $\mathcal{S}$, which is a family of signals, each of them defined on a bounded interval, whose countable concatenations stay in $\mathcal{S}$.

Definition 6.1 For every $t \geq 0$ let $\mathcal{F}_t$ be a subset of $L^\infty([0,t],\mathcal{M})$ and denote by $\mathcal{F}$ the union

$$\mathcal{F} = \cup_{t\geq 0}\mathcal{F}_t.$$

Set

$$\mathcal{F}_\infty = \left\{A_1 * A_2 * \cdots * A_k * \cdots \mid A_k \in \mathcal{F}_{t_k} \text{ for } k \in \mathbb{N},\ \sum_{k\in\mathbb{N}} t_k = \infty\right\}$$

and

$$\Phi(\mathcal{F}_t) = \{\Phi_A(t,0) \mid A \in \mathcal{F}_t\}, \quad \text{for every } t \geq 0, \qquad \Phi(\mathcal{F}) = \cup_{t\geq 0}\Phi(\mathcal{F}_t).$$

We say that $\mathcal{F}$ is a *skeleton* of $\mathcal{S}$ if

$$\mathcal{F}_s * \mathcal{F}_t \subset \mathcal{F}_{s+t}, \qquad \text{for every } s,t \geq 0, \tag{6.1}$$

and

$$\mathcal{F}_\infty \subset \mathcal{S}.$$

Moreover, we associate with a skeleton $\mathcal{F}$ the quantity

$$\lambda_\#(\mathcal{F}) = \limsup_{t\to+\infty} \sup_{R_t\in\Phi(\mathcal{F}_t)} \frac{\log\|R_t\|}{t},$$

with the usual convention that $\sup(\emptyset) = -\infty$. For $t \geq 0$ we denote by $\mathcal{F}_t^\star$ the set of signals $A \in L^\infty([0,t],\overline{\mathrm{co}}(\mathcal{M}))$ such that there exist a sequence $(t_n)_{n\in\mathbb{N}}$ in

$[0, +\infty)$ converging to t and a sequence $(A_n)_{n\in\mathbb{N}}$ with $A_n \in \mathcal{F}_{t_n}$ such that A_n weak-$\star$ converges to A.[1] Finally, we set $\mathcal{F}^\star = \cup_{t\geq 0}\mathcal{F}_t^\star$.

Remark 6.2 Let $\mathcal{F}$ be a skeleton of $\mathcal{S}$. Condition (6.1) implies that $\Phi(\mathcal{F})$ is a semigroup. Moreover, if $A \in \mathcal{F}$ and $B \in \mathcal{F}_\infty$, then $A * B \in \mathcal{F}_\infty$.

Notice that, if $\mathcal{F}$ is a skeleton of $\mathcal{S}$, then

$$\lambda_\#(\mathcal{F}) \leq \lambda(\mathcal{S}). \tag{6.2}$$

We look for conditions ensuring that $\mathcal{F}$ carries enough information about $(\Sigma_\mathcal{S})$, and in particular yielding $\lambda_\#(\mathcal{F}) = \lambda(\mathcal{S})$.

Definition 6.3 Let $\mathcal{F}$ be a skeleton of $\mathcal{S}$. We say that $\mathcal{F}$ is *extensive* if there exists $t_* \geq 0$ such that for every $t \geq 0$ and $A \in \mathcal{S}$, there exist $\tau_j \in [0, t_*]$ and $A_j \in L^\infty([0, \tau_j], \mathcal{M})$, $j = 1, 2$, such that $A_1 * A|_{[0,t]} * A_2$ is in $\mathcal{F}$.

The intuitive idea motivating the definition of extensive skeleton is that it is large enough to capture the asymptotic behavior of the system. More precisely, the fact that every restriction $A|_{[0,t]}$ can be extended to a signal in the extensive skeleton by gluing "short" pieces before and after it, guarantees that the uniform asymptotic behavior of the trajectories of $(\Sigma_\mathcal{S})$ is not affected by the "short" patches. The latter can therefore be characterized by looking only at the trajectories corresponding to signals in $\mathcal{F}_\infty$.

It is useful to associate with an extensive skeleton $\mathcal{F}$ the quantity

$$\kappa_* = \max\{\|\Phi_B(0, \tau)\|^2 \mid B \in L^\infty([0, \tau], \overline{\mathrm{co}}(\mathcal{M})),\ \tau \in [0, t_*]\} < +\infty. \tag{6.3}$$

By definition of κ_* and the fact that $\mathcal{F}$ is extensive, for every $A \in \mathcal{S}$ and $t \geq 0$ there exist $t' \in [t, t + 2t_*]$ and $A' \in \mathcal{F}_{t'}$ such that

$$\|\Phi_A(t, 0)\| \leq \kappa_* \|\Phi_{A'}(t', 0)\|. \tag{6.4}$$

Several examples of extensive skeletons will be presented in the next section.

Remark 6.4 Recall that $\mathcal{S}^\star$ denotes the weak-$\star$ closure of $\mathcal{S}$. By a simple compactness argument, we deduce that if $\mathcal{F}$ is an extensive skeleton of $\mathcal{S}$ then $\mathcal{F}^\star$ is an extensive skeleton of $\mathcal{S}^\star$ (with values in $\overline{\mathrm{co}}(\mathcal{M})$).

We next investigate the equality case in (6.2) and we provide a version of Fenichel's lemma for skeletons.

Lemma 6.5 *If $\mathcal{F}$ is an extensive skeleton of $\mathcal{S}$ then $\lambda_\#(\mathcal{F}) = \lambda(\mathcal{S})$. If, moreover, the semigroup $\Phi(\mathcal{F})$ is irreducible, then $\mathcal{F}$ satisfies the following property: if*

---

[1] More rigorously: $A_n|_{[0,t]}$ weak-$\star$ converges to A up to the extension of A_n on $(t_n, t]$ by any element of $\mathcal{M}$ in the case $t_n < t$.

$\lim_{j\to\infty} R_j = 0$ *for every sequence* $R_j \in \Phi(\mathcal{F}_{t_j})$ *with* t_j *tending to* $+\infty$, *then* $\lambda_\#(\mathcal{F}) < 0$.

Proof We already noticed that, by definition, $\lambda_\#(\mathcal{F}) \le \lambda(\mathcal{S})$. The converse inequality follows from the fact that $\mathcal{F}$ is extensive and (6.4).

As for the last part of the statement, assume that $\lim_{j\to\infty} R_j = 0$ for every sequence $R_j \in \Phi(\mathcal{F}_{t_j})$ with $t_j \to +\infty$, and let us first show that $(\Sigma_{\mathcal{S}^\star})$ is attractive. Fix $A \in \mathcal{S}^\star$, $x \in \mathbb{R}^n$, and a sequence t_j tending to infinity such that $\lim_{j\to\infty} \|\Phi_A(t_j,0)x\| = \limsup_{t\to\infty} \|\Phi_A(t,0)x\|$. By Remark 6.4 we can apply (6.4) to $\mathcal{F}^\star$ and deduce that for every $j \in \mathbb{N}$ there exists $Q_j \in \cup_{\tau\in[t_j,t_j+2t_*]}\Phi(\mathcal{F}_\tau^\star)$ such that

$$\|\Phi_A(t_j,0)x\| \le \kappa_* \|Q_j\| \|x\|.$$

In order to prove that $(\Sigma_{\mathcal{S}^\star})$ is attractive, we are going to show that Q_j goes to zero as $j \to \infty$. Notice that, for every $t > 0$,

$$\sup\{\|R\| \mid R \in \Phi(\mathcal{F}_t^\star)\} = \limsup_{\delta\downarrow 0} \sup\{\|U\| \mid \tau \in (-\delta,\delta),\ U \in \Phi(\mathcal{F}_{t+\tau})\}, \tag{6.5}$$

as it follows from Proposition 1.3 in the case $\Phi(\mathcal{F}_t^\star) \neq \emptyset$, while in the case $\Phi(\mathcal{F}_t^\star) = \emptyset$ both terms in (6.5) are equal to $-\infty$ by definition of $\mathcal{F}_t^\star$ (Definition 6.1). By assumption,

$$\limsup_{t\to\infty} \sup\{\|U\| \mid \tau \in (-\delta,\delta),\ U \in \Phi(\mathcal{F}_{t+\tau})\} = 0, \qquad \forall \delta > 0,$$

concluding the proof that $(\Sigma_{\mathcal{S}^\star})$ is attractive.

Fenichel's lemma (Lemma 1.25) applied to $\mathcal{S}^\star$ then implies that $\lambda_\#(\mathcal{F}) = \lambda(\mathcal{S}) \le \lambda(\mathcal{S}^\star) < 0$. □

At the light of Lemma 6.5, a crucial assumption on skeletons, also appearing in the following sections, is that $\Phi(\mathcal{F})$ is irreducible. The purpose of what follows is to provide a sufficient condition for the irreducibility of $\Phi(\mathcal{F})$ not relying on the integration of any flow.

In the following, we denote by $G_{\mathcal{M}}$ the matrix group generated by $\mathcal{M}$ (i.e, the smallest subgroup of $\mathrm{GL}(n)$ containing e^{tM} for every $t \in \mathbb{R}$ and $M \in \mathcal{M}$). Recall that, by the orbit theorem (Theorem A.19), $G_{\mathcal{M}}$, endowed with its corresponding orbit topology, is an immersed analytic submanifold of $\mathrm{GL}(n)$.

Definition 6.6 We say that a skeleton $\mathcal{F}$ is *rich* if $\Phi(\mathcal{F})$ has nonempty interior in $G_{\mathcal{M}}$. If there exists $\tau > 0$ such that $\cup_{s\in[0,\tau]} \Phi(\mathcal{F}_s)$ has nonempty interior in $G_{\mathcal{M}}$, then we say that $\mathcal{F}$ is *finite-time rich*.

Remark 6.7 The skeleton of $\mathcal{S}_{\text{arb}}(\mathcal{M})$ defined by $\mathcal{F}^t_{\text{arb}}(\mathcal{M}) = L^\infty([0,t],\mathcal{M})$, $t \ge 0$, is finite-time rich, as it follows from Krener's theorem (Theorem A.18) applied to the manifold $G_{\mathcal{M}}$.

Remark 6.8 Let $\mathcal{F}$ be a skeleton and $\mathcal{M}$ be reducible. Let $\mathcal{F}_1, \dots, \mathcal{F}_r$ be obtained from $\mathcal{F}$ based on the maximal invariant flag decomposition of Proposition 1.31. For $i = 1, \dots, r$, let $\pi_{ii} \colon M_n(\mathbb{R}) \to M_{n_i}(\mathbb{R})$ be the application introduced in (1.14). Then, the projection by π_{ii} of an open set (for the orbit topology) of the matrix group generated by $\mathcal{M}$ is an open set of the matrix group generated by $\mathcal{M}_i$ (for its own orbit topology). As a consequence, if $\mathcal{F}$ is rich (respectively, finite-time rich) then each $\mathcal{F}_i$ is a rich (respectively, finite-time rich) skeleton of the corresponding class $\mathcal{S}_i$.

Proposition 6.9 *If $\mathcal{F}$ is a rich skeleton and $\mathcal{M}$ is irreducible, then $\Phi(\mathcal{F})$ is irreducible.*

Proof Let $x, z \in \mathbb{R}^n \setminus \{0\}$. We have to prove that there exists $R \in \Phi(\mathcal{F})$ such that $z^\top R x \neq 0$. Since $\mathcal{M}$ is irreducible, it follows from Lemma 1.29 that there exists a piecewise constant signal $A \colon [0, t] \to \mathcal{M}$ such that $z^\top \Phi_A(t, 0)x \neq 0$. Consider an analytic path $[0, 1] \ni \xi \mapsto M_\xi$ in the matrix group $G_{\mathcal{M}}$ connecting $\Phi_A(t, 0)$ to a matrix in the interior of $\Phi(\mathcal{F})$, whose existence follows from the connectedness of $G_{\mathcal{M}}$ (see Lemma A.16). The function $\xi \mapsto z^\top M_\xi x$ is analytic and not identically equal to zero. Its zeros are therefore isolated (Lemma A.13), whence the conclusion with R of the form M_ξ. □

Proposition 6.10 *Let $\mathcal{F}$ be a skeleton of $\mathcal{S}$. Let $A \colon [0, t] \to \mathcal{M}$ be in $\mathcal{F}$. Assume that there exist $t_0 = 0 < t_1 < \cdots < t_\ell < t = t_{\ell+1}$ such that A is analytic on $[t_j, t_{j+1})$ for $0 \leq j \leq \ell$ and* $\operatorname{span}\{A(t_1), \dots, A(t_\ell)\} = \operatorname{span}(\mathcal{M})$. *Assume, moreover, that $A \circ \psi \in \mathcal{F}$ for every reparameterization ψ in a $C^{1,1}$-neighborhood of the identity (that is, there exists $\varepsilon \in (0, 1)$ such that $A \circ \psi \in \mathcal{F}$ for every $\psi \colon [0, T'] \to [0, T]$ Lipschitz continuous with $\psi(0) = 0$, $\psi(T') = T$, and $|\psi'(s) - 1| < \varepsilon$ for almost every $s \in [0, T']$). Then $\mathcal{F}$ is finite-time rich.*

Proof Denote by $G_{\mathcal{M}}$ the matrix group generated by $\mathcal{M}$. For every $k \in \mathbb{N}$, $\tau = (\tau_1, \dots, \tau_k) \in (0, +\infty)^k$, and $M = (M_1, \dots, M_k) \in \{A(t_1), \dots, A(t_\ell)\}^k$, set $T_\tau = \sum_{j=1}^k \tau_j$ and denote by $B_{\tau,M} \colon [0, T_\tau] \to \mathcal{M}$ the piecewise constant signal

$$B_{\tau,M}(s) = M_j, \quad \text{if } \sum_{i=1}^{j-1} \tau_i < s < \sum_{i=1}^{j} \tau_i, \ 1 \leq j \leq k.$$

Since $\operatorname{span}\{A(s) \mid s \in [0, t)\} = \operatorname{span}(\mathcal{M})$ and thanks to Krener's theorem (Theorem A.18 applied on the manifold $G_{\mathcal{M}}$), denoting by k the dimension of $G_{\mathcal{M}}$, there exist τ and M such that the partial derivatives $\frac{\partial}{\partial s_j} \Phi_{B_{s,M}}(T_s, 0)\Big|_{s=\tau}$, $j = 1, \dots, k$, span the entire tangent to $G_{\mathcal{M}}$ at $\Phi_{B_{\tau,M}}(T_\tau, 0)$, where $s = (s_1, \dots, s_k)$.

Up to replacing A by a concatenation of A with itself (finitely many times), we can choose $\ell = k$ and $t_1, \dots, t_k$ such that $A(t_j) = M_j$ for $j = 1, \dots, k$. Let us now introduce a suitable one-parameter deformation of A into $B_{\tau,M}$. For every

$\eta \in (0, 1]$ and $j = 0, \dots, k$ set $\tau_{j,\eta} = \tau_j + \eta(t_{j+1} - t_j - \tau_j)$ (with $\tau_0 = 0$) and define $A_{j,\eta} \colon [0, \tau_{j,\eta}] \to \mathcal{M}$ by

$$A_{j,\eta}(s) = A\left(t_j + \left(\frac{s}{\tau_{j,\eta}}\right)^{\frac{1}{\eta}} \tau_{j,\eta}\right), \qquad s \in [0, \tau_{j,\eta}].$$

Let, moreover, $t(\eta) = \sum_{j=0}^{k} \tau_{j,\eta}$ and $A_\eta = A_{0,\eta} * \cdots * A_{k,\eta} \colon [0, t(\eta)] \to \mathcal{M}$. In particular, $A_1 = A$ and $\lim_{\eta \downarrow 0} A_\eta = B_{\tau,M}$ almost everywhere on $[0, T_\tau]$, where we use that

$$\lim_{\eta \downarrow 0} \left(\frac{s}{\tau_{j,\eta}}\right)^{\frac{1}{\eta}} = 0 \quad \text{for every } s \in [0, \tau_j).$$

For $\eta > 0$, $j = 0, \dots, k$, and $\varepsilon \in (-\tau_{j,\eta}, \tau_{j,\eta})$, let $A^{\varepsilon}_{j,\eta} \colon [0, \tau_{j,\eta} + \varepsilon] \to \mathcal{M}$ be defined by

$$A^{\varepsilon}_{j,\eta}(s) = A\left(t_j + \left(\frac{s}{\tau_{j,\eta} + \varepsilon}\right)^{\frac{1}{\eta}} \tau_{j,\eta}\right), \qquad s \in [0, \tau_{j,\eta} + \varepsilon],$$

and define $A^{(j),\varepsilon}_{\eta} = A_{0,\eta} * \cdots A_{j-1,\eta} * A^{\varepsilon}_{j,\eta} * A_{j+1,\eta} * \cdots * A_{k,\eta}$. Set $v_{j,\eta} = \frac{d}{d\varepsilon}\Big|_{\varepsilon=0} \Phi_{A^{(j),\varepsilon}_{\eta}}(t(\eta) + \varepsilon, 0)$.

We claim that

$$\lim_{\eta \to 0} v_{j,\eta} = v_j, \qquad i = 1, \dots, k. \tag{6.6}$$

In order to prove the claim, notice that, for $j = 1, \dots, k$ and $\eta \in (0, 1]$,

$$\begin{aligned} \frac{d}{d\varepsilon}\Big|_{\varepsilon=0} \Phi_{A^{\varepsilon}_{j,\eta}}(\tau_{j,\eta} + \varepsilon, 0) &= \frac{d}{d\varepsilon}\Big|_{\varepsilon=0} \Phi_{(1+\frac{\varepsilon}{\tau_{j,\eta}})A_{j,\eta}}(\tau_{j,\eta}, 0) \\ = \frac{1}{\tau_{j,\eta}} \int_0^{\tau_{j,\eta}} \Phi_{A_{j,\eta}}(\tau_{j,\eta}, s) &A\left(t_j + \tau_{j,\eta}\left(\frac{s}{\tau_{j,\eta}}\right)^{\frac{1}{\eta}}\right) \Phi_{A_{j,\eta}}(s, 0) ds, \end{aligned}$$

where we used the variation of constants formula (Lemma A.6)

$$\begin{aligned} \Phi_{(1+\frac{\varepsilon}{\tau_{j,\eta}})A_{j,\eta}}(\tau_{j,\eta}, 0) &= \Phi_{A_{j,\eta} + \frac{\varepsilon}{\tau_{j,\eta}} A_{j,\eta}}(\tau_{j,\eta}, 0) \\ &= \Phi_{A_{j,\eta}}(\tau_{j,\eta}, 0) \\ &\quad + \frac{\varepsilon}{\tau_{j,\eta}} \int_0^{\tau_{j,\eta}} \Phi_{A_{j,\eta}}(\tau_{j,\eta}, s) \\ &\quad \times A\left(t_j + \tau_{j,\eta}\left(\frac{s}{\tau_{j,\eta}}\right)^{\frac{1}{\eta}}\right) \Phi_{(1+\frac{\varepsilon}{\tau_{j,\eta}})A_{j,\eta}}(s, 0) ds. \end{aligned}$$

Hence, for $j = 1, \ldots, k$, by the L^1 convergence of $A_{j,\eta}$ to the constant $A(t_j) = M_j$ on $[0, \tau_j]$,

$$\lim_{\eta \to 0} \frac{d}{d\varepsilon}\bigg|_{\varepsilon=0} \Phi_{A^{\varepsilon}_{j,\eta}}(\tau_{j,\eta} + \varepsilon, 0) = M_j e^{\tau_j M_j},$$

yielding (6.6).

The construction of $B_{\tau,M}$ implies that for η close to zero the directions $v_{1,\eta}, \ldots, v_{k,\eta}$ span the entire tangent to $G_{\mathcal{M}}$ at $\Phi_{A_\eta}(t(\eta), 0)$. The analytic dependence of $\Phi_{A_\eta}(t(\eta), 0)$ and $v_{j,\eta}$ with respect to η then guarantee that, for almost all $\eta \in (0, 1]$, $v_{1,\eta}, \ldots, v_{k,\eta}$ span the entire tangent space to $G_{\mathcal{M}}$ at $\Phi_{A_\eta}(t(\eta), 0)$ (Lemma A.13). Since the assumptions made on A imply that for $j \in \{1, \ldots, k\}$, η close to one, and ε close to zero the signal $A_\eta^{(j),\varepsilon}$ is in $\mathcal{F}$, then for every $t' > t$ the set $\cup_{s \in [0,t']} \Phi(\mathcal{F}_s)$ covers a neighborhood of $\Phi_{A_\eta}(t(\eta), 0)$ in $G_{\mathcal{M}}$ for almost every η close to one. □

6.1.2 *Extensive and Rich Skeletons for Relevant Classes of Switching Signals*

We give in this section some examples of skeleton for the classes of switching signals introduced in Sect. 1.1.1.

- A skeleton for the class $\mathcal{S}_{\mathrm{dw},\tau}(\mathcal{M})$ of signals with dwell-time τ is given by the family $\mathcal{F}_{\mathrm{dw},\tau}(\mathcal{M})$ of all piecewise constant signals on finite intervals with dwell-time τ and such that the first and last subintervals on which the signal is constant have length at least τ. In particular, each element of $\mathcal{F}_{\mathrm{dw},\tau}(\mathcal{M})$ is defined on an interval of length at least τ.
- Given $\mathcal{M} = \{M_\ell = N_0 + \sum_{i=1}^m \ell_i N_i \mid \ell = (\ell_1, \ldots, \ell_m) \in [0, 1]^m\}$, a skeleton for $\mathcal{S}_{\mathrm{pe},T,\mu}(\mathcal{M})$ is given by the family $\mathcal{F}_{\mathrm{pe},T,\mu}(\mathcal{M})$ of signals $[0, t] \ni s \mapsto M_{\alpha(s)}$ with $t \geq T - \mu$, $\alpha \in L^\infty([0, t], [0, 1]^m)$, $\int_s^{s+T} \alpha_i(\tau) d\tau \geq \mu$ for every $s \in [0, t - T]$ and $i = 1, \ldots, m$, and $\alpha \equiv (1, \ldots, 1)$ on $[t - T + \mu, t]$.

For the classes $\mathcal{S}_{\mathrm{cont}}(\mathcal{M})$, $\mathcal{S}_{\mathrm{lip},L}(\mathcal{M})$, and $\mathcal{S}_{\mathrm{bv},T,\nu}(\mathcal{M})$ we assume that $\mathcal{M}$ is *star-shaped*, i.e., there exists $\overline{M} \in \mathcal{M}$ such that every segment connecting a point of $\mathcal{M}$ to $\overline{M}$ is contained in $\mathcal{M}$, and we define

- $\mathcal{F}_{\mathrm{cont}}(\mathcal{M})$ as the set of continuous loops starting and ending at $\overline{M}$;
- $\mathcal{F}_{\mathrm{lip},L}(\mathcal{M})$ as the set of L-Lipschitz loops starting and ending at $\overline{M}$;
- $\mathcal{F}_{\mathrm{bv},T,\nu}(\mathcal{M})$ as the set of signals on finite intervals $[0, t]$, $t \geq T$, starting and ending at $\overline{M}$, constant on $[t - T, t]$, and whose restriction to every interval $[s, s + T]$ with $s \in [0, t - T]$ has total variation at most ν.

It is easy to check that all these classes of finite-interval signals define an extensive skeleton. Moreover, as summarized in the next lemma, it follows easily from Proposition 6.10 that they are also (finite-time) rich.

Lemma 6.11

1. *$\mathcal{F}_{\mathrm{dw},\tau}(\mathcal{M})$ is an extensive and finite-time rich skeleton of $\mathcal{S}_{\mathrm{dw},\tau}(\mathcal{M})$.*
2. *Given $\mathcal{M} = \{N_0 + \sum_{i=1}^{m} \ell_i N_i \mid \ell = (\ell_1, \ldots, \ell_m) \in [0, 1]^m\}$, $\mathcal{F}_{\mathrm{pe},T,\mu}(\mathcal{M})$ is an extensive and finite-time rich skeleton of $\mathcal{S}_{\mathrm{pe},T,\mu}(\mathcal{M})$.*
3. *Let $\mathcal{M}$ be star-shaped. Then $\mathcal{F}_{\mathrm{cont}}(\mathcal{M})$, $\mathcal{F}_{\mathrm{lip},L}(\mathcal{M})$, and $\mathcal{F}_{\mathrm{bv},T,\nu}(\mathcal{M})$ are extensive and finite-time rich skeletons of, respectively, $\mathcal{S}_{\mathrm{cont}}(\mathcal{M})$, $\mathcal{S}_{\mathrm{lip},L}(\mathcal{M})$, and $\mathcal{S}_{\mathrm{bv},T,\nu}(\mathcal{M})$.*

Remark 6.12 In the discussion above the hypothesis on the star-shapedness of $\mathcal{M}$ can be replaced by a weaker one, preserving the fact that the proposed skeleton is extensive and finite-time rich. For instance, if we consider $\mathcal{S}_{\mathrm{cont}}(\mathcal{M})$, we could simply ask $\mathcal{M}$ to be pathwise connected by analytic paths (and fix any $\overline{M}$ in $\mathcal{M}$).

In the case of $\mathcal{S}_{\mathrm{lip},L}(\mathcal{M})$, one should add to the previous weaker assumption an additional constraint on the length of paths that render $\mathcal{M}$ pathwise connected; for instance, one could assume that there exists $C > 0$ such that every two points of $\mathcal{M}$ can be connected by a analytic curve lying in $\mathcal{M}$ of length smaller than C.

Finally, for $\mathcal{S}_{\mathrm{bv},T,\nu}(\mathcal{M})$ we could ask that for every $M, M' \in \mathcal{M}$ there exists a finite sequence of points in $\mathcal{M}$ whose first and last elements are M and M' and such that the distance between two subsequent elements is smaller than ν.

6.1.3 Barabanov Skeletons and Quasi-Extremal Trajectories

The following definition extends to skeletons the notions of extremal and Barabanov norms.

Definition 6.13 (Extremal and Barabanov Skeletons) Let $\mathcal{F}$ be a skeleton. We say that $\mathcal{F}$ is *extremal* if there exists a norm w on $\mathbb{R}^n$ (which is said to be an *extremal norm for $\mathcal{F}$*) such that the induced matrix norm $\|\cdot\|_w$ satisfies, for every $t \geq 0$ and $R \in \Phi(\mathcal{F}_t)$,

$$\|R\|_w \leq e^{\lambda_{\#}(\mathcal{F})t}. \tag{6.7}$$

We say that $\mathcal{F}$ is a *Barabanov skeleton* if it is extremal and there exists $C_{\mathrm{b}} > 0$ such that for every $x \in \mathbb{R}^n$ and $t \geq 0$ there exist $t' \geq t$ and $R \in \Phi(\mathcal{F}_{t'})$ such that

$$\|Rx\| \geq C_{\mathrm{b}} e^{\lambda_{\#}(\mathcal{F})t'} \|x\|. \tag{6.8}$$

Let us give some sufficient conditions for a skeleton to be Barabanov.

Proposition 6.14 *Let $\mathcal{F}$ be an extensive skeleton of $\mathcal{S}$ with $\Phi(\mathcal{F})$ irreducible. Then $\mathcal{F}$ is a Barabanov skeleton. Moreover, there exists $C > 0$ such that $\|\Phi_A(t,0)\| \leq Ce^{t\lambda(\mathcal{S})}$ for every $A \in \mathcal{S}$ and $t \geq 0$.*

Proof First recall that $\lambda_{\#}(\mathcal{F}) = \lambda(\mathcal{S})$, since $\mathcal{F}$ is extensive (Lemma 6.5). Set

$$\mathscr{R}_\infty = \{R \in M_n(\mathbb{R}) \mid \exists t_k \to \infty,\ R_k \in \Phi(\mathcal{F}_{t_k}) \text{ such that } e^{-t_k\lambda(\mathcal{S})}R_k \to R\}.$$

By the semigroup property of $\Phi(\mathcal{F})$, it follows that $\mathscr{R}_\infty$ is a semigroup of matrices. Moreover, for the same reason,

$$e^{-t\lambda(\mathcal{S})}TS,\, e^{-t\lambda(\mathcal{S})}ST \in \mathscr{R}_\infty, \qquad \forall t \geq 0,\ \forall T \in \Phi(\mathcal{F}_t),\ \forall S \in \mathscr{R}_\infty. \tag{6.9}$$

Notice that $\mathscr{R}_\infty$ is closed since it coincides with $\bigcap_{s>0}\overline{\bigcup_{t>s}e^{-t\lambda(\mathcal{S})}\Phi(\mathcal{F}_t)}$. It is also bounded, as it can be checked by following the same arguments as those in the proof of Lemma 4.7.

Define

$$\hat{v}(x) = \max_{R\in\mathscr{R}_\infty}\|Rx\|.$$

We claim that $\hat{v}$ is a norm. The homogeneity is obvious and subadditivity follows directly from the definition. As for the positive-definiteness of $\hat{v}$, assume that $\hat{v}(x) = 0$ and let us prove that $x = 0$. Property (6.9) implies that $\hat{v}(Mx) = 0$ for every $M \in \Phi(\mathcal{F})$. This means that $\{y \mid \hat{v}(y) = 0\}$ is invariant for $\Phi(\mathcal{F})$. By irreducibility of $\Phi(\mathcal{F})$ we conclude that $x = 0$, as required, or that $\mathscr{R}_\infty = \{0\}$. The latter is impossible due to Lemma 6.5.

The norm $\hat{v}$ is actually extremal, as it follows from (6.9). In particular, the skeleton $\mathcal{F}$ is extremal. In order to show that $\mathcal{F}$ is a Barabanov skeleton, let $c > 0$ be such that

$$c\|x\| \leq \hat{v}(x) \qquad \text{for all } x \in \mathbb{R}^n.$$

For every $x_0 \in \mathbb{R}^n$, there exists a sequence $(R_k)_{k\in\mathbb{N}}$ so that $R_k \in \Phi(\mathcal{F}_{t_k})$ with $\lim_{k\to\infty} t_k = \infty$ and $\hat{v}(x_0) = \lim_{k\to\infty} e^{-\lambda(\mathcal{S})t_k}\|R_kx_0\|$. For every $k \in \mathbb{N}$, let $A_k \in \mathcal{F}_{t_k}$ be such that $R_k = \Phi_{A_k}(t_k, 0)$. Fix now $t \geq 0$. Let $k \in \mathbb{N}$ be such that $t_k \geq t$ and $e^{-\lambda(\mathcal{S})t_k}\|R_kx_0\| \geq \hat{v}(x_0)/2$. Then

$$\|R_kx_0\| \geq \frac{c}{2}e^{\lambda(\mathcal{S})t_k}\|x_0\|,$$

proving (6.8) with $t' = t_k$, $R = R_k$, and $C_b = \frac{c}{2}$.

Concerning the last part of the statement, let $A \in \mathcal{S}$ and $t \geq 0$. Since $\mathcal{F}$ is extensive and thanks to (6.4), there exist $t_* \geq 0$ and $\kappa_* \geq 1$ (independent of A and t) such that there exist $\tau \in [t, t + 2t_*]$ and $B \in \mathcal{F}_\tau$ with

$$\|\Phi_A(t, 0)\| \leq \kappa_* \|\Phi_B(\tau, 0)\|.$$

By the extremality of $\mathcal{F}$ and the equivalence of all norms on $\mathbb{R}^n$, there exists $\tilde{C} > 0$ independent of A such that $\|\Phi_B(\tau, 0)\| \leq \tilde{C} e^{\tau\lambda(\mathcal{S})}$. Hence,

$$\|\Phi_A(t, 0)\| \leq C e^{t\lambda(\mathcal{S})},$$

where $C = \tilde{C}\kappa_* \max(1, e^{2t_*\lambda(\mathcal{S})})$. □

Definition 6.15 (Quasi-Extremal Trajectory) A trajectory $x\colon t \mapsto \Phi_A(t, 0)x_0$ with $x_0 \neq 0$ and $A \in \mathcal{S}^\star$ is said to be *quasi-extremal for* $\mathcal{S}$ if there exists $C_{\mathrm{qx}} \geq 1$ such that

$$\frac{1}{C_{\mathrm{qx}}} e^{\lambda(\mathcal{S})t} \|x_0\| \leq \|x(t)\| \leq C_{\mathrm{qx}} e^{\lambda(\mathcal{S})t} \|x_0\|, \qquad \forall t \geq 0.$$

The following result ensures the existence of quasi-extremal trajectories under the hypotheses of Proposition 6.14.

Theorem 6.16 *Let $\mathcal{F}$ be an extensive skeleton of $\mathcal{S}$ and assume that $\Phi(\mathcal{F})$ is irreducible. Then, for every $x_0 \in \mathbb{R}^n \setminus \{0\}$, $\mathcal{S}$ admits a quasi-extremal trajectory with initial condition x_0 corresponding to a signal in the weak-$\star$ closure of $\mathcal{F}_\infty$.*

Proof Let t_* be as in Definition 6.3 and $\kappa_* \geq 1$ be defined as in (6.3). Let C be as in the statement of Proposition 6.14. Moreover, since $\mathcal{F}$ is a Barabanov skeleton, there exists $C_{\mathrm{q}} > 0$ as in (6.8). Recall, moreover, that $\lambda_\#(\mathcal{F}) = \lambda(\mathcal{S})$ by Lemma 6.5.

Fix $x_0 \neq 0$. There exist an increasing sequence of times t_k going to infinity and a sequence of signals A_k in $\mathcal{F}_{t_k}$ such that $\|\Phi_{A_k}(t_k, 0)x_0\| \geq C_{\mathrm{b}} e^{\lambda(\mathcal{S})t_k} \|x_0\|$. We now claim that there exists $C_0 > 0$ independent of x_0 such that, for every $k \in \mathbb{N}$ and every $s \in [0, t_k]$,

$$\left\|\Phi_{A_k}(s, 0)x_0\right\| \geq \frac{1}{C_0} e^{\lambda(\mathcal{S})s} \|x_0\|. \tag{6.10}$$

Indeed, since $\mathcal{F}$ is extensive, one gets that $\Phi_{A_k}(t_k, s) = MRM'$ with $\|M\|, \|M'\| \leq \sqrt{\kappa_*}$ and $R \in \Phi(\mathcal{F}_{t_k - s + \delta})$, with $\delta \in [0, 2t_*]$. One can therefore write $\Phi_{A_k}(t_k, 0) = MRM'\Phi_{A_k}(s, 0)$. It follows that

$$\|\Phi_{A_k}(t_k, 0)x_0\| \leq \kappa_* C e^{\lambda(\mathcal{S})(t_k - s + \delta)} \left\|\Phi_{A_k}(s, 0)x_0\right\|.$$

On the other hand $\|\Phi_{A_k}(t_k, 0)x_0\| \geq C_{\mathrm{b}} e^{\lambda(\mathcal{S})t_k} \|x_0\|$, which proves (6.10) with

$$C_0 = \frac{\kappa_* C \max(1, e^{2\lambda(\mathcal{S})t_*})}{C_{\mathrm{b}}}.$$

For $k \in \mathbb{N}$, let $B_k \in \mathcal{F}_\infty$ be such that A_k is the restriction of B_k to $[0, t_k]$. Up to a subsequence we can assume that B_k weak-$\star$ converges to some $B_\star$ in the weak-$\star$ closure of $\mathcal{F}_\infty$. Passing to the limit in (6.10) using Proposition 1.3, we deduce that for every $s \geq 0$,

$$\left\| \Phi_{B_\star}(s, 0)x_0 \right\| \geq \frac{1}{C_0} e^{\lambda(\mathcal{S})s} \|x_0\|. \tag{6.11}$$

Similarly, for every $s \geq 0$, $x_0 \in \mathbb{R}^n$ and B in $\mathcal{S}^\star$, it holds

$$\|\Phi_B(s, 0)x_0\| \leq Ce^{\lambda(\mathcal{S})s} \|x_0\|,$$

where C is again the constant appearing in the statement of Proposition 6.14. This conclude the proof. □

Another consequence of Proposition 6.14 is the equality between the maximal Lyapunov exponent and the uniform exponential rate. We recall that such an equality was obtained in Proposition 1.26 in the case where $\mathcal{S}$ long-time approximates $\mathcal{S}^\star$.

Lemma 6.17 *Let $\mathcal{S}$ admit an extensive skeleton $\mathcal{F}$ and assume that $\Phi(\mathcal{F})$ is irreducible. Then $\lambda(\mathcal{S}) = \hat{\lambda}(\mathcal{S})$.*

Proof Recall that $\hat{\lambda}(\mathcal{S}) \leq \lambda(\mathcal{S})$ (Proposition 1.20). Fix $x_0 \neq 0$. Proposition 6.14 implies that $\mathcal{F}$ is a Barabanov skeleton. By (6.8), there exist two sequences $(t_k)_{k\in\mathbb{N}}$ and $(R_k)_{k\in\mathbb{N}}$ with $t_k \geq k$ and $R_k \in \Phi(\mathcal{F}_{t_k})$ for every k such that

$$\|R_k \cdots R_1 x_0\| \geq (C_b)^k e^{\lambda(\mathcal{S})(t_1+\cdots+t_k)} \|x_0\|, \qquad k \in \mathbb{N}.$$

In particular, using that $t_1 + \cdots + t_k \geq k(k+1)/2$ for every $k \in \mathbb{N}$, we have

$$\hat{\lambda}(\mathcal{S}) \geq \limsup_{k\to\infty} \frac{\log \|R_k \cdots R_1\|}{t_1 + \cdots + t_k} \geq \lambda(\mathcal{S}) + \liminf_{k\to\infty} \frac{k \log(C_b)}{t_1 + \cdots + t_k} = \lambda(\mathcal{S}),$$

concluding the proof. □

We now extend the previous result to the case where $\mathcal{M}$ may be reducible, under the assumption that $\mathcal{S}$ admits a rich skeleton.

Corollary 6.18 *Let $\mathcal{S}$ admit an extensive and rich skeleton $\mathcal{F}$. Then $\lambda(\mathcal{S}) = \hat{\lambda}(\mathcal{S})$.*

Proof Let us apply the block decomposition of Proposition 1.31. By Proposition 1.32, we have that $\lambda(\mathcal{S}) = \max_{i=1,\ldots,r} \lambda(\mathcal{S}_i)$, where $\mathcal{S}_1, \ldots, \mathcal{S}_r$ correspond to a maximal invariant flag decomposition of $\mathcal{M}$. Let $\mathcal{F}_1, \ldots, \mathcal{F}_r$ be obtained from $\mathcal{F}$ based on the same maximal invariant flag decomposition. Then, for every $j = 1, \ldots, k$, $\mathcal{F}_j$ is trivially an extensive skeleton of $\mathcal{S}_j$. Moreover, each $\mathcal{F}_j$ is also rich, as noticed in Remark 6.8. Then, by Proposition 6.9 and Lemma 6.17, $\hat{\lambda}(\mathcal{S}_i) = \lambda(\mathcal{S}_i)$ for $i = 1, \ldots, r$. Notice now that $\hat{\lambda}(\mathcal{S}) \geq \hat{\lambda}(\mathcal{S}_i)$ for every $i = 1, \ldots, r$, since each solution of $(\Sigma_{\mathcal{S}_i})$ is the projection of a solution of $(\Sigma_{\mathcal{S}})$.

Hence, $\hat{\lambda}(\mathcal{S}) \geq \max_{i=1,\dots,r} \hat{\lambda}(\mathcal{S}_i) = \max_{i=1,\dots,r} \lambda(\mathcal{S}_i) = \lambda(\mathcal{S})$. Since, on the other hand, $\hat{\lambda}(\mathcal{S}) \leq \lambda(\mathcal{S})$ (Proposition 1.20) then the proof of the corollary is complete. □

Remark 6.19 Since all the skeletons introduced in Sect. 6.1.2 are extensive and rich, $\lambda(\mathcal{S}) = \hat{\lambda}(\mathcal{S})$ for the corresponding classes of signals.

Let us conclude the section by deducing a result concerning the equality between $\lambda(\mathcal{S})$ and the joint spectral exponent $\mu(\mathcal{S})$ introduced in Sect. 1.3.3, which generalizes Theorem 4.40.

Proposition 6.20 *Let $\mathcal{S}$ admit a rich and extensive skeleton $\mathcal{F}$. Then $\mu(\mathcal{S}) = \lambda(\mathcal{S})$.*

Proof Recall that we always have $\mu(\mathcal{S}) \leq \lambda(\mathcal{S})$. Up to a spectral shift as in (1.8), we can assume that $\lambda(\mathcal{S}) = 0$ and we are left to prove that $\mu(\mathcal{S}) \geq 0$.

Reasoning as in Corollary 6.18, we can restrict to the case where $\mathcal{M}$, and hence $\Phi(\mathcal{F})$, are irreducible. By Proposition 6.14, there exists $C > 0$ such that $\|\Phi_A(t, 0)\| \leq C$ for every $A \in \mathcal{S}$ and $t \geq 0$. By Theorem 6.16, moreover, $\mathcal{S}$ admits a quasi-extremal trajectory, that is, there exist $C_{\mathrm{qx}} \geq 1$, $A \in \mathcal{S}^\star$, and a unit vector $x_0 \in \mathbb{R}^n$ such that

$$\frac{1}{C_{\mathrm{qx}}} \leq \|x(t)\| \leq C_{\mathrm{qx}}, \qquad \forall t \geq 0.$$

Let us select an increasing unbounded sequence $(t_j)_{j\in\mathbb{N}}$ in $[0, +\infty)$ such that $x(t_j)$ converges to some $x_* \in \mathbb{R}^n$ as $j \to \infty$, with $\|x_*\| \geq 1/C_{\mathrm{qx}} > 0$. We can assume without loss of generality that $t_{j+1} - t_j \to +\infty$ as $j \to \infty$. Define $M_j = \Phi_A(t_{j+1}, t_j)$ for every $j \in \mathbb{N}$ and notice that $\|M_j\| \leq C$. We can then extract a subsequence $(M_{j_k})_{k\in\mathbb{N}}$ converging to some $M_* \in M_n(\mathbb{R})$. By taking the limit as $j \to \infty$ in the relation $x(t_{j+1}) = M_j x(t_j)$, we deduce that $x_* = M_* x_*$. Hence $\rho(M_*) \geq 1$. By continuity of the spectral radius, we deduce that $\rho(M_j) > 1/2$ for j large enough. Moreover, for every $j \in \mathbb{N}$ there exists $B_j \in \mathcal{S}$ such that $\rho(\Phi_{B_j}(t_{j+1} - t_j, 0)) > \rho(M_j)/2$, because $A(t_j + \cdot) \in \mathcal{S}^\star$.

Hence,

$$\mu(\mathcal{S}) \geq \limsup_{j\to\infty} \frac{\log(\rho(\Phi_{B_j}(t_{j+1} - t_j, 0)))}{t_{j+1} - t_j} = 0,$$

concluding the proof. □

Remark 6.21 Reasoning as in Remark 6.19, we can deduce that for all the classes $\mathcal{S}_{\mathrm{label}}(\mathcal{M})$ of signals corresponding to the skeletons $\mathcal{F}_{\mathrm{label}}(\mathcal{M})$ introduced in Sect. 6.1.2 it holds $\mu(\mathcal{S}_{\mathrm{label}}(\mathcal{M})) = \lambda(\mathcal{S}_{\mathrm{label}}(\mathcal{M}))$.

In particular, the proof of Theorem 4.41 can be easily adapted to deduce that $\mathcal{M} \mapsto \lambda(\mathcal{S}_{\mathrm{label}}(\mathcal{M}))$ is lower-semicontinuous on the set of compact subsets of $M_n(\mathbb{R})$ satisfying the constraints imposed on the class (e.g., simplexes of matrices for $\mathcal{S}_{\mathrm{pe},T,\mu}(\mathcal{M})$).

Together with the upper-semicontinuity observed in Remark 1.23, this yields the continuity of $\mathcal{M} \mapsto \lambda(\mathcal{S}_{\text{label}}(\mathcal{M}))$.

6.2 Periodization

Given a class $\mathcal{S}$ of signals with values in $\mathcal{M} \subset M_n(\mathbb{R})$, let us define

$$\lambda^{\text{per}}(\mathcal{S}) = \sup_{(A,x_0)\in\mathfrak{S}_{\text{per}}} \limsup_{t\to\infty} \frac{\log\left(\|\Phi_A(t,0)x_0\|\right)}{t}, \tag{6.12}$$

where $\mathfrak{S}_{\text{per}}$ is the set of pairs (A, x_0) such that $A \in \mathcal{S} \cap \mathcal{S}_{\text{per}}(\mathcal{M})$, $x_0 \in \mathbb{R}^n \setminus \{0\}$, and there exists $T > 0$ such that both A and $t \mapsto \Pi(\Phi_A(t,0)x_0)$ are T-periodic, where we recall that $\Pi\colon \mathbb{R}^n \setminus \{0\} \to \mathbb{RP}^{n-1}$ denotes the canonical projection. In other words, $\mathfrak{S}_{\text{per}}$ corresponds to the periodic trajectories of the projected switched system $(\Pi\Sigma_{\mathcal{S}})$ introduced in Sect. 1.5. Recall also that for every $M \in GL(n)$ and $q \in \mathbb{RP}^{n-1}$, Mq denotes the element $\Pi(Mx) \in \mathbb{RP}^{n-1}$ where $x \in \mathbb{R}^n$ is such that $q = \Pi(x)$.

The goal of this section is to prove the following result.

Theorem 6.22 *Let $\mathcal{S}$ admit an extensive and finite-time rich skeleton $\mathcal{F}$. Then*

$$\lambda^{\text{per}}(\mathcal{S}) = \lambda(\mathcal{S}).$$

The result applies, in particular, to the case where $\mathcal{S} = \mathcal{S}_{\text{arb}}(\mathcal{M})$ and to all the classes of signals discussed in Sect. 6.1.2.

The proof of Theorem 6.22 requires several preliminary results. Let us start by introducing some useful notation.

Given a skeleton $\mathcal{F}$, consider the time-reverse $\mathcal{F}^-$ of $\mathcal{F}$, i.e.,

$$\mathcal{F}^- = \cup_{t\geq 0}\mathcal{F}_t^-, \qquad \mathcal{F}_t^- = \{-A(t-\cdot) \mid A \in \mathcal{F}_t\}.$$

Notice that $\mathcal{F}^-$ is itself a skeleton of a family of signals with values in $-\mathcal{M}$ (e.g., the family of all time-shifts of signals in $\mathcal{F}_\infty^-$, where the latter is obtained from $\mathcal{F}^-$ in analogy to how $\mathcal{F}_\infty$ is obtained from $\mathcal{F}$). Given $T > 0$, we also write

$$\mathcal{F}_{\leq T} = \cup_{t\in[0,T]}\mathcal{F}_t, \qquad \mathcal{F}_{\leq T}^- = \cup_{t\in[0,T]}\mathcal{F}_t^-.$$

Notice that for every $A \in \mathcal{F}_t$ the flow of $-A(t-\cdot)$ from time 0 to t is $\Phi_A(0,t) = \Phi_A(t,0)^{-1}$. In particular,

$$q_1 \in \Phi(\mathcal{F}_{\leq T})q_0 \quad \text{if and only if} \quad q_0 \in \Phi(\mathcal{F}_{\leq T}^-)q_1.$$

In addition to the previous notation, for every $q_0 \in \mathbb{RP}^{n-1}$ we denote the *orbit of* $G_{\mathcal{M}}$ *through* q_0 by

$$\mathcal{O}(q_0) = \{Mq_0 \mid M \in G_{\mathcal{M}}\}, \tag{6.13}$$

i.e., $\mathcal{O}(q_0) = G_{\mathcal{M}} q_0$, where we recall that $G_{\mathcal{M}}$ denotes the matrix group generated by $\mathcal{M}$. Let us list some useful properties of the orbits of $G_{\mathcal{M}}$. Recall that a C^1 map between two manifolds is a *submersion* if its differential is everywhere surjective. As a consequence, if $F\colon \mathcal{M}_1 \to \mathcal{M}_2$ is a submersion and ω is a neighborhood of a point x_1 in $\mathcal{M}_1$, then $F(\omega)$ is a neighborhood of $F(x_1)$ in $\mathcal{M}_2$.

Proposition 6.23 *The following holds true:*

(i) For every $q_0 \in \mathbb{RP}^{n-1}$, $\mathcal{O}(q_0)$ is an immersed analytic submanifold of $\mathbb{RP}^{n-1}$ and $M \mapsto Mq_0$ is a submersion from $G_{\mathcal{M}}$ to $\mathcal{O}(q_0)$;
(ii) There exists $q_0 \in \mathbb{RP}^{n-1}$ for which $\mathcal{O}(q_0)$ is an embedded compact analytic submanifold of $\mathbb{RP}^{n-1}$;
(iii) If $\mathcal{M}$ is irreducible then for every $q_0 \in \mathbb{RP}^{n-1}$ and every proper subspace $V \subset \mathbb{R}^d$ the orbit $\mathcal{O}(q_0)$ is not contained in $\{\Pi(x) \mid x \in V,\ x \neq 0\}$.

Proof Let us first consider Item **(i)**. Fix $q_0 \in \mathbb{RP}^{n-1}$. The fact that $\mathcal{O}(q_0)$ is an immersed analytic submanifold of $\mathbb{RP}^{n-1}$ is a direct consequence of the orbit theorem (Theorem A.19 in the appendix). The submersion property follows from the linearity of the map $M \mapsto Mq_0$ and the characterization of the tangent space to an orbit given in the orbit theorem, which guarantees that, given any $M \in G_{\mathcal{M}}$ and denoting $q = Mq_0$,

$$T_M G_{\mathcal{M}} = \{NM \mid N \in \mathrm{Lie}(\mathcal{M})\}, \qquad T_q\mathcal{O}(q_0) = \{Nq \mid N \in \mathrm{Lie}(\mathcal{M})\},$$

where $\mathrm{Lie}(\mathcal{M})$ is the Lie algebra of matrices generated by $\mathcal{M}$ (see the appendix for details).

Item **(ii)** follows directly from Theorem A.20.

As for Item **(iii)**, assume by contradiction that there exists $q_0 = \Pi(x_0) \in \mathbb{RP}^{n-1}$ and a proper subspace $V \subset \mathbb{R}^d$ such that $\mathcal{O}(q_0) \subset \{\Pi(x) \mid x \in V,\ x \neq 0\}$. Let

$$W = \mathrm{span}\{Mx_0 \mid M \in G_{\mathcal{M}}\}.$$

Then W is contained in V and is invariant for all matrices in $\mathcal{M}$. Since V is a proper subspace and $W \neq \{0\}$, this contradicts the irreducibility of $\mathcal{M}$. □

From now on, we say that an orbit $\mathcal{O}(q_0)$ as in Item **(ii)** of Proposition 6.23 is a *compact orbit for* $(\Sigma_{\mathcal{M}})$.

Lemma 6.24 *Let $M \in \mathrm{GL}(n)$ and $T > 0$ be such that $M \in \mathrm{int}_{G_{\mathcal{M}}}(\Phi(\mathcal{F}_{\leq T}))$. Then $M^{-1} \in \mathrm{int}_{G_{\mathcal{M}}}(\Phi(\mathcal{F}^-_{\leq T}))$. In particular, if $\mathcal{F}$ is a finite-time rich skeleton then for every $q_0 \in \mathbb{RP}^{n-1}$ there exist $q_1 \in \mathcal{O}(q_0)$ and $T > 0$ such that $q_0 \in \mathrm{int}_{\mathcal{O}(q_0)}(\Phi(\mathcal{F}^-_{\leq T})q_1)$.*

Proof The first part of the statement follows from the equality $\Phi(\mathcal{F}^-_{\le T}) = \{N^{-1} \mid N \in \Phi(\mathcal{F}_{\le T})\}$ and by noticing that $N \mapsto N^{-1}$ is an automorphism of $\mathrm{GL}(n)$ sending $G_{\mathcal{M}}$ into itself.

In order to prove the second part of the statement, pick $T > 0$ and $M \in \mathrm{int}_{G_{\mathcal{M}}}(\Phi(\mathcal{F}_{\le T}))$. Fix $q_0 \in \mathbb{RP}^{n-1}$ and let $q_1 = Mq_0$. Using the first part of the lemma, there exists a neighborhood ω of M^{-1} in $G_{\mathcal{M}}$ contained in $\Phi(\mathcal{F}^-_{\le T})$. By the submersion property in Item **(i)** of Proposition 6.23, $\omega q_1 = \{Nq_1 \mid N \in \omega\}$ is a neighborhood of q_0 in $O(q_0)$. The conclusion then follows, since $\omega q_1 \subset \Phi(\mathcal{F}^-_{\le T})q_1$.
□

A crucial intermediate result for the proof of Theorem 6.22 is the following.

Proposition 6.25 *Let O be a compact orbit for $(\Sigma_{\mathcal{M}})$ and let $\mathcal{F}$ be a finite-time rich skeleton of a class of signals with values in $\mathcal{M}$. Then there exists a nonempty subset $\mathfrak{C}$ of O, open for the induced topology and such that:*

(i) $\Phi(\mathcal{F})q = \mathfrak{C}$ *for every* $q \in \mathfrak{C}$;
(ii) $\overline{\Phi(\mathcal{F})q} = \overline{\mathfrak{C}}$ *for every* $q \in \overline{\mathfrak{C}}$;
(iii) *for every* $q \in \mathfrak{C}$ *there exists* $T > 0$ *such that* $\overline{\mathfrak{C}} \subset \Phi(\mathcal{F}^-_{\le T})q$.

Proof Endow the set O with the partial order defined by $q_1 \le q_2$ if $q_2 \in \Phi(\mathcal{F})q_1$. Let $\mathcal{K}$ be a totally ordered subset of O and let us prove that it admits an upper bound. Since $\overline{\Phi(\mathcal{F})q}$ is compact for every $q \in \mathcal{K}$, we deduce from Cantor's intersection theorem (Theorem A.24) that the set $\cap_{q\in\mathcal{K}} \overline{\Phi(\mathcal{F})q}$ is nonempty. Pick now $z \in \cap_{q\in\mathcal{K}}\overline{\Phi(\mathcal{F})q}$. Since $\mathcal{F}$ is finite-time rich, we deduce from Lemma 6.24 that there exists $\bar{z} \in O$ such that $z \in \mathrm{int}_O(\Phi(\mathcal{F}^-)\bar{z})$. Then, for every $q \in \mathcal{K}$, one has by construction that $\Phi(\mathcal{F})q$ intersects $\Phi(\mathcal{F}^-)\bar{z}$, which implies that $\bar{z} \in \Phi(\mathcal{F})q$. Therefore, $\bar{z}$ is an upper bound for $\mathcal{K}$.

By Zorn's lemma (Lemma A.23), there exists a maximal element $\bar{q}$ for $(O, \le)$, meaning that

$$\Phi(\mathcal{F})q = \Phi(\mathcal{F})\bar{q}, \qquad \text{if } q \in \Phi(\mathcal{F})\bar{q}. \tag{6.14}$$

Set $\mathfrak{C} = \Phi(\mathcal{F})\bar{q}$ and let us prove that $\mathfrak{C}$ is open in O. Let $q \in \mathfrak{C}$ and pick $q^+ \in \mathrm{int}_O(\Phi(\mathcal{F})q)$, whose existence follows from the fact that $\mathcal{F}$ is rich and the submersion property in Item **(i)** of Proposition 6.23. Then (6.14) implies that q^+ is in $\mathfrak{C}$ and that $q \in \Phi(\mathcal{F})q^+$. Then there exist $t > 0$ and $A \in \mathcal{F}_t$ such that $q = \Phi_A(t, 0)q^+$. Since $\Phi_A(t, 0)$ sends $\mathrm{int}_O(\Phi(\mathcal{F})q)$ into itself and $\mathrm{int}_O(\Phi(\mathcal{F})q)$ is contained in $\mathfrak{C}$, we deduce that $\mathfrak{C}$ contains a neighborhood of q for the topology of O. This completes the proof of the openness of $\mathfrak{C}$ and (6.14) proves Item **(i)**.

In order to prove Item **(ii)**, consider $q \in \overline{\mathfrak{C}} = \overline{\Phi(\mathcal{F})\bar{q}}$. Since $\Phi(\mathcal{F})$ is a semigroup, one immediately gets that $\Phi(\mathcal{F})q$, and hence $\overline{\Phi(\mathcal{F})q}$, are contained in $\overline{\mathfrak{C}}$. We are left to prove that $\overline{\mathfrak{C}} \subset \overline{\Phi(\mathcal{F})q}$. By Lemma 6.24 there exists $z \in O$ such that $q \in \mathrm{int}_O(\Phi(\mathcal{F}^-)q)$. Since $\mathrm{int}_O(\Phi(\mathcal{F}^-)q)$ intersects $\mathfrak{C}$, we deduce from (6.14) that z is in $\mathfrak{C}$. Applying again (6.14), $\mathfrak{C} \subset \Phi(\mathcal{F})q$, which yields the conclusion up to taking closures.

Let us conclude by proving Item **(iii)**. Fix $q \in \mathfrak{C}$ and $q_0 \in \overline{\mathfrak{C}}$. Let us prove that there exist $T > 0$ and a neighborhood Ω of q_0 (in $\mathbb{RP}^{n-1}$ or, equivalently, in O) such that $q \in \Phi(\mathcal{F}_{\leq T})q_1$ for every $q_1 \in \Omega \cap \mathfrak{C}$. The conclusion then follows by compactness of $\overline{\mathfrak{C}}$. Applying Lemma 6.24, there exist $\tau > 0$ and $q_2 \in O$ such that $q_0 \in \mathrm{int}_O(\Phi(\mathcal{F}^-_{\leq \tau})q_2)$. According to Item **(ii)**, there exist $t \geq 0$ and $M \in \Phi(\mathcal{F}_t)$ such that $q = Mq_2$. Hence the claim is proved with $T = \tau + t$ and $\Omega = \mathrm{int}_O(\Phi(\mathcal{F}^-_{\leq \tau})q_2)$. □

We can now conclude the proof of the main result of the section.

Proof of Theorem 6.22 First notice that it is sufficient to prove that $\lambda(\mathcal{S}) \leq \lambda^{\mathrm{per}}(\mathcal{S})$, the other inequality being obvious by definition.

Let us first prove the inequality when $\mathcal{M}$ is irreducible. Let O be a compact orbit of $(\Sigma_{\mathcal{M}})$ and consider $\mathfrak{C}$ as in Proposition 6.25. We claim that there exist d linearly independent unit vectors $v_1, \ldots, v_d \in \mathbb{R}^d$ such that their projections $q_i = \Pi(v_i)$, $i = 1, \ldots, d$, belong to $\mathfrak{C}$. Indeed, since O is an analytic orbit and because of the openness of $\mathfrak{C}$, then

$$\mathrm{span}\{x \in \mathbb{R}^d \setminus \{0\} \mid \Pi(x) \in \mathfrak{C}\} = \mathrm{span}\{x \in \mathbb{R}^d \setminus \{0\} \mid \Pi(x) \in O\},$$

as it follows from Lemma A.13 applied to each function $t \mapsto v^\top \gamma(t)$ with v orthogonal to $\mathrm{span}\{x \in \mathbb{R}^d \setminus \{0\} \mid \Pi(x) \in \mathfrak{C}\}$ and γ analytic curve in O crossing $\mathfrak{C}$ (since every point of $\mathfrak{C}$ can be joined to any point of O by an analytic curve, cf. Lemma A.16). Moreover, $\mathrm{span}\{x \in \mathbb{R}^d \setminus \{0\} \mid \Pi(x) \in O\} = \mathbb{R}^d$, as it follows from Item **(iii)** of Proposition 6.23.

Since the map $B \mapsto \max_{i=1,\ldots,d} \|Bv_i\|$ is a norm in $M_n(\mathbb{R})$, Lemma 6.5 implies that for every $\varepsilon > 0$ there exist $t > 0$, $A \in \mathcal{F}_t$, and $\ell \in \{1, \ldots, d\}$ such that

$$\frac{\log(\|\Phi_A(t,0)v_\ell\|)}{t} > \lambda(\mathcal{S}) - \varepsilon.$$

Moreover, t can be taken arbitrarily large. Let T be as in Item **(iii)** of Proposition 6.25 with $q = q_\ell$. Then there exist $t' \in [0, T]$ and a signal $A' \in \mathcal{F}_{t'}$ such that

$$q_\ell = \Phi_{A'}(t',0)\Phi_A(t,0)q_\ell.$$

Let then $A''(\cdot)$ be the $(t+t')$-periodic signal in $\mathcal{F}_\infty \subset \mathcal{S}$ defined by the concatenation of A and A' on each period. By construction, the pair (A'', v_ℓ) is in $\mathfrak{S}_{\mathrm{per}}$. Since T is independent of ε, we can moreover assume that

$$\lambda^{\mathrm{per}}(\mathcal{S}) \geq \frac{\log(\|\Phi_{A''}(t+t',0)v_\ell\|)}{t+t'} > \lambda(\mathcal{S}) - 2\varepsilon.$$

One then concludes by letting ε tend to zero.

Let us consider now the case where $\mathcal{M}$ is reducible. By Proposition 1.32, we can identify from $(\Sigma_{\mathcal{S}})$ finitely many irreducible subsystems $(\Sigma_{\mathcal{S}_1}), \dots, (\Sigma_{\mathcal{S}_r})$ with $\lambda(\mathcal{S}) = \max_{i=1,\dots,r} \lambda(\mathcal{S}_i)$. The skeletons of $\mathcal{S}_i$ induced by that of $\mathcal{S}$ are clearly extensive and they are also finite-time rich as noticed in Remark 6.8. If a subsystem $\mathcal{S}_i$ with $\lambda(\mathcal{S}) = \lambda(\mathcal{S}_i)$ happens to correspond to an invariant subspace for $\mathcal{M}$, we have

$$\lambda(\mathcal{S}) = \lambda(\mathcal{S}_i) = \lambda(\mathcal{S}_i^{\mathrm{per}}) \leq \lambda(\mathcal{S}^{\mathrm{per}}),$$

concluding the argument. Otherwise, fix i minimal such that $\lambda(\mathcal{S}) = \lambda(\mathcal{S}_i)$. Using the notation of Proposition 1.32 and (1.14), write the restriction of every $M \in \mathcal{M}$ and $A \in \mathcal{S}$ to the invariant space E_i as

$$M|_{E_i} = \begin{pmatrix} \hat{M}_{11} & \hat{M}_{12} \\ 0 & \hat{M}_{22} \end{pmatrix} \quad \text{and} \quad A|_{E_i} = \begin{pmatrix} \hat{A}_{11} & \hat{A}_{12} \\ 0 & \hat{A}_{22} \end{pmatrix},$$

with $\hat{M}_{22} = \pi_{ii}(M)$ and $\hat{A}_{22} = \pi_{ii}(A)$.

Then, $\{\hat{M}_{22} \mid M \in \mathcal{M}\}$ is irreducible and

$$\lambda(\hat{\mathcal{S}}_1) = \max_{j=1,\dots,i-1} \lambda(\mathcal{S}_j) < \lambda(\hat{\mathcal{S}}_2) = \lambda(\mathcal{S}),$$

where we use $\hat{\mathcal{S}}_j$ to denote the class of signals $\{\hat{A}_{jj}(\cdot) \mid A(\cdot) \in \mathcal{S}\}$, $j = 1, 2$. Notice that

$$\lambda^{\mathrm{per}}(\hat{\mathcal{S}}_1) \leq \lambda(\hat{\mathcal{S}}_1) < \lambda(\hat{\mathcal{S}}_2) = \lambda^{\mathrm{per}}(\hat{\mathcal{S}}_2), \tag{6.15}$$

where the last equality results from the first part of the argument. Setting $\mathcal{S}|_{E_i} = \{A|_{E_i} \mid A \in \mathcal{S}\}$, we have by invariance of E_i that $\lambda^{\mathrm{per}}(\mathcal{S}|_{E_i}) \leq \lambda^{\mathrm{per}}(\mathcal{S})$. Let us conclude the argument by showing that $\lambda^{\mathrm{per}}(\hat{\mathcal{S}}_2) \leq \lambda^{\mathrm{per}}(\mathcal{S}|_{E_i})$. This is done by proving that a trajectory of $(\Sigma_{\hat{\mathcal{S}}_2})$ whose projection by Π is periodic can be lifted to a trajectory of $(\Sigma_{\mathcal{S}|_{E_i}})$ whose projection by Π is also periodic. By definition of $\lambda^{\mathrm{per}}(\hat{\mathcal{S}}_2)$, for every $\varepsilon > 0$ there exist $\tau > 0$, a τ-periodic signal $A \in \mathcal{S}$, $\mu \in \mathbb{R}$, and $x_2 \neq 0$ such that

$$\Phi_{\hat{A}_{22}}(\tau, 0)x_2 = \mu x_2, \qquad \frac{\log|\mu|}{\tau} > \lambda^{\mathrm{per}}(\hat{\mathcal{S}}_2) - \varepsilon.$$

Let $\varepsilon > 0$ be taken such that $\lambda^{\mathrm{per}}(\hat{\mathcal{S}}_2) - \varepsilon > \lambda^{\mathrm{per}}(\hat{\mathcal{S}}_1)$, whose existence is guaranteed by (6.15). In particular $\frac{\log|\mu|}{\tau} > \lambda^{\mathrm{per}}(\hat{\mathcal{S}}_1)$, implying that μ is not in the spectrum of $\Phi_{\hat{A}_{11}}(\tau, 0)$.

For every $x \in E_i$ of the form $x = (x_1, x_2)$, one has

$$\Phi_{A|_{E_i}}(\tau, 0)x = \begin{pmatrix} \Phi_{\hat{A}_{11}}(\tau, 0)x_1 + \int_0^\tau \Phi_{\hat{A}_{11}}(\tau, s)\hat{A}_{12}(s)\Phi_{\hat{A}_{22}}(s, 0)x_2 ds \\ \mu x_2 \end{pmatrix}.$$

The required lift is obtained by choosing x_1 as

$$x_1 = (\mu \, \mathrm{Id}_n - \Phi_{\hat{A}_{11}}(\tau, 0))^{-1} \left(\int_0^\tau \Phi_{\hat{A}_{11}}(\tau, s)\hat{A}_{12}(s)\Phi_{\hat{A}_{22}}(s, 0)x_2 ds \right),$$

so that $\Phi_{A|_{E_i}}(\tau, 0)x = \mu x$. This concludes the proof of the theorem. □

6.3 Stability of Systems Having a Nonstrict Common Quadratic Lyapunov Function

In this section, we investigate the asymptotic properties of trajectories of a switched linear system $(\Sigma_{\mathcal{S}})$ under the assumption that $x \mapsto \|x\|^2$ is a nonstrict common quadratic Lyapunov function, that is,

(H) for every $M \in \mathcal{M}$, one has $M^\top + M \preceq 0$.

In particular, by Theorem 3.3 one has $\lambda(\mathcal{S}) \leq \lambda(\mathcal{M}) \leq 0$.

Remark 6.26 Hypothesis (H) is a particular instance of a more general condition on $\mathcal{M}$, namely the existence of a real symmetric positive-definite matrix P such that

$$M^\top P + PM \preceq 0, \quad \forall M \in \mathcal{M}. \tag{6.16}$$

However, it is immediate to reduce, with no loss of generality as regards the asymptotic properties of $(\Sigma_{\mathcal{S}})$, (6.16) to Hypothesis (H) by considering the square root $P^{1/2}$ of P (i.e., the unique real symmetric positive-definite matrix such that $P^{1/2}P^{1/2} = P$) and performing the linear change of variable $x \mapsto P^{1/2}x$, which amounts to replace $\mathcal{M}$ by $P^{1/2}\mathcal{M}P^{-1/2}$.

The purpose of this section consists in identifying a class of signals $\mathcal{S}$ for which the ω-limit sets of the corresponding trajectories can be characterized in terms of linear subspaces $\mathcal{N}_M$ of $\mathbb{R}^n$ associated with every element of $M \in \mathcal{M}$. Once this is performed, sufficient conditions on these subspaces are proposed which guarantee asymptotic stability of $(\Sigma_{\mathcal{S}})$, in the case where $\mathcal{M}$ is finite. The subspaces $\mathcal{N}_M$ are defined in the following lemma, which also gathers their basic properties.

Lemma 6.27 *For $M \in M_n(\mathbb{R})$, denote by $s_M = (M + M^\top)/2$ and $a_M = (M - M^\top)/2$ the symmetric and skew-symmetric parts of M, respectively. Consider*

$$\mathcal{N}_M = \cap_{k\geq 0}\mathrm{Ker}\Big(s_M M^k\Big). \tag{6.17}$$

If $s_M \preceq 0$ then the following properties hold:

1. *$x \in \mathbb{R}^n$ belongs to $\mathcal{N}_M$ if and only if $\|e^{tM}x\| = \|x\|$ for all $t \in \mathbb{R}$;*
2. *$x \in \mathbb{R}^n$ belongs to $\mathcal{N}_M$ if and only if there exists $\tau \neq 0$ such that $\|e^{\tau M}x\| = \|x\|$;*
3. *Both $\mathcal{N}_M$ and its orthogonal complement $\mathcal{N}_M^\perp$ are invariant by M and the restriction of M to $\mathcal{N}_M$ (respectively, $\mathcal{N}_M^\perp$) is skew-symmetric (respectively, Hurwitz) in any orthonormal basis of $\mathcal{N}_M$ (respectively, $\mathcal{N}_M^\perp$).*

Proof For $x \in \mathbb{R}^n$, the vector $s_M M^k x$ with $k \geq 0$ is the evaluation at $t = 0$ of the k-th time derivative of the analytic function $t \mapsto s_M e^{tM}x$. Hence $x \in \mathcal{N}_M$ if and only if $s_M e^{tM}x = 0$ for all $t \in \mathbb{R}$, i.e., if and only if $x^\top e^{tM^\top} s_M e^{tM}x = 0$ for all $t \in \mathbb{R}$, since $s_M \preceq 0$. The latter being the time derivative of the analytic function $t \mapsto \|e^{tM}x\|^2/2$, we get that $x \in \mathcal{N}_M$ if and only if $\|e^{tM}x\| = \|x\|$ for all $t \in \mathbb{R}$, proving Item 1.

Since $t \mapsto \|e^{tM}x\|$ is nonincreasing and $t \mapsto e^{tM}x$ is analytic, one obtains that $\|e^{tM}x\| = \|x\|$ for $t \in \mathbb{R}$ if and only if there exists $\tau \neq 0$ such that $\|e^{\tau M}x\| = \|x\|$. This proves Item 2.

To get Item 3, we only show the invariance of $\mathcal{N}_M^\perp$ by M, the rest being elementary. This follows from the fact that $M^\top x = -Mx \in \mathcal{N}_M$ for $x \in \mathcal{N}_M$. One deduces that for $y \in \mathcal{N}_M^\perp$ then $x^\top My = -y^\top Mx = 0$ for every $x \in \mathcal{N}_M$, hence the conclusion. □

We need the following preliminary description of ω-limit sets of trajectories of $(\Sigma_{\mathcal{S}})$ and $(\Sigma_{\mathcal{S}}^{\mathrm{lift}})$.

Lemma 6.28 *Assume that Hypothesis (H) holds true. For $A(\cdot) \in \mathcal{S}$ and $x \in \mathbb{R}^n$, let $\omega_A(x)$ and Ω_A be the ω-limit sets of the trajectories $\Phi_A(\cdot, 0)x$ in $\mathbb{R}^n$ and $\Phi_A(\cdot, 0)$ in $M_n(\mathbb{R})$. The following properties hold true:*

1. *$\omega_A(x) = \Omega_A x$ and there exists $r \geq 0$ such that $\omega_A(x)$ is a nonempty, compact, and connected subset of the sphere of radius r centered at the origin;*
2. *the map $t \mapsto \Phi_A(t, 0)^\top \Phi_A(t, 0)$ taking values in the set of positive-definite real symmetric matrices is nondecreasing and admits a limit as t tends to infinity;*
3. *there exists a compact and connected subset O_A of the orthogonal group $\mathrm{O}(n)$ such that $\Omega_A = O_A S_A$, where S_A is the square root of $\lim_{t\to+\infty} \Phi_A(t, 0)^\top \Phi_A(t, 0)$.*

Proof Item 1 is standard since $t \mapsto \|\Phi_A(t,0)x\|$ is nondecreasing, hence converging at infinity to some $r \geq 0$. Item 2 is obtained by a direct computation using Hypothesis (H):

$$\frac{d}{dt}\Big(\Phi_A(t,0)^\top \Phi_A(t,0)\Big) = \Phi_A(t,0)^\top \Big(A(t)^\top + A(t)\Big)\Phi_A(t,0) \preceq 0,$$

for a.e. $t \geq 0$.

Hence the map $t \mapsto \Phi_A(t,0)^\top \Phi_A(t,0)$ being monotone non-increasing and bounded below by zero converges to some $S \succeq 0$ as t tends to infinity.

To derive Item 3, first define, for $t \geq 0$, the positive-definite real symmetric matrix $S(t)$ as the square root of $\Phi_A(t,0)^\top \Phi_A(t,0)$. Then $S(t) \succ 0$ for $t \geq 0$ and $t \mapsto S(t)$ is absolutely continuous. Define $\Gamma(t) = \Phi_A(t,0)S^{-1}(t)$ for $t \geq 0$ and notice that $\Gamma(t) \in O(n)$. The set O_A is then the ω-limit set of the absolutely continuous trajectory $t \mapsto \Gamma(t)$. □

Remark 6.29 System $(\Sigma_{\mathcal{S}})$ is attractive if and only if for every $A(\cdot) \in \mathcal{S}$ one has $\Omega_A = \{0\}$, i.e., $S_A = 0$.

We now provide the basic relationship between the previously defined ω-limit sets and the linear subspaces N_M defined in Lemma 6.27. For that purpose we restrict our attention to piecewise constant signals and, with $A \in \mathcal{S}_{\mathrm{pc}}(\mathcal{M})$ we associate the increasing sequence (possibly finite) of its discontinuity times $(a_k)_{k\in\mathbb{N}}$. We refer to each a_k as a *switching time* of A. We next introduce the restrained classes of signals needed in the sequel.

Definition 6.30 A signal $A \in \mathcal{S}_{\mathrm{pc}}(\mathcal{M})$ is said to be *mildly switching* if, for every $\tau > 0$ and every sequence $(a_k)_{k\in\mathbb{N}}$ in $\mathbb{R}_+$ tending to infinity, it holds

$$\limsup_{k\to\infty} \Big(\sup\{t-s \mid a_k \leq s \leq t \leq a_k+\tau,\ A_{|[s,t]} \text{ is constant}\}\Big) > 0.$$

Example 6.31 Simple examples of mildly switching signals are the elements of the classes $\mathcal{S}_{\mathrm{dw},\tau}(\mathcal{M})$ and $\mathcal{S}_{\mathrm{av-dw},\tau,N_0}(\mathcal{M})$, with $\tau > 0$ and N_0 a positive integer, introduced in Sect. 1.1.1.

We also associate with $A \in \mathcal{S}_{\mathrm{pc}}(\mathcal{M})$ the set Ω_A^{st} of all limit points of $t \mapsto \Phi_A(t,0)$ along sequences in the set of switching times $(a_k)_{k\in\mathbb{N}}$ of A.

Proposition 6.32 *Let $\mathcal{M}$ be compact and assume that Hypothesis (H) holds true. For every $A(\cdot) \in \mathcal{S}_{\mathrm{pc}}(\mathcal{M})$ and $x \in \mathbb{R}^n$, one has*

$$\omega_A(x) \subset \Omega_A^{\mathrm{st}}x \cup \Big(\cup_{M\in\mathcal{M}} N_M\Big). \tag{6.18}$$

If, moreover, $A(\cdot)$ is mildly switching, then $\omega_A(x) \subset \cup_{M\in\mathcal{M}} N_M$ for every $x \in \mathbb{R}^n$.

Proof Let $l \in \omega_A(x)$ and $(t_k)_{k\in\mathbb{N}}$ be an unbounded sequence such that $l = \lim_{k\to\infty} \Phi_A(t_k, 0)x$. Consider the sequence $(a_{n_k})_{k\in\mathbb{N}}$ where a_{n_k} is the unique switching time such that $t_k \in [a_{n_k}, a_{n_k+1})$. Denote by M_k the matrix in $\mathcal{M}$ such that $A(\cdot) = M_k$ on the interval $[a_{n_k}, a_{n_k+1})$, $k \in \mathbb{N}$. Up to extracting a subsequence, we can assume that the limits as k tends to infinity of $\Phi_A(a_{n_k}, 0)x$ and M_k exist and are denoted by l_0 and M, respectively.

Assume first that there is a subsequence of the durations $t_k - a_{n_k}$, $k \in \mathbb{N}$, which tends to zero. Hence $(a_{n_k})_{k\in\mathbb{N}}$ is unbounded and $l_0 \in \Omega_A^{\mathrm{st}}$. Moreover, $\Phi_A(t_k, a_{n_k})$ tends to Id_n as k tends to infinity, implying that $l = l_0$. Otherwise, there exists $\delta > 0$ such that $t_k - a_{n_k} \geq \delta$ for k large enough, and hence $e^{-\delta M} l = \lim_{k\to\infty} \Phi_A(t_k - \delta, 0)x$ also belongs to $\omega_A(x)$, in addition to l. By Item 1 of Lemma 6.28, $e^{-\delta M} l$ and l have same norm and, by Item 2 of Lemma 6.27, one concludes that $l \in \mathcal{N}_M$.

Assume now that A is mildly switching. Let $(a_{n_k})_{k\in\mathbb{N}}$ be an unbounded sequence of switching times such that $l = \lim_{k\to\infty} \Phi_A(a_{n_k}, 0)x$. We should prove that $l \in \mathcal{N}_M$ for some $M \in \mathcal{M}$. Since A is mildly switching, for every $\tau > 0$ there exists $\varepsilon > 0$ and a subsequence $(a_{n_{k_j}})_{j\in\mathbb{N}}$ of $(a_{n_k})_{k\in\mathbb{N}}$ such that A is constant on a subinterval $[b_j^\tau, b_j^\tau + \varepsilon]$ of $[a_{n_{k_j}}, a_{n_{k_j}} + \tau]$. Reasoning as in the first part of the proof, up to possibly extracting a further subsequence of $(b_j^\tau)_{j\in\mathbb{N}}$, the limit $\lim_{j\to\infty} \Phi_A(b_j^\tau, 0)x =: l_\tau$ exists and it belongs to $\mathcal{N}_{M_\tau}$ for some $M_\tau \in \mathcal{M}$. The proof is concluded by letting τ goes to zero and by extracting a converging sequence of M_τ, since $l_\tau \to l$ as τ goes to zero. □

Let us introduce a suitable notion of regularity for piecewise constant signals.

Definition 6.33

1. For $M \in \mathcal{M}$, a signal $A(\cdot) \in \mathcal{S}_{\mathrm{pc}}(\mathcal{M})$ is said to satisfy *Hypothesis* (H_M) if there exist a sequence $(t_k)_{k\in\mathbb{N}}$ tending to infinity and $\tau > 0$ such that $A \equiv M$ on $[t_k, t_k + \tau]$ for every $k \in \mathbb{N}$.
2. When $\mathcal{M}$ is finite, a signal $A(\cdot) \in \mathcal{S}_{\mathrm{pc}}(\mathcal{M})$ is said to be *regular* if it is mildly switching and satisfies Hypothesis (H_M) for every $M \in \mathcal{M}$. We denote by $\mathcal{S}_{\mathrm{reg}}(\mathcal{M})$ the sequence of all regular signals.

Notice that $\mathcal{S}_{\mathrm{reg}}(\mathcal{M})$ is shift-invariant and that it is not weak-$\star$ compact, unless $\mathcal{M}$ is a singleton.

Remark 6.34 Let $A(\cdot)$ in $\mathcal{S}_{\mathrm{pc}}(\mathcal{M})$ be mildly switching. Recalling that $\omega_A(x) = \Omega_A x$, it follows from the last part of Proposition 6.32 that $\mathrm{Im}(J) \subset \cup_{M\in\mathcal{M}} \mathcal{N}_M$ for every $J \in \Omega_A$. Since $\mathrm{Im}(J)$ is a linear subspace, when $\mathcal{M}$ is finite we even get that, for every $J \in \Omega_A$, there exists $M \in \mathcal{M}$ such that $\mathrm{Im}(J) \subset \mathcal{N}_M$.

Assume now that a signal $A(\cdot) \in \mathcal{S}_{\mathrm{pc}}(\mathcal{M})$ satisfies Hypothesis (H_M). Then $\omega_A(x) \cap \mathcal{N}_M \neq \emptyset$ for every $x \in \mathbb{R}^n$. Hence, for $\mathcal{M}$ finite and $A \in \mathcal{S}_{\mathrm{reg}}(\mathcal{M})$, one has

$$\dim \mathrm{Im}(S_A) \leq \min_{M\in\mathcal{M}} \dim \mathcal{N}_M.$$

We now provide a stability result for regular signals.

Proposition 6.35 *Assume that $\mathcal{M}$ is finite and that Hypothesis (H) hold true. Assume, moreover, that*

$$\textbf{(C)} \quad S^{n-1} \cap \left(\cup_{M \in \mathcal{M}} \mathcal{N}_M \right) \text{ is not connected or } \mathcal{N}_M = \{0\} \text{ for some } M \in \mathcal{M}. \tag{6.19}$$

Then $(\Sigma_{\mathcal{S}_{\mathrm{reg}}(\mathcal{M})})$ is attractive.

Proof Arguing by contradiction, there exist $x \in \mathbb{R}^n$ and $r > 0$ such that $\omega_A(x) \subset rS^{n-1}$. By Proposition 6.32 and Remark 6.34, the connected set $\omega_A(x)$ is included in $\cup_{M \in \mathcal{M}} \mathcal{N}_M$, since A is mildly switching, and, for each $M \in \mathcal{M}$, it intersects the space $\mathcal{N}_M$, since A satisfies hypothesis (H_M). This contradicts **(C)** and concludes the proof of the proposition. □

As a consequence of the proposition and of Fenichel's lemma (Lemma 1.25), for a weak-$\star$ closed and shift-invariant subclass $\mathcal{S}$ of $\mathcal{S}_{\mathrm{reg}}(\mathcal{M})$ we have $\lambda(\mathcal{S}) < 0$.

One notices that Condition **(C)** is not easy to check in practice and it also implies that $\cap_{M \in \mathcal{M}} \mathcal{N}_M = \{0\}$. This motivates to find easily checkable conditions sufficient to derive Condition **(C)**. In that direction, we have the following result.

Theorem 6.36 *Assume that $\mathcal{M}$ is finite and that Hypothesis (H) holds true. Denote by p the cardinality of $\mathcal{M}$. Then $(\Sigma_{\mathcal{S}_{\mathrm{reg}}(\mathcal{M})})$ is attractive if one of the following conditions holds true:*

(1) $\mathcal{N}_M = \{0\}$ *for some* $M \in \mathcal{M}$;
(2) $\dim \mathcal{N}_M = 1$ *for some* $M \in \mathcal{M}$ *and* $\mathcal{N}_M \not\subseteq \mathcal{N}_{\hat{M}}$ *for* $\hat{M} \in \mathcal{M} \setminus \{M\}$;
(3) $p = 2$ *and* $\cap_{M \in \mathcal{M}} \mathcal{N}_M = \{0\}$;
(4) $p > 2$ *and* $\dim (\sum_{M \in \mathcal{M}} \mathcal{N}_M) > \sum_{M \in \mathcal{M}} \dim \mathcal{N}_M - p + 1$.

Proof By Proposition 6.35, it is enough to show that any of the above conditions imply Condition **(C)**. This is clearly true for the first three conditions and we are left to provide an argument for the fourth one. The latter goes by contradiction and assume that Condition **(C)** does not hold. There exists a connected component Γ of $S^{n-1} \cap \left(\cup_{M \in \mathcal{M}} \mathcal{N}_M \right)$ intersecting each space $\mathcal{N}_M$, $M \in \mathcal{M}$. This implies that for every $\bar{M} \in \mathcal{M}$ the linear subspace $\mathcal{N}_{\bar{M}} \cap \left(\sum_{M \in \mathcal{M} \setminus \{\bar{M}\}} \mathcal{N}_M \right)$ contains a nonzero vector. One deduces that

$$\begin{aligned} \dim \left(\textstyle\sum_{M \in \mathcal{M}} \mathcal{N}_M\right) &= \dim \mathcal{N}_{\bar{M}} + \dim \left(\textstyle\sum_{M \in \mathcal{M} \setminus \{\bar{M}\}} \mathcal{N}_M\right) \\ &\quad - \dim \left(\mathcal{N}_{\bar{M}} \cap \left(\textstyle\sum_{M \in \mathcal{M} \setminus \{\bar{M}\}} \mathcal{N}_M\right)\right) \\ &\leq \dim \mathcal{N}_{\bar{M}} + \dim \left(\textstyle\sum_{M \in \mathcal{M} \setminus \{\bar{M}\}} \mathcal{N}_M\right) - 1. \end{aligned}$$

One concludes by an easy inductive argument on p. □

6.4 Notes and References

The main results of Sect. 6.1 can be found in [CMS17a], with several arguments simplified in the present exposition. The notion of skeleton corresponds to *concatenable subfamilies* in [CMS17a], while *extensiveness* replaces *fatness*. The proof of Proposition 6.14 follows basically the Barabanov norm construction of [Wir02, Proposition 3.2].

The periodization results of Sect. 6.2 (and, in particular, Theorem 6.22) are inspired by Colonius and Kliemann [CK93], where the approach has been proposed in the case of arbitrary switching, under a suitable Lie bracket generating assumption. Such an approach has been extended in [BS20] to the case of dwell-time signals, removing the Lie bracket generating assumption. A key step is the existence of a set $\mathfrak{C}$ as in Proposition 6.25, a result originally proved in [AKO86, Theorem 3.1], from the point of view of *invariant control sets* (see also [Kli87, Lemma 2.1] and [SMT95]). An earlier periodization result in the case of arbitrary switching may be found in [PR91b], where the authors assume that $\mathcal{M}$ is a segment of matrices and the switched system admits a (nontrivial) invariant cone.

The results presented in Sect. 6.3 are mainly taken from [JN13]. The main stability results of Sect. 6.3 (Theorem 6.36) generalize those in [BJ11, RSD10, SVR11] and can be seen as versions of LaSalle's principle for switched systems. Notice that mildly switching signals are called *nonchaotic* in [BJ11, JN13]. Former occurrences of properties in the same spirit are [BM05, GST08, Hes04, MAG06, SGT07].

Chapter 7
Switched Controlled Dynamics

In this chapter, we consider switched linear control systems $(C_{\mathcal{T}})$ of the type

$$\dot{x}(t) = A(t)x(t) + B(t)u(t), \tag{7.1}$$

$$y(t) = C(t)x(t), \tag{7.2}$$

with state x, control u, output y and switching signal $\big(A(\cdot), B(\cdot), C(\cdot)\big)$. Note that the output map defined in (7.2) is often given in the more general form $y(t) = C(t)x(t) + D(t)u(t)$. For simplicity of presentation, we only consider the case $D(\cdot) \equiv 0$ since addressing the general situation increases the notational burden without any extra technical difficulty.

As mentioned in the introduction, the switching signal can be seen either as a parameter for which the user has no or only partial information or as a design parameter with which the user can tune in order to achieve a preassigned goal. Till now, the main issue has been to address the (asymptotic) stability of the system and the switching signal was considered to be a perturbation possibly inducing more instability. Instead, in the present chapter, the switching signal will be envisioned from both points of view, namely as a perturbation or a tunable parameter. In the first two sections of the chapter, emphasis is made on the second point of view in order to first understand the reachable set associated with a switched linear control system, and then study its minimal realizations. In the third section we address L^2-gain issues, while in the fourth section, we undertake a stabilization purpose for a class of restrained switching signals where the required feedback must be designed in order to stabilize, for each switching signal, the corresponding nonautonomous linear control system. Hence, in both cases, a switching signal has to be envisioned by its destabilizing effect. Instead, in the fifth section, we aim at stabilizing a switched linear control system under the hypothesis that the switching signal is part of the feedback design, so that it can be considered as a tunable parameter.

Y. Chitour et al., *Dynamics and Stability of Continuous-Time Switched Linear Systems*, PNLDE Subseries in Control 105,
https://doi.org/10.1007/978-3-031-82982-6_7

7.1 Controllability and Observability

In this section, we address standard control theoretical issues for switched linear control systems, namely, controllability and observability. In particular, we aim at establishing a necessary and sufficient condition for controllability (and also for the dual notion of observability) in terms of a Kalman-type algebraic criterion, only depending on $\mathcal{N}$ under mild conditions on the class $\mathcal{T}$ of switching signals.

We first provide some basic notation used in the sequel. Let $\pi_A, \pi_{AB}, \ldots,$ be the projections from $M_n(\mathbb{R}) \times M_{n,m}(\mathbb{R}) \times M_{p,n}(\mathbb{R})$ onto the corresponding factors. We extend them to any functions taking values in $\mathcal{N}$. Given a switching signal $\sigma(\cdot) = (A, B, C)$, we use $\sigma_A, \sigma_{AB}, \ldots,$ to denote the projections of σ by $\pi_A, \pi_{AB}, \ldots,$ respectively. We also use $\mathcal{T}_A, \mathcal{N}_A, \mathcal{T}_{AB}, \mathcal{N}_{AB}, \ldots$ to denote the corresponding projections of $\mathcal{T}$ and $\mathcal{N}$.

We next state properties for trajectories of (7.1). Given $t_0 \geq 0$, $x_0 \in \mathbb{R}^n$, $\sigma_{AB} \in \mathcal{T}_{AB}$, and $u \in \mathcal{U}$, the trajectory of (7.1) starting at x_0 at $t = t_0$ and associated with σ_{AB} and u is well-defined for all times $t \geq t_0$ and we use $x(\cdot; t_0, x_0, \sigma_{AB}, u)$ to denote it. By the variation of constants formula (Lemma A.6), one has

$$x(t; t_0, x_0, \sigma_{AB}, u) = \Phi_A(t, t_0)x_0 + \int_{t_0}^{t} \Phi_A(t, s)B(s)u(s)ds. \tag{7.3}$$

Similarly, we use $y(\cdot; t_0, x_0, \sigma, u)$ to denote the output associated with $\sigma \in \mathcal{T}$ and corresponding to the trajectory of (7.1) starting at $x_0 \in \mathbb{R}^n$ at $t = t_0$ and associated with σ_{AB} and u. In particular, for $t \geq t_0$,

$$\begin{aligned} y(t; t_0, x_0, \sigma, u) &= C(t)x(t; t_0, x_0, \sigma_{AB}, u) \\ &= C(t)\Phi_A(t, t_0)x_0 + C(t) \int_{t_0}^{t} \Phi_A(t, s)B(s)u(s)ds. \end{aligned} \tag{7.4}$$

A consequence of Proposition 1.3 and Eq. (7.3) is the following proposition.

Proposition 7.1 *Let $(\sigma_{AB}^{(k)})_{k\in\mathbb{N}}$ be a sequence in $\mathcal{T}_{AB}$ weak-$\star$ converging to $\sigma_{AB} \in \mathcal{S}_{\rm arb}(\mathcal{N}_{AB})$, $(x_k)_{k\in\mathbb{N}}$ a sequence in $\mathbb{R}^n$ converging to some $x \in \mathbb{R}^n$ and $(u_k)_{k\in\mathbb{N}}$ be a sequence in $\mathcal{U}$ converging (in the L^1-norm on every bounded interval of $[0, +\infty)$) to some locally integrable u. Then the sequence $(x(\cdot; 0, x_k, \sigma_{AB}^{(k)}, u_k))_{k\in\mathbb{N}}$ converges to $x(\cdot; 0, x, \sigma_{AB}, u)$ uniformly on compact intervals.*

The basic definitions needed to characterize our findings are given below.

Definition 7.2 Let $(C_{\mathcal{T}})$ be a switched linear control system defined by (7.1), (7.2), $\mathcal{T}$, and $\mathcal{U}$.

- A state $x_1 \in \mathbb{R}^n$ is *reachable from a state $x_0 \in \mathbb{R}^n$ in time $T \geq 0$* if there exists $(\sigma_{AB}, u) \in \mathcal{T}_{AB} \times \mathcal{U}$ such that $x_1 = x(T; 0, x_0, \sigma_{AB}, u)$. The *reachable set from $x_0 \in \mathbb{R}^n$ in time $T \geq 0$* is the set $R_{\mathcal{T}}(x_0, T)$ made of the reachable points from x_0

in time $T \geq 0$. Finally, the *reachable set from* x_0 is the set $R_{\mathcal{T}}(x_0)$ defined as the union of all the $R_{\mathcal{T}}(x_0, T)$, $T \geq 0$.

- The switched linear control system $(C_{\mathcal{T}})$ is *controllable from* $x_0 \in \mathbb{R}^n$ if $R_{\mathcal{T}}(x_0) = \mathbb{R}^n$ and *completely controllable* if it is controllable from every $x_0 \in \mathbb{R}^n$.
- A state $x_1 \in \mathbb{R}^n$ is *unobservable from a state* $x_0 \in \mathbb{R}^n$ *in time* $T > 0$ if for every $(\sigma, u) \in \mathcal{T} \times \mathcal{U}$ the output maps $[0, T] \ni t \mapsto y(t; 0, x_0, \sigma, u)$ and $[0, T] \ni t \mapsto y(\cdot; 0, x_1, \sigma, u)$ are equal. The *unobservable set from* $x_0 \in \mathbb{R}^n$ *in time* $T > 0$ is the set $I_{\mathcal{T}}(x_0, T)$ made of x_0 and all the unobservable points from x_0 in time $T > 0$. Finally, the *unobservable set from* x_0 is the set $I_{\mathcal{T}}(x_0)$ defined as the intersection of all the $I_{\mathcal{T}}(x_0, T)$, $T > 0$.
- The switched linear control system $(C_{\mathcal{T}})$ is *observable from* $x_0 \in \mathbb{R}^n$ if $I_{\mathcal{T}}(x_0) = \{x_0\}$ and *completely observable* if it is observable from every $x_0 \in \mathbb{R}^n$.

From the above definitions, we make the following standard remark.

Remark 7.3 From the definition of observability and Eq. (7.4), a state $x_1 \in \mathbb{R}^n$ is unobservable from a state $x_0 \in \mathbb{R}^n$ in time $T > 0$ if and only if, for every $\sigma_{AC} \in \mathcal{T}_{AC}$, $x_1 - x_0$ belongs to all the kernels of the linear mappings $\omega(t, \sigma_{AC}) := C(t)\phi_A(t, 0)$, $t \in [0, T]$. One deduces that the unobservable sets $I_{\mathcal{T}}(x_0, T)$, $T \geq 0$, are equal to the affine subspaces $x_0 + I_{\mathcal{T}}(0, T)$ where $I_{\mathcal{T}}(0, T)$ is equal to the intersection of all Ker $\omega(t, \sigma_{AC})$ with $0 \leq t \leq T$ and $\sigma_{AC} \in \mathcal{T}_{AC}$. One is led to define the *observable space* $O_{\mathcal{T}}(0, T)$ *at time* T and the *observable space* $O_{\mathcal{T}}(0)$ associated with a switched linear control system $(C_{\mathcal{T}})$ by $O_{\mathcal{T}}(0, T) = I_{\mathcal{T}}(0, T)^{\perp}$ and $O_{\mathcal{T}}(0) = I_{\mathcal{T}}(0)^{\perp}$. Then

$$O_{\mathcal{T}}(0, T) = I_{\mathcal{T}}(0, T)^{\perp} = \sum_{\substack{0 \leq t \leq T \\ \sigma_{AC} \in \mathcal{T}_{AC}}} \operatorname{Im} \omega(t, \sigma_{AC})^{\top}, \quad T \geq 0,$$

with a similar characterization of $O_{\mathcal{T}}(0)$.

Hence, a switched linear control system $(C_{\mathcal{T}})$ is completely observable if and only if $O_{\mathcal{T}}(0) = \mathbb{R}^n$.

We now aim at characterising algebraically the reachable and observable sets in terms of $\mathcal{N}$ and it is worthwhile recalling what happens for the unswitched case, i.e., when $\mathcal{N}$ reduces to a single triple (A, B, C). It is a standard fact that, for every $x_0 \in \mathbb{R}^n$ and $T > 0$, $R_{\mathcal{T}}(x_0, T) = e^{AT}x_0 + R_{\mathcal{T}}(0, T)$ and $R_{\mathcal{T}}(0, T)$ is equal to the image of the controllability matrix $C(A, B) = [B\ AB\ \cdots\ A^{n-1}B]$ or equivalently

$$R_{\mathcal{T}}(0, T) = \operatorname{Span}\{A^i b \mid 0 \leq i \leq n-1,\ b \text{ column of } B\}. \tag{7.5}$$

We extend in what follows the characterization given by (7.5) to the switched system case. More precisely, we show that the appropriate extension relies on the next space

$$\mathcal{R}(\mathcal{N}) := \operatorname{Span}\left\{A_1 \cdots A_i b \;\middle|\; \begin{array}{l} 0 \leq i \leq n-1,\ A_1, \ldots, A_i \in \mathcal{N}_A \\ b \text{ column of some } B \in \mathcal{N}_B \end{array}\right\}. \tag{7.6}$$

We first have the following observation.

Lemma 7.4 *Set*

$$\bar{\mathcal{R}} := \operatorname{Span}\{A_1 \cdots A_i b \mid i \geq 0,\ A_1, \ldots, A_i \in \mathcal{N}_A,\ b \text{ column of some } B \in \mathcal{N}_B\}.$$

Then $\bar{\mathcal{R}} = \mathcal{R}(\mathcal{N})$.

Proof Define $\mathcal{V}_0 = \sum_{B \in \mathcal{N}_B} \operatorname{Im} B$ and, recursively for $k \geq 0$, $\mathcal{V}_{k+1} = \mathcal{V}_k + \sum_{A \in \mathcal{N}_A} A\mathcal{V}_k$. Then $(\mathcal{V}_k)_{k \geq 0}$ is an increasing sequence of subspaces of $\bar{\mathcal{R}}$. Hence there exists $k_* \leq n-1$ so that $\mathcal{V}_{k_*+1} = \mathcal{V}_{k_*}$ and it is immediate to see that $\mathcal{V}_k = \mathcal{V}_{k_*}$ for every $k \geq k_*$. Hence, by construction, $\mathcal{V}_{k_*} = \bar{\mathcal{R}}$ and since $\mathcal{V}_{k_*} \subset \mathcal{R}(\mathcal{N}) \subset \bar{\mathcal{R}}$, one gets the lemma. □

Remark 7.5 As regards a characterization of observability in the unswitched case, first recall that, for every $T > 0$, $I_{\mathcal{T}}(0, T)$ is equal to the kernel of the observability matrix $\mathcal{O}(A, C)$ whose lines are given by CA^i, $0 \leq i \leq n-1$, i.e., $\mathcal{O}_{\mathcal{T}}(0, T)$ is equal to the image of $\mathcal{O}(A, C)^\top$ or equivalently $\mathcal{O}_{\mathcal{T}}(0, T) = \operatorname{Span}\{(A^\top)^i c^\top \mid 0 \leq i \leq n-1,\ c \text{ line of } C\}$. Note that $\mathcal{O}(A, C)^\top = \mathcal{C}(A^\top, C^\top)$ establishing a (linear) duality between the properties of controllability and observability. The extension to the switched system case is also based on a duality principle and leads one to consider the space

$$\mathcal{O}(\mathcal{N}) := \operatorname{Span}\left\{A_1^\top \cdots A_i^\top c^\top \;\middle|\; \begin{array}{l} 0 \leq i \leq n-1,\ A_1, \ldots, A_i \in \mathcal{N}_A \\ c \text{ line of some } C \in \mathcal{N}_C \end{array}\right\}. \tag{7.7}$$

The duality principle above mentioned can now be established in the switched system case by defining $\mathcal{N}^\top$ as the subset of $M_n(\mathbb{R}) \times M_{n,p}(\mathbb{R}) \times M_{m,n}(\mathbb{R})$ such that $(A^\top, C^\top, B^\top) \in \mathcal{N}^\top$ if and only if $(A, B, C) \in \mathcal{N}$. Then $\mathcal{O}(\mathcal{N}) = \mathcal{R}(\mathcal{N}^\top)^\top$. Such a result will enable us to deduce at once observability results from controllability ones.

7.1.1 Case of Piecewise Constant Switching Signals with Piecewise Constant Controls

In this section, we assume that $\mathcal{T} = \mathcal{S}_{\mathrm{pc}}(\mathcal{N})$ and $\mathcal{U}$ is the set $\mathcal{PC}_m$. We will often make a slight abuse of notation by identifying a mode in $\mathcal{N}$ with the corresponding constant signal in $\mathcal{T}$.

To provide the main result of this section, we need the following definitions.

Definition 7.6 For $N \geq 1$ and $\tau = (t_1, \ldots, t_N) \in \mathbb{R}^N$, let $T_\tau = \sum_{j=1}^{N} t_j$. For every $\beta = ((A_1, B_1), \ldots, (A_N, B_N)) \in \mathcal{N}_{AB}^N$, let σ_β be the piecewise constant switching signal defined on $[0, T_\tau]$ by $\sigma_\beta = (A_j, B_j)$ on each interval $[\sum_{l=1}^{j-1} t_l, \sum_{l=1}^{j} t_l)$, for $1 \leq j \leq N$. Then $\Phi_{\tau,\beta}$ is defined as the linear mapping from $\mathcal{PC}_m$ to $\mathbb{R}^n$ given by

$$\Phi_{\tau,\beta}(u) = x(T_\tau; 0, 0, \sigma_\beta, u) = \sum_{j=1}^{N}\Big(\prod_{l=j+1}^{N} e^{A_l t_l}\Big)\Big(\int_0^{t_j} e^{sA_j} B_j u\Big(s + \sum_{l=1}^{j-1} t_l\Big) ds\Big). \tag{7.8}$$

If moreover $u \in \mathcal{PC}_m$ is constantly equal to some $u_j \in \mathbb{R}^m$ on each interval $[\sum_{l=1}^{j-1} t_l, \sum_{l=1}^{j} t_l)$, (7.8) reduces to

$$\Phi_{\tau,\beta}(u) = \sum_{j=1}^{N}\Big(\prod_{l=j+1}^{N} e^{A_l t_l}\Big)\Big(\int_0^{t_j} e^{sA_j} ds\Big) B_j u_j. \tag{7.9}$$

We next provide the main result of the section.

Theorem 7.7 *Let $\mathcal{N}$ be a bounded subset of $M_n(\mathbb{R}) \times M_{n,m}(\mathbb{R}) \times M_{p,n}(\mathbb{R})$. Consider the switched linear control system $(C_\mathcal{T})$ defined by (7.1) and (7.2) with $\mathcal{T} = \mathcal{S}_{\mathrm{pc}}(\mathcal{N})$ and $\mathcal{U} = \mathcal{PC}_m$. Then, for every $T > 0$, one has that*

$$R_\mathcal{T}(0) = R_\mathcal{T}(0, T) = \mathcal{R}(\mathcal{N}),$$

and there exist a positive integer $\bar{N}$ and $(\bar{\tau}, \bar{\beta}) \in (0, \infty)^{\bar{N}} \times \mathcal{N}_{AB}^{\bar{N}}$ with $T_{\bar{\tau}} = T$ such that $\mathrm{Im}\Phi_{\bar{\tau},\bar{\beta}} = \mathcal{R}(\mathcal{N})$. In particular, $(C_\mathcal{T})$ is completely controllable if and only if $\mathcal{R}(\mathcal{N}) = \mathbb{R}^n$.

Proof If N is a positive integer and $\beta = ((A_1, B_1), \ldots, (A_N, B_N)) \in \mathcal{N}_{AB}^N$, denote by E_β the sum of the spaces $\mathrm{Im}\, B_i$, $1 \leq i \leq N$. For every $T > 0$, recall that $R_\mathcal{T}(0, T)$ is defined as the union of the subspaces $\mathrm{Im}\Phi_{\tau,\beta}$, where N is a positive integer, $\tau = (t_1, \ldots, t_N) \in (0, \infty)^N$ with $T_\tau = T$ and $\beta \in \mathcal{N}_{AB}^N$. Moreover, one has that $R_\mathcal{T}(0, t) \subset R_\mathcal{T}(0, T)$ for every $t \leq T$, as it follows by considering among the controls on $[0, T]$ defining $R_\mathcal{T}(0, T)$ those that are zero on $[0, T - t]$.

Consider next $\bar{N} \geq 1$, $\bar{\tau} = (\bar{t}_1, \ldots, \bar{t}_{\bar{N}}) \in (0, \infty)^{\bar{N}}$, and $\bar{\beta} \in \mathcal{N}_{AB}^{\bar{N}}$ so that the rank of $\Phi_{\bar{\tau},\bar{\beta}}$ is maximal. We first want to prove that

$$R_\mathcal{T}(0, T) = \mathrm{Im}\Phi_{\bar{\tau},\bar{\beta}}, \quad \text{for every } T \geq \bar{T}. \tag{7.10}$$

For $N' \geq 1$, $\tau' \in (0,\infty)^{N'}$, and $\beta' \in \mathcal{N}_{AB}^{N'}$, consider the concatenations $\bar{\tau}\tau' \in (0,\infty)^{\bar{N}+N'}$ and $\bar{\beta}\beta' \in \mathcal{N}_{AB}^{\bar{N}+N'}$ by first applying (τ',β'). Using (7.3), one gets that

$$\operatorname{Im}\Phi_{\bar{\tau}\tau',\bar{\beta}\beta'} = \Big(\prod_{i=1}^{\bar{N}} e^{\bar{t}_i\bar{A}_i}\Big)\operatorname{Im}\Phi_{\tau',\beta'} + \operatorname{Im}\Phi_{\bar{\tau},\bar{\beta}},$$

where $(\bar{A}_1,\dots,\bar{A}_{\bar{N}})$ is the A-component of $\bar{\beta}$. Since $\operatorname{Im}\Phi_{\bar{\tau},\bar{\beta}}$ is included in $\operatorname{Im}\Phi_{\bar{\tau}\tau',\bar{\beta}\beta'}$, it follows from the maximality of the rank of $\Phi_{\bar{\tau},\bar{\beta}}$ that these two subspaces of $\mathbb{R}^n$ are actually equal. Therefore,

$$\operatorname{Im}\Phi_{\tau',\beta'} \subset \Big(\prod_{i=1}^{\bar{N}} e^{\bar{t}_i\bar{A}_i}\Big)^{-1}\operatorname{Im}\Phi_{\bar{\tau},\bar{\beta}}, \qquad \text{for every } \tau' \in (0,\infty)^{N'} \text{ and } \beta' \in \mathcal{N}_{AB}^{N'}. \tag{7.11}$$

By first choosing $(\tau',\beta') = (\bar{\tau},\bar{\beta})$ in (7.11), one gets that $\Big(\prod_{i=1}^{\bar{N}} e^{\bar{t}_i\bar{A}_i}\Big)^{-1}\operatorname{Im}\Phi_{\bar{\tau},\bar{\beta}} = \operatorname{Im}\Phi_{\bar{\tau},\bar{\beta}}$, since the two spaces have the same dimension. Hence, again by (7.11), for every $T > 0$, $R_{\mathcal{T}}(0,T)$ is included in $\operatorname{Im}\Phi_{\bar{\tau},\bar{\beta}}$, which is itself included in $R_{\mathcal{T}}(0,\bar{T})$. This yields (7.10).

We now prove that for every $T > 0$, $R_{\mathcal{T}}(0,T) \subset \mathcal{R}(\mathcal{N})$. This follows by Lemma 7.4 and from the fact that, for every $\tau = (t_1,\dots,t_N) \in (0,\infty)^N$ with $T_\tau = T$ and $\beta \in \mathcal{N}_{AB}^N$, every element $\Phi_{\tau,\beta}(u)$ of $\operatorname{Im}\Phi_{\tau,\beta}$ with $u \in \mathcal{PC}_m$, as given in (7.8) can be expressed only using powers A_i^j with $0 \leq j \leq n-1$, thanks to Cayley-Hamilton theorem (cf. Theorem A.26), and hence belongs to $\mathcal{R}(\mathcal{N})$.

We next establish that $R_{\mathcal{T}}(0,T)^\perp \subset \mathcal{R}(\mathcal{N})^\perp$ for every $T > 0$. Pick any $z \in \mathbb{R}^n$ orthogonal to $R_{\mathcal{T}}(0,T)$. Hence, for every $N \geq 1$, $\tau = (t_1,\dots,t_N) \in (0,\infty)^N$ with $T_\tau < T$, and $\beta \in \mathcal{N}_{AB}^N$, one has $z^\top\Phi_A(T_\tau,0)B(0) = 0$. Since $B(0)$ can be taken in an arbitrary way in $\mathcal{N}_B$, one deduces that the map $(t_1,\dots,t_N) \mapsto z^\top\prod_{i=1}^N e^{t_iA_i}b$, where $\tau = (t_1,\dots,t_N) \in (0,\infty)^N$ with $T_\tau < T$ and b is any column b of an arbitrary element $B \in \mathcal{N}_B$, is identically equal to zero. Recalling that $(t_1,\dots,t_N) \mapsto z^\top\prod_{i=1}^N e^{t_iA_i}b$ is expandable in power series, it follows that all the coefficients of that power series must also be equal to zero and one gets the conclusion.

From what precedes, it follows that $R_{\mathcal{T}}(0,T) = \mathcal{R}(\mathcal{N})$ for $T \geq \bar{T}$. To conclude the proof of Theorem 7.7, it remains to prove that the same is true for $T \in (0,\bar{T})$. To show the latter, let r be the dimension of $\mathcal{R}(\mathcal{N})$ and we can assume with no loss of generality that there exist $u^{(1)},\dots,u^{(r)} \in (\mathbb{R}^m)^{\bar{N}}$ so that $\Phi_{\bar{\tau},\bar{\beta}}(u_{\bar{\tau}}^{(l)})$, $1 \leq l \leq r$, are linearly independent in $\mathbb{R}^n$, where for every $\tau \in (0,+\infty)^{\bar{N}}$ we denote by $u_\tau^{(l)}$ the piecewise constant control whose values are the elements of the $\bar{N}$-tuple $u^{(l)}$, taken on time intervals whose lengths are the elements of the $\bar{N}$-tuple τ.

Finally, fix $n-r$ vectors $v_{r+1}, \ldots, v_n$ in $\mathbb{R}^n$ so that

$$\Phi_{\bar{\tau},\bar{\beta}}(u_{\bar{\tau}}^{(1)}), \ldots, \Phi_{\bar{\tau},\bar{\beta}}(u_{\bar{\tau}}^{(r)}), v_{r+1}, \ldots, v_n$$

is a basis of $\mathbb{R}^n$. Consider the map $\Psi\colon (0,\infty)^{\bar{N}} \to \mathbb{R}$ defined by

$$\Psi(\tau) = \det(\Phi_{\tau,\bar{\beta}}(u_{\tau}^{(1)}), \ldots, \Phi_{\tau,\bar{\beta}}(u_{\tau}^{(r)}), v_{r+1}, \ldots, v_n).$$

Notice that Ψ is a nontrivial analytic map. It follows from Lemma A.13 that there exists a sequence $(\tau_k)_{k\in\mathbb{N}}$ in $(0,\infty)^{\bar{N}}$ with T_{τ_k} tending to zero as k tends to infinity so that, for every $k \in \mathbb{N}$, $\Psi(\tau_k) \neq 0$, so that the rank of $\Phi_{\tau_k,\bar{\beta}}$ is equal to r. Since moreover,

$$\mathrm{Im}\Phi_{\tau_k,\bar{\beta}} \subset R_{\mathcal{T}}(0, T_{\tau_k}) \subset R_{\mathcal{T}}(0, \bar{T}) = \mathcal{R}(\mathcal{N})$$

it follows that $R_{\mathcal{T}}(0, T_{\tau_k}) = \mathcal{R}(\mathcal{N})$. One concludes by recalling that the mapping $T \mapsto R_{\mathcal{T}}(0, T)$ is monotone. □

By the duality principle presented in Remark 7.5, one derives the following result on the observability of switched linear control systems.

Theorem 7.8 *Let $\mathcal{N}$ be a bounded subset of $M_n(\mathbb{R}) \times M_{n,m}(\mathbb{R}) \times M_{p,n}(\mathbb{R})$. Consider the switched linear control system $(C_{\mathcal{T}})$ defined by* (7.1) *and* (7.2) *with $\mathcal{T} = \mathcal{S}_{\mathrm{pc}}(\mathcal{N})$ and $\mathcal{U} = \mathcal{PC}_m$. Then, for every $T > 0$,*

$$O_{\mathcal{T}}(0) = O_{\mathcal{T}}(0, T) = \mathcal{O}(\mathcal{N})$$

and there exists a positive integer $\bar{N}$ and $((t_1, \ldots, t_{\bar{N}}), (\gamma_1, \ldots, \gamma_{\bar{N}}))) \in \mathbb{R}^{\bar{N}} \times \mathcal{N}_{AC}^{\bar{N}}$ such that $T_{\bar{t}} = T$ and $\mathrm{Im}\,\omega(T, \sigma_{AC})^\top = \mathcal{O}(\mathcal{N})$, where ω is defined as in Remark 7.3 and σ_{AC} is the piecewise constant switching signal defined on $[0, T_{\bar{t}}]$ by $\sigma_{AC}(t) = \gamma_i$ on each interval $[\sum_{l=1}^{i-1} t_l, \sum_{l=1}^{i} t_l)$ for $1 \le i \le \bar{N}$. In particular, $(C_{\mathcal{T}})$ is completely observable if and only if $\mathcal{O}(\mathcal{N}) = \mathbb{R}^n$.

7.1.2 General Case

In this section, we extend the results of the previous section to a general a switched linear control system $(C_{\mathcal{T}})$. For $t \ge 0$ and $\sigma \in \mathcal{S}_{\mathrm{arb}}(\mathcal{N})$, we define the controllability and observability Gramians at time t given, respectively, by

$$W^c(t, \sigma_{AB}) = \int_0^t \Phi_A(t,s)B(s)B(s)^\top \Phi_A(t,s)^\top ds, \tag{7.12}$$

$$W^o(t, \sigma_{AC}) = \int_0^t \Phi_A(t, s)^\top C(s)^\top C(s)\Phi_A(t, s)ds. \tag{7.13}$$

We next introduce the following assumption for a class of measurable signals.

(T) A class $\mathcal{T}$ of measurable signals verifies Assumption **(T)** if, for every $\varepsilon > 0$ and $\sigma \in \mathcal{S}_{\mathrm{pc}}(\mathcal{N})$, there exists a path $(0, 1] \ni \lambda \mapsto (t_\lambda, \sigma^\lambda) \in (0, \infty) \times \mathcal{S}_{\mathrm{arb}}(\mathcal{N})$ such that:

- the map $\lambda \mapsto (W^c(t_\lambda, \sigma^\lambda_{AB}), W^o(t_\lambda, \sigma^\lambda_{AC}))$ is analytic on $(0, 1]$,
- $\|W^c(t_1, \sigma^1_{AB}) - W^c(t, \sigma_{AB})\| \leq \varepsilon$ and $\|W^o(t_1, \sigma^1_{AC}) - W^o(t, \sigma_{AC})\| \leq \varepsilon$,
- the set $\{\lambda \in (0, 1) \mid \sigma^\lambda \in \mathcal{T}\}$ has nonempty interior.

In the next proposition, we show that most of the classes of switching signals considered previously do satisfy Assumption **(T)**.

Proposition 7.9 *Let $\mathcal{T}$ be a class of measurable signals taking values in $\mathcal{N} \subset M_n(\mathbb{R}) \times M_{n,m}(\mathbb{R}) \times M_{p,n}(\mathbb{R})$ and equal to one of the classes introduced in Sect. 1.1.1. Assume moreover that, if $\mathcal{T} = \mathcal{S}_{\mathrm{lip},L}(\mathcal{N})$ or $\mathcal{T} = \mathcal{S}_{\mathrm{bv},T,\nu}(\mathcal{N})$, then $\mathcal{N}$ is star-shaped, i.e., there exists $(A_*, B_*, C_*) \in \mathcal{N}$ such that for every $(A, B, C) \in \mathcal{N}$ the segment between (A, B, C) and (A_*, B_*, C_*) is contained in $\mathcal{N}$. Then $\mathcal{T}$ satisfies Assumption **(T)**.*

Proof We only provide arguments relative to the controllability Gramian since those relative to the observability Gramian are entirely similar.

Take $t, \varepsilon > 0$ and $\sigma = (A, B, C) \in \mathcal{S}_{\mathrm{pc}}(\mathcal{N})$.

If $\mathcal{T} = \mathcal{S}_{\mathrm{per}}(\mathcal{N})$, we take $\sigma^\lambda = \sigma^\#$ and $t_\lambda = t$ for every $\lambda = (0, 1]$, where $\sigma^\#$ is the t-periodic signal coinciding with σ on $[0, t)$.

For the cases $\mathcal{T} = \mathcal{S}_{\mathrm{dw},\tau}(\mathcal{N})$ and $\mathcal{T} = \mathcal{S}_{\mathrm{av-dw},\tau,N_0}(\mathcal{N})$ one can find the path $\lambda \mapsto W^c(t_\lambda, \sigma^\lambda_{AB})$ as follows. For every $\lambda > 0$, take $t_\lambda = t/\lambda$ and consider the time-reparameterized signal $\sigma^\lambda = \Big(A(\lambda\,\cdot), B(\lambda\,\cdot), C(\lambda\,\cdot)\Big)$. Then $\sigma^\lambda \in \mathcal{T}$ for λ small enough and, after a rescaling of time, one has

$$W^c(t_\lambda, \sigma^\lambda_{AB}) = \frac{1}{\lambda}\int_0^t \Phi_{\frac{A(\cdot)}{\lambda}}(t, s)B(s)B(s)^\top \Phi_{\frac{A(\cdot)}{\lambda}}(t, s)^\top ds.$$

This is due to the fact that, for every $s \geq 0$, $\Phi_{A(\lambda\,\cdot)}(t/\lambda, s/\lambda) = \Phi_{A(\cdot)/\lambda}(t, s)$. In addition, by expanding $\Phi_{A(\cdot)/\lambda}(t, s)$ in Volterra series, we deduce that there exists $C > 0$ such that

$$\Phi_{A(\cdot)/\lambda}(t, s) = \sum_{\ell \geq 0} \frac{\varphi_\ell(s)}{\lambda^\ell},$$

with $|\varphi_\ell(s)| \leq C$ for every $\ell \geq 0$ and $s \in [0, t]$. This implies that $(0, \infty) \ni \lambda \mapsto W^c(\frac{t}{\lambda}, \sigma^\lambda_{AB})$ is analytic and equal to $W^c(t, \sigma_{AB})$ for $\lambda = 1$.

For the cases $\mathcal{T} = \mathcal{S}_{\mathrm{PE},T,\mu}(\mathcal{N})$ and $\mathcal{T} = \mathcal{S}_{\mathrm{BV},T,\mu}(\mathcal{N})$, the previous construction can be modified as follows. First notice that, up to adding some short intervals on which σ is constant we can assume that

- Case $\mathcal{T} = \mathcal{S}_{\mathrm{PE},T,\mu}(\mathcal{N})$: the value of σ on the last interval on which it is constant is equal to (A_1, B_1, C_1) (i.e., corresponding to the value one for the switching parameter α);
- Case $\mathcal{T} = \mathcal{S}_{\mathrm{BV},T,\nu}(\mathcal{N})$: the distance between two subsequent values of σ is smaller than ν.

The modification of σ can be taken so that the corresponding variation of $W^c(t, \sigma_{AB})$ is smaller than ε. The argument now works as before in the case $\mathcal{T} = \mathcal{S}_{\mathrm{BV},T,\nu}(\mathcal{N})$. In the case $\mathcal{T} = \mathcal{S}_{\mathrm{PE},T,\mu}(\mathcal{N})$ one can take $\sigma^\lambda = \Big(A(\cdot/\lambda), B(\cdot/\lambda), C(\cdot/\lambda)\Big)$ on $[0, \lambda t_0]$ and $\sigma^\lambda = (A_1, B_1, C_1)$ on $[\lambda t_0, t]$, where t_0 is such that $\sigma|_{[t_0,t]} \equiv (A_1, B_1, C_1)$.

We are left to discuss the case $\mathcal{T} = \mathcal{S}_{\mathrm{lip},L}(\mathcal{N})$. Then we modify σ by adding, at each switching time, a Lipschitz continuous arc defined on a small time interval and bridging the discontinuity (one may take a Lipschitz continuous arc whose graph is the union of two segments joining at (A_*, B_*, C_*)). These modifications can be done while keeping the corresponding variation of $W^c(t, \sigma_{AB})$ smaller than ε. By a time-reparameterization $\lambda \mapsto \sigma^\lambda = \Big(A(\lambda\,\cdot), B(\lambda\,\cdot), C(\lambda\,\cdot)\Big)$ we can lower the Lipschitz constant and complete the proof as above. □

In the following theorem, we provide a complete characterization of the controllability and observability of a switched linear control system subject to Assumption **(T)**.

Theorem 7.10 *Consider the switched linear control system $(C_{\mathcal{T}})$. Assume that $\mathcal{T}$ satisfies Assumption **(T)**. Then $R_{\mathcal{T}}(0) = \mathcal{R}(\mathcal{N})$ and there exist $t^* > 0$ and a switching law $\sigma \in \mathcal{T}$ such that the range of the controllability Gramian $W^c(t^*, \sigma_{AB})$ is equal to $\mathcal{R}(\mathcal{N})$.*

Similarly, $O_{\mathcal{T}}(0) = O(\mathcal{N})$ and there exist $t^* > 0$ and a switching law $\sigma \in \mathcal{T}$ such that the range of the observability Gramian $W^o(t^*, \sigma_{AC})$ is equal to $O(\mathcal{N})$.

Proof The observability results are deduced from the duality principle put forward in Remark 7.5 and we therefore focus on proving the controllability part.

We first consider the case where $\mathcal{T} = \mathcal{S}_{\mathrm{arb}}(\mathcal{N})$ and the set $\mathcal{U}$ of admissible controls u is made of all locally integrable functions $u\colon [0, +\infty) \to \mathbb{R}^m$. Since $\mathcal{S}_{\mathrm{arb}}(\mathcal{N})$ and $\mathcal{U}$ contain $\mathcal{S}_{\mathrm{pc}}(\mathcal{N})$ and $\mathcal{U} = \mathcal{PC}_m$ respectively, one deduces from Theorem 7.7 that $R_{\mathcal{S}_{\mathrm{arb}}(\mathcal{N})}(0)$ contains $\mathcal{R}(\mathcal{N})$.

To prove the converse inclusion, notice that every element $\sigma \in \mathcal{S}_{\mathrm{arb}}(\mathcal{N})$ (respectively, $u \in \mathcal{U}$) is the limit point of a sequence $(\sigma^k)_{k\geq 0}$ in $\mathcal{S}_{\mathrm{pc}}(\mathcal{N})$ (respectively, $(u^k(\cdot))_{k\geq 0}$ in $\mathcal{PC}_m$) for the L^1_{loc} topology. By applying Proposition 7.1, one deduces that $R_{\mathcal{S}_{\mathrm{arb}}(\mathcal{N})}(0)$ is contained in the closure of the reachable set from 0 for the switched linear control system associated with $\mathcal{T} = \mathcal{S}_{\mathrm{pc}}(\mathcal{N})$ and admissible controls in $\mathcal{PC}_m$. The latter being equal to the closed set $\mathcal{R}(\mathcal{N})$, the equality $R_{\mathcal{S}_{\text{arb}}(\mathcal{N})}(0) = \mathcal{R}(\mathcal{N})$ is proved.

Consider now an arbitrary class $\mathcal{T}$ of signals satisfying property **(T)** together with a set $\mathcal{U}$ of admissible controls that contains $\mathcal{PC}_m$. Since $\mathcal{T}$ is a subset of $\mathcal{S}_{\text{arb}}(\mathcal{N})$, one deduces that $R_{\mathcal{T}}(0) \subset \mathcal{R}(\mathcal{N})$.

Let us prove the opposite inclusion. According to Theorem 7.7, there exist $N \geq 1$, $\bar{\tau} = (\bar{t}_1, \ldots, \bar{t}_N) \in (0, +\infty)^N$, and $\bar{\beta} \in \mathcal{N}_{AB}^N$ such that $\text{Im}\Phi_{\bar{\tau},\bar{\beta}} = \mathcal{R}(\mathcal{N})$. Let $\bar{T} = \sum_i^N \bar{t}_i$. Let $\bar{\sigma} \in \mathcal{S}_{\text{pc}}(\mathcal{N})$ be such that $\bar{\sigma}_{AB}$ on the interval $[0, \bar{T}]$ is the concatenation of the elements in $\bar{\beta}$ taken on intervals of lengths $\bar{t}_1, \ldots, \bar{t}_N$. Then the range of $W^c(\bar{T}, \bar{\sigma})$ is equal to $\mathcal{R}(\mathcal{N})$. This is a consequence of the fact that $W^c(\bar{T}, \bar{\sigma}) = \Phi_{\bar{\tau},\bar{\alpha}}\Phi^*_{\bar{\tau},\bar{\beta}}$, where $\Phi^*_{\bar{\tau},\bar{\beta}}$ is the adjoint of $\Phi_{\bar{\tau},\bar{\beta}} : \mathcal{PC}_m \to \mathbb{R}^n$.

Fix $\varepsilon > 0$ such that the rank of W is larger than or equal to the rank of $W^c(\bar{T}, \bar{\sigma})$ for every $n \times n$ matrix W satisfying $\|W - W^c(\bar{T}, \bar{\sigma})\| < \varepsilon$. Let $\lambda \mapsto (t_\lambda, \sigma^\lambda)$ be the analytic path provided by Assumption **(T)**. It then follows from Lemma A.13 that there exists $\lambda^* \in (0, 1]$ such that $\sigma^{\lambda^*} \in \mathcal{T}$ and the rank of the corresponding controllability Gramian $W^c(t_{\lambda^*}, \sigma^{\lambda^*})$ is larger than or equal to the rank of $W^c(\bar{T}, \bar{\sigma})$, itself equal to $\dim \mathcal{R}(\mathcal{N})$. Since the range of $W^c(t_{\lambda^*}, \sigma^{\lambda^*})$ is included in $\mathcal{R}(\mathcal{N})$ this concludes the proof. □

We next provide a result of controllability and observability when signals belong to a skeleton. For that purpose, consider a switched linear control system $(C_{\mathcal{T}})$ and consider a skeleton $\mathcal{G} = \cup_{t \geq 0} \mathcal{G}_t$ of $\mathcal{T}$. We introduce the following definitions.

Definition 7.11

1. A point $x \in \mathbb{R}^n$ is *$\mathcal{G}$-reachable for* $(C_{\mathcal{T}})$ if there exist $t \geq 0$, a switching law $(A, B, C) \in \mathcal{G}_t$ and an input $u \in \mathcal{U}$ such that the corresponding trajectory x_u starting at 0 reaches x at time t. The *reachable set* $\mathfrak{R}(\mathcal{G})$ is the set of all $\mathcal{G}$-reachable points. System $(C_{\mathcal{T}})$ is said to be *$\mathcal{G}$-controllable* if $\mathfrak{R}(\mathcal{G}) = \mathbb{R}^n$.
2. A point $x \in \mathbb{R}^n$ is *$\mathcal{G}$-observable for* $(C_{\mathcal{T}})$ if there exist $t \geq 0$ and a switching law $(A, B, C) \in \mathcal{G}_t$ such that the trajectory x_0 associated with the zero input and starting at x gives rise to an output y satisfying $y|_{[0,t]} \neq 0$. The *observable set* $\mathfrak{O}(\mathcal{G})$ is the set of all $\mathcal{G}$-observable points together with the origin $0 \in \mathbb{R}^n$. System $(C_{\mathcal{T}})$ is said to be *$\mathcal{G}$-observable* if $\mathfrak{O}(\mathcal{G}) = \mathbb{R}^n$.

We need the following assumption on $\mathcal{G}$.

(T-Sk) The skeleton $\mathcal{G}$ satisfies Assumption **(T-Sk)** if, for every $t^*, \varepsilon > 0$ and $(A, B, C) \in \mathcal{S}_{\text{pc}}(\mathcal{N})$, there exists a path $(0, 1] \ni \lambda \mapsto (t_\lambda, A_\lambda, B_\lambda, C_\lambda) \in (0, \infty) \times \mathcal{S}_{\text{arb}}(\mathcal{N})$ such that

- $\lambda \mapsto (W^c_\lambda(t_\lambda), W^o_\lambda(t_\lambda))$ is analytic,
- $\|W^c_1(t_1) - W^c(t^*)\| \leq \varepsilon$ and $\|W^o_1(t_1) - W^o(t^*)\| \leq \varepsilon$,
- the set $\{\lambda \in (0, 1) \mid (A_\lambda, B_\lambda, C_\lambda) \in \mathcal{G}_\infty\}$ has nonempty interior.

An adaptation of the proof of Proposition 7.9 combined with the tools introduced in the proof of Proposition 6.10 yields the following result.

Lemma 7.12 *Let $\mathcal{T}$ be one of the classes introduced in Sect. 1.1.1 with corresponding $\mathcal{G}$ defined in Sect. 6.1.2. Assume moreover that if $\mathcal{T} = \mathcal{S}_{\text{lip},L}(\mathcal{N})$ or $\mathcal{T} = \mathcal{S}_{\text{bv},T,\nu}(\mathcal{N})$ then $\mathcal{N}$ is star-shaped. Then the skeleton $\mathcal{G}$ satisfies Assumption* **(T-Sk)**.

We can now state the following proposition.

Proposition 7.13 *Consider a switched linear control system $(C_\mathcal{T})$ and a skeleton $\mathcal{G}$ of the class $\mathcal{T}$ satisfying Assumption* **(T-Sk)**. *Then $\mathfrak{R}(\mathcal{G}) = \mathcal{R}(\mathcal{N})$ and there exist $t^* > 0$ and a switching law $(A, B, C) \in \mathcal{G}_{t^*}$ such that the range of the controllability and observability Gramians $W^c(t^*)$ and $W^o(t^*)$ are equal to $\mathcal{R}(\mathcal{N})$ and $\mathcal{O}(\mathcal{N})$, respectively.*

Proof The inclusions $\mathfrak{R}(\mathcal{G}) \subset \mathcal{R}(\mathcal{N})$ and $\mathfrak{O}(\mathcal{G}) \subset \mathcal{O}(\mathcal{N})$ follow from Theorems 7.7 and 7.8, respectively. The opposite inclusions and the existence of t_* and (A, B, C) can be proved following the arguments in the corresponding part of the proof of Theorem 7.10. □

7.2 Minimal Realization

In this section we extend to switched linear control systems the classical notions of controllable and observable parts, as well as that of minimal realization.

We start with the controllable part. Let $(C_\mathcal{T})$ be a switched linear control system satisfying the hypotheses of Theorem 7.10. Consider a direct sum decomposition $\mathbb{R}^n = R_\mathcal{T}(0) \oplus E$ of the state space and set $r = \dim R_\mathcal{T}(0)$. Recall that $R_\mathcal{T}(0) = \mathcal{R}(\mathcal{N})$ and, according to Eq. (7.6), $R_\mathcal{T}(0)$ is invariant by every matrix $A \in \mathcal{N}_A$ and the range of every $B \in \mathcal{N}_B$ is included in $R_\mathcal{T}(0)$. Therefore, there exists $P \in \mathrm{GL}(n)$ such that $P^{-1}x = \begin{pmatrix} x^c \\ x^u \end{pmatrix}$ with $x^c \in R_\mathcal{T}(0)$ and $x^u \in E$, where $R_\mathcal{T}(0)$ is identified with $\mathbb{R}^r$ and E with $\mathbb{R}^{n-r}$. Then, for every $(A, B, C) \in \mathcal{N}$,

$$P^{-1}AP = \begin{pmatrix} A^c & * \\ 0 & A^u \end{pmatrix}, \quad P^{-1}B = \begin{pmatrix} B^c \\ 0 \end{pmatrix}, \quad CP = \begin{pmatrix} C^c * \end{pmatrix},$$

where A^c and B^c belong to $M_r(\mathbb{R})$ and $M_{r,m}(\mathbb{R})$ respectively. Moreover, the switched linear control system defined on $\mathbb{R}^r$ and associated with $\mathcal{T}(\mathcal{N}^c)$, the class of measurable signals obtained from $\mathcal{T}$ and taking values in

$$\mathcal{N}^c = \{(A^c, B^c, C^c) \mid (A, B, C) \in \mathcal{N}\} \subset M_r(\mathbb{R}) \times M_{r,m}(\mathbb{R}) \times M_{p,r}(\mathbb{R})$$

is completely controllable. We refer to $\mathcal{N}^c$ as the *controllable part* of $\mathcal{N}$. Notice that $\mathcal{N}^c = \Pi^c(\mathcal{N})$ where

$$\Pi^c(A, B, C) = (UP^{-1}APU^\top, UP^{-1}B, CPU^\top),$$

with $U = (\mathrm{Id}_r \ 0_{r,n-r})$. Also notice that the output y corresponding to the original system satisfies $y = C^c x^c$.

We proceed similarly to define the observable part of $(C_{\mathcal{T}})$. Consider a direct sum decomposition $\mathbb{R}^n = O_{\mathcal{T}}(0) \oplus F$ of the state space and set $s = \dim O_{\mathcal{T}}(0)$. As previously, one deduces that there exists $Q \in \mathrm{GL}(n)$ such that $Q^{-1}x = \begin{pmatrix} x^o \\ x^u \end{pmatrix}$ with $x^o \in O_{\mathcal{T}}(0)$ and $x^u \in F$. For every $(A, B, C) \in \mathcal{N}$,

$$Q^{-1}AQ = \begin{pmatrix} A^o & 0 \\ * & A^u \end{pmatrix}, \quad Q^{-1}B = \begin{pmatrix} B^o \\ * \end{pmatrix}, \quad CQ = \begin{pmatrix} C^o & 0 \end{pmatrix},$$

where A^o and C^o belong to $M_s(\mathbb{R})$ and $M_{p,s}(\mathbb{R})$ respectively. Moreover, the switched linear control system defined on $\mathbb{R}^s$ associated with $\mathcal{T}(\mathcal{N}^o)$, the class of measurable signals obtained from $\mathcal{T}$ and taking values in

$$\mathcal{N}^o = \{(A^o, B^o, C^o) \mid (A, B, C) \in \mathcal{N}\} \subset M_s(\mathbb{R}) \times M_{s,m}(\mathbb{R}) \times M_{p,s}(\mathbb{R})$$

is completely observable. We refer to $\mathcal{N}^o$ as the *observable part* of $\mathcal{M}$. Notice that $\mathcal{N}^o = \Pi^o(\mathcal{N})$ where

$$\Pi^o(A, B, C) = (VQ^{-1}AQV^\top, VQ^{-1}B, CQV^\top),$$

and $V = (\mathrm{Id}_s \ 0_{s,n-s})$. The corresponding output y satisfies $y = C^o x^o$.

Combining the above two constructions, one can associate with every switched linear control system $(C_{\mathcal{T}})$ a reduced controllable and observable switched linear control system $(C_{\mathcal{T}(\mathcal{N}_r)})$ with same output map. One first reduces $(C_{\mathcal{T}})$ to its reachable space $R_{\mathcal{T}}(0)$ and get a controllable system $(C_{\mathcal{T}(\mathcal{N}^c)})$ with same output map. Then one reduces the latter system to its observable space $O_{\mathcal{T}(\mathcal{N}^c)}(0)$ to finally obtain a switched linear control system $(C_{\mathcal{T}(\mathcal{N}_r)})$, where $\mathcal{N}_r = \Pi^o(\Pi^c(\mathcal{N}))$, with the required properties. We say that $(C_{\mathcal{T}(\mathcal{N}_r)})$ is a *canonical reduction of* $(C_{\mathcal{T}})$.

Remark 7.14 Note that the canonical reduction $(C_{\mathcal{T}(\mathcal{N}_r)})$ is uniquely defined by $(C_{\mathcal{T}})$ up to a linear change of coordinates, even if its construction depends on the choice of subspaces supplementary to $R_{\mathcal{T}}(0)$ in the original state space and to $O_{\mathcal{T}(\mathcal{N}^c)}(0)$ in $R_{\mathcal{T}}(0)$.

Let us provide some definitions relative to the classical concept of minimal realization, adapted to the framework of switched systems.

Definition 7.15

- (*Input-output map.*) Let $(C_{\mathcal{T}})$ be a switched linear control system. The *input-output map* $y_{C_{\mathcal{T}}}$ of $(C_{\mathcal{T}})$ is defined as the mapping which associates with every $(\sigma, u) \in \mathcal{T} \times \mathcal{U}$ the output $y(\cdot; 0, 0, \sigma, u)$.
- (*Switched linear system realization of an input-output map.*) An input-output map $y_{C_{\mathcal{T}}}$ is said to be *realized by a switched linear control system* $(C_{\hat{\mathcal{T}}})$, with the

switching signals in $\hat{\mathcal{T}}$ taking values in $\hat{\mathcal{N}}$, if the set $\mathcal{U}$ of admissible controls is the same for $(C_{\mathcal{T}})$ and $(C_{\hat{\mathcal{T}}})$ and there exists $\varphi\colon \mathcal{N} \to \hat{\mathcal{N}}$ such that $\{\varphi \circ \sigma \mid \sigma \in \mathcal{T}\} = \hat{\mathcal{T}}$ and $y_{C_{\hat{\mathcal{T}}}}(\varphi \circ \sigma, u) = y_{C_{\mathcal{T}}}(\sigma, u)$ for every $(\sigma, u) \in \mathcal{T} \times \mathcal{U}$. In that case, $(C_{\hat{\mathcal{T}}})$ is said to be a *realization of* $y_{C_{\mathcal{T}}}$ and one says that the dimension $\dim C_{\hat{\mathcal{T}}}$ of the realization is the dimension of its state space.

- (*Minimality.*) A switched linear control system $(C_{\hat{\mathcal{T}}})$ is a *minimal realization of an input-output map* $y_{C_{\mathcal{T}}}$ if it is a realization of $y_{C_{\mathcal{T}}}$ of minimal dimension.
- (*Switched linear system morphism between realizations.*) Let $(C_{\hat{\mathcal{T}}})$ be a realization of $y_{C_{\mathcal{T}}}$ with φ as above. Let $\hat{n}$ be the dimension of $(C_{\hat{\mathcal{T}}})$ and assume that there exists a linear mapping $S\colon \mathbb{R}^{\hat{n}} \to \mathbb{R}^n$ such that

$$AS = S\hat{A}, \quad B = S\hat{B}, \quad CS = \hat{C} \tag{7.14}$$

for every $(A, B, C) \in \mathcal{N}$ with $(\hat{A}, \hat{B}, \hat{C}) = \varphi(A, B, C)$. Then we say that S is a *switched linear system morphism* from $(C_{\hat{\mathcal{T}}})$ to $(C_{\mathcal{T}})$. If S is an isomorphism we say that the realizations $(C_{\mathcal{T}})$ and $(C_{\hat{\mathcal{T}}})$ are algebraically similar.

Remark 7.16 Given an input-output map $y_{C_{\mathcal{T}}}$, it is by definition realized by $(C_{\mathcal{T}})$, hence there always exists a minimal realization for $y_{C_{\mathcal{T}}}$. Moreover, all minimal realizations of $y_{C_{\mathcal{T}}}$ clearly have the same dimension.

Remark 7.17 Consider a switched linear system $(C_{\mathcal{T}})$ and its canonical reductions $(C_{\mathcal{T}(\mathcal{N}_r)})$ defined as in the first part of the section. By construction, $(C_{\mathcal{T}(\mathcal{N}_r)})$ are controllable and observable realizations of $y_{C_{\mathcal{T}}}$. They are all algebraically similar as realizations, and Theorem 7.18 below guarantees that, for some relevant classes of signals, they are actually minimal.

The main result of this section consists in a characterization of minimal realizations of input-output maps.

Theorem 7.18 *Let* $y_{C_{\mathcal{T}}}$ *be an input-output map of a switched linear control system* $(C_{\mathcal{T}})$. *Assume that* $\mathcal{T}$ *is one of the following classes (introduced in Sect. 1.1.1):* $\mathcal{S}_{\mathrm{arb}}(\mathcal{N})$, $\mathcal{S}_{\mathrm{pc}}(\mathcal{N})$, $\mathcal{S}_{\mathrm{per}}(\mathcal{N})$, $\mathcal{S}_{\mathrm{dw},\delta}(\mathcal{N})$, $\mathcal{S}_{\mathrm{av-dw},\delta,N_0}(\mathcal{N})$, $\mathcal{S}_{\mathrm{pe},T,\mu}(\mathcal{N})$. *Then the following statements are equivalent:*

(i) $(C_{\mathcal{T}})$ *is a minimal realization of* $y_{C_{\mathcal{T}}}$*;*
(ii) $(C_{\mathcal{T}})$ *is controllable and observable;*
(iii) If $(C_{\hat{\mathcal{T}}})$ *is controllable and it is a realization of* $y_{C_{\mathcal{T}}}$, *then there exists a surjective switched linear system morphism from* $(C_{\hat{\mathcal{T}}})$ *to* $(C_{\mathcal{T}})$.

In addition, all minimal realizations of $y_{C_{\mathcal{T}}}$ *are algebraically similar.*

Proof Assuming that the equivalence between (i), (ii), and (iii) is established, let us prove the last part of the theorem. Let $(C_{\mathcal{T}_1})$ and $(C_{\mathcal{T}_2})$ be two minimal realizations of some input-output map $y_{C_{\mathcal{T}}}$. Note that they have same dimension n, thanks to Remark 7.16. According to (ii) and (iii), there exists a surjective switched linear system morphism from $(C_{\mathcal{T}_2})$ to $(C_{\mathcal{T}_1})$. The latter must therefore be an isomorphism of $\mathbb{R}^n$, yielding the thesis.

The implication $(i) \Rightarrow (ii)$ is immediate from the construction of a reduced switched linear control system $(C_{\mathcal{T}(\mathcal{N}_r)})$ defined above. Indeed, a switched linear system and its reduced system have the same input-output map. Moreover, the dimension of any reduced system is strictly smaller than that of the original switched linear control system if the latter is either not controllable or not observable. Therefore a minimal realization must be controllable and observable.

As for the implication $(iii) \Rightarrow (i)$, consider $y_{C_\mathcal{T}}$ as in (iii) and a minimal realization $(C_{\hat{\mathcal{T}}})$. As proved previously, $(C_{\hat{\mathcal{T}}})$ is controllable and then (iii) implies that $\dim C_{\hat{\mathcal{T}}} \geq \dim C_\mathcal{T}$. By minimality of $(C_{\hat{\mathcal{T}}})$, (i) follows.

To conclude the proof of the theorem, it remains to prove the implication $(ii) \Rightarrow (iii)$. For that purpose, assume that $(C_{\hat{\mathcal{T}}})$ is a controllable realization of $y_{C_\mathcal{T}}$, where $(C_\mathcal{T})$ is controllable and observable. Let $\varphi : \mathcal{N} \to \hat{\mathcal{N}}$ be such that $y_{C_{\hat{\mathcal{T}}}}(\varphi \circ \sigma, u) = y_{C_\mathcal{T}}(\sigma, u)$ for every $(\sigma, u) \in \mathcal{T} \times \mathcal{U}$.

We first assume that the classes of signals $\mathcal{T}$ and $\hat{\mathcal{T}}$ contain $\mathcal{S}_{\mathrm{pc}}(\mathcal{N})$ and $\mathcal{S}_{\mathrm{pc}}(\hat{\mathcal{N}})$ respectively. Let us associate with each matrix product $A_w = A_1 \cdots A_\ell$ of elements in $\mathcal{N}_A$ a matrix product $\hat{A}_w$ of elements in $\hat{\mathcal{N}}_A$ as follows: choose $(A_j, B_j, C_j) \in \mathcal{N}$ for $1 \leq i \leq \ell$ and set $\hat{A}_w = \hat{A}_1 \cdots \hat{A}_\ell$ with $(\hat{A}_j, \hat{B}_j, \hat{C}_j) = \varphi(A_j, B_j, C_j)$, for $1 \leq j \leq \ell$. It is useful to associate with A_w the element $\beta_w = ((A_1, B_1), \ldots, (A_\ell, B_\ell))$ of $\mathcal{N}_{AB}^\ell$.

We claim the following: for every $(A, B, C) \in \mathcal{N}$ and every matrix product $A_w = A_1 \cdots A_\ell$, setting $(\hat{A}, \hat{B}, \hat{C}) = \varphi(A, B, C)$ one has

$$\hat{C}\hat{A}_w\hat{B} = CA_wB. \tag{7.15}$$

To see that, let

$$\beta = ((A_1, B_1), \ldots, (A_\ell, B_\ell), (A, B)), \qquad \text{where } ((A_1, B_1), \ldots, (A_\ell, B_\ell)) = \beta_w.$$

We use (7.9) to identify $\Phi_{\tau,\beta}$, by an abuse of notation, with a function on $(\mathbb{R}^m)^{\ell+1}$. Consider the linear mapping $C\Phi_{\tau,\beta}$, which is obtained as the output map $y(\sum_{i=1}^{\ell+1} t_i; 0, 0, \cdot)$ and is analytic with respect to $\tau = (t_1, \ldots, t_{\ell+1})$. Since this output map is realized by $(C_\mathcal{T})$ and $(C_{\hat{\mathcal{T}}})$, one derives two equal values for every coefficient of the power series $\tau \mapsto y(\sum_{i=1}^{\ell+1} t_i; 0, 0, \cdot)$ expressed first with elements of $\mathcal{N}$ and then with elements of $\hat{\mathcal{N}}$. Equation (7.15) corresponds for the equality of the coefficients associated with the monomial $\prod_{i=1}^{\ell+1} t_i$.

We next choose $k \in \mathbb{N}$, a subset $\mathcal{N}_k$ of $\mathcal{N}$, and a subset $\hat{\mathcal{N}}_k$ of $\hat{\mathcal{N}}$, each of them with k elements (A_i, B_i, C_i), $1 \leq i \leq k$, and $(\hat{A}_i, \hat{B}_i, \hat{C}_i)$, $1 \leq i \leq k$, such that

(a) $\mathcal{R}(\mathcal{N}_k) = \mathbb{R}^n$ and $\mathcal{O}(\mathcal{N}_k) = \mathbb{R}^n$;
(b) $\mathcal{R}(\hat{\mathcal{N}}_k) = \mathbb{R}^{\hat{n}}$;
(c) $\varphi(A_i, B_i, C_i) = (\hat{A}_i, \hat{B}_i, \hat{C}_i)$ for every $1 \leq i \leq k$.

From Item (b), there exist $\hat{n}$ matrix products $\hat{A}_{w_1}, \ldots, \hat{A}_{w_{\hat{n}}}$ with elements in $\hat{\mathcal{N}_k}$ and $\hat{n}$ vectors $\hat{b}_1, \ldots, \hat{b}_{\hat{n}}$, with $\hat{b}_j$ column vector of some $\hat{B}_i$, $1 \le i \le k$, such that $\hat{A}_{w_j}\hat{b}_j$ is a basis of $\mathbb{R}^{\hat{n}}$. By a slight abuse of notation, we say that $\hat{b}_j$ corresponds to a column b_j of B_i. We next define a switched linear system morphism S from $(C_{\hat{\mathcal{T}}})$ to $(C_{\mathcal{T}})$ as follows. Let $S\colon \mathbb{R}^{\hat{n}} \to \mathbb{R}^n$ be the linear mapping defined by $S(\hat{A}_{w_j}\hat{b}_j) = A_{w_j}b_j$, $1 \le j \le \hat{n}$. For every $\hat{A} \in \hat{\mathcal{N}_A}$, there exist $a_{ij} \in \mathbb{R}$, $1 \le i, j \le \hat{n}$, such that $\hat{A}\hat{A}_{w_j}\hat{b}_j = \sum_{i=1}^{\hat{n}} a_{ij}\hat{A}_{w_i}\hat{b}_i$ for every $1 \le j \le \hat{n}$. Therefore, for every $C \in \mathcal{N}_C$ and matrix product A_w of elements in $\mathcal{N}_A$, one has

$$
\begin{aligned}
CA_wS(\hat{A}\hat{A}_{w_j}\hat{b}_j) &= \sum_{i=1}^{\hat{n}} a_{ij}CA_wS(\hat{A}_{w_i}\hat{b}_i) = \sum_{i=1}^{\hat{n}} a_{ij}CA_wA_{w_i}b_i \\
&= \sum_{i=1}^{\hat{n}} a_{ij}\hat{C}\hat{A}_w\hat{A}_{w_i}\hat{b}_i \\
&= \hat{C}\hat{A}_w\hat{A}\hat{A}_{w_j}\hat{b}_j = CA_wAA_{w_j}b_j = CA_wAS(\hat{A}_{w_j}\hat{b}_j),
\end{aligned}
$$

where we have used twice Eq. (7.15). Since $(C_{\mathcal{T}})$ is observable, the rows of all matrices of the form CA_w span $\mathbb{R}^n$ (Theorem 7.8), and one deduces that $S(\hat{A}\hat{A}_{w_j}\hat{b}_j) = AS(\hat{A}_{w_j}\hat{b}_j)$ for every $1 \le j \le \hat{n}$, hence $S\hat{A} = AS$. One derives by similar computations the two other equalities of Eq. (7.14). Therefore S defines a morphism from $(C_{\hat{\mathcal{T}}})$ to $(C_{\mathcal{T}})$. It remains to prove that it is surjective. From Item (a), there exist n matrix products $A_{z_1}, \ldots, A_{z_n}$ of elements in $\mathcal{N}_k$ and n vectors $v_1, \ldots, v_n$, with v_l column vector of some B_i, $1 \le i \le k$, such that $A_{z_l}v_l$ is a basis of $\mathbb{R}^n$. According to Item (c), $A_{z_l}v_l$ corresponds to some $\hat{A}_{z_l}\hat{v}_l$ which is equal to $\sum_{i=1}^{\hat{n}} b_{il}\hat{A}_{w_i}\hat{b}_i$ for every $1 \le l \le n$ for some $b_{il} \in \mathbb{R}$, $1 \le i \le \hat{n}$, $1 \le l \le n$.

For every $C \in \mathcal{N}_C$ and matrix product A_w of elements in $\mathcal{N}_A$, one has

$$
\begin{aligned}
CA_wA_{z_l}v_l &= \hat{C}\hat{A}_w\hat{A}_{z_l}\hat{v}_l = \sum_{i=1}^{\hat{n}} b_{il}\hat{C}\hat{A}_w\hat{A}_{w_i}\hat{b}_i = \sum_{i=1}^{\hat{n}} b_{il}CA_wA_{w_i}b_i \\
&= CA_w\left(\sum_{i=1}^{\hat{n}} b_{il}A_{w_i}b_i\right) = CA_wS\left(\sum_{i=1}^{\hat{n}} b_{il}\hat{A}_{w_i}\hat{b}_i\right).
\end{aligned}
$$

By observability of $(C_{\mathcal{T}})$, one gets that $A_{z_l}v_l$ belongs to the range of S for every $1 \le l \le n$, hence the surjectivity of S.

Notice that the assumption on $\mathcal{T}$ and $\hat{\mathcal{T}}$ to contain their respective piecewise constant signals was used to derive Eq. (7.15). Since every piecewise constant signal on a finite time-interval is the restriction to that time-interval of a signal in $\mathcal{S}_{\text{per}}(\mathcal{N})$, moreover, Eq. (7.15) also holds in the case $\mathcal{T} = \mathcal{S}_{\text{per}}(\mathcal{N})$.

Assume now that $\mathcal{T} = \mathcal{S}_{\mathrm{PE},T,\mu}(\cdot)(\mathcal{N})$ with $\mathcal{N} = \{N_\delta = (1-\delta)N_0 + \delta N_1 \mid \delta \in [0,1]\}$. Since every piecewise constant controls defined on $[0,\mu]$ is the restriction of an element of $\mathcal{T}$, once again (7.15) holds true.

Consider now the case where $\mathcal{T}$ is equal to $\mathcal{S}_{\mathrm{dw},\delta}(\mathcal{N})$ or $\mathcal{S}_{\mathrm{av-dw},\delta,N_0}(\mathcal{N})$. Notice that the latter class contains the former one. Then, one can consider the linear mappings $C\Phi_{\tau,\beta}$ (on $(\mathbb{R}^m)^{\ell+1}$), which are equal to the output maps $y(\sum_{i=1}^{\ell+1} t_i; 0, 0, \cdot)$, for values of t_i larger than the dwell-time δ. By analyticity of $\tau \mapsto C\Phi_{\tau,\beta}$, Eq. (7.15) still holds true. □

7.3 L^2-Gain for Switched Linear Control Systems

In this section, we address robust stabilization issues for switched linear control systems of the type $(C_{\mathcal{T}})$. Here, the class of admissible controls $\mathcal{U}$ is chosen to be equal to $L^2_{\mathrm{loc}}([0,+\infty), \mathbb{R}^m)$ (that contains $\mathcal{P}C_m$). We denote in the sequel $L^2 = L^2([0,+\infty), \mathbb{R}^m)$.

Let us first recall the notion L^2-gain for unswitched systems and how it arises in the context of robust stabilization. Given a linear plant $\dot{x} = Ax + Bu,\ y = Cx$, consider the problem of stabilizing it with a linear feedback. This means taking $u = Kx$, i.e., finding a matrix K such that $A_K := A + BK$ is Hurwitz. Usually the choice of such a K is not unique and one wants to compare the performances of such stabilizing feedbacks. For that purpose, one considers the perturbed feedback $u = Kx + d$, yielding the dynamics $\dot{x} = A_K x + Bd$ and the output map $d \mapsto y(\cdot; 0, 0, d)$ for the trajectory starting at the origin and associated with the disturbance $d \in L^2$. Then, one defines the L^2-gain $\gamma_2(K)$ associated with the linear plant (A_K, B, C) as

$$\gamma_2(K) = \sup\left\{\frac{\|y(\cdot; 0, 0, d)\|_2}{\|d\|_2} \mid d \in L^2 \setminus \{0\}\right\}. \tag{7.16}$$

Then one says that a stabilizing linear feedback $u = Kx$ exhibits better performances than another stabilizing linear feedback $u = K'x$ if $\gamma_2(K) < \gamma_2(K')$.

In the context of switched linear control systems, we define the *L^2-gain of $\mathcal{T}$* and denote it by $\gamma_2(\mathcal{T})$, as follows

$$\gamma_2(\mathcal{T}) = \sup\left\{\frac{\|y(\cdot; 0, 0, \sigma, d)\|_2}{\|d\|_2} \mid d \in L^2 \setminus \{0\},\ \sigma \in \mathcal{T}\right\}, \tag{7.17}$$

where $y(\cdot; 0, 0, \sigma, u)$ is the output map of (7.2) and defined in (7.4). We investigate qualitative properties of $\gamma_2(\mathcal{T})$ and, in particular, we are interested in finding conditions on $\mathcal{T}$ ensuring its finiteness.

To state the main result of this section, we need the following definition.

Definition 7.19 We say that a switched linear control system $(C_{\mathcal{T}})$ is *uniformly observable* (or, equivalently, that $\mathcal{T}$ is *uniformly observable*) if there exist $T, \gamma > 0$

such that, for every $\sigma \in \mathcal{T}$ and every $t \geq 0$, one has

$$W^o(t+T, \sigma_{AC}) - W^o(t, \sigma_{AC}) \geq \gamma \mathrm{Id}_n, \tag{7.18}$$

where $W^o(t, \sigma_{AC})$ is the observability Gramian at time t associated with σ and defined in (7.13).

Notice that if a switched linear control system is uniformly observable and T, γ are as in Definition 7.19, the observability Gramian at time T associated with a switching law belonging to the weak-$\star$ closure of $\mathcal{T}$ still satisfies (7.18).

Remark 7.20 Consider the case where $\mathcal{T}$ contains all the constant $\mathcal{N}$-valued switching signals. (Notice that this is the case for all classes of switching signals introduced in Sect. 1.1.1 except that of persistently exciting signals.) It is then easy to see that uniform observability implies that (A, C) is observable for every $(A, B, C) \in \mathcal{N}$. In the case where $\mathcal{N}$ is compact and the class of signals under consideration is $\mathcal{S}_{\mathrm{dw},\tau}(\mathcal{N})$ for some $\tau > 0$, one easily shows that the converse is also true, namely, the observability of each pair (A, C) implies uniform observability.

Recall that for the unswitched case, the L^2-gain associated with a linear plant $\dot{x} = Ax + Bu$, $y = Cx$, is finite if and only if $A^{\min}$ is Hurwitz, where $(A^{\min}, B^{\min}, C^{\min})$ is any minimal realization of (A, B, C). We aim at deriving a counterpart of this classical result in the switched case. In that direction, we provide the following theorem, which is the main result of this section.

Theorem 7.21 *Let $(C_{\mathcal{T}})$ be a switched linear control system. Suppose that $\mathcal{T}$ admits a rich and extensive skeleton $\mathcal{G}$ satisfying assumption* **(T-Sk)**. *Let $(C_{\mathcal{T}(\mathcal{N}^r)})$ be a canonical reduction of $(C_{\mathcal{T}})$ as defined in Sect. 7.2.Then $\gamma_2(\mathcal{T})$ is finite if $\lambda(\mathcal{T}_A(\mathcal{N}^r)) < 0$ and infinite if either $\lambda(\mathcal{T}_A(\mathcal{N}^r)) > 0$ or $\lambda(\mathcal{T}_A(\mathcal{N}^r)) = 0$ and $\mathcal{T}(\mathcal{N}^r)$ is uniformly observable.*

Recall that the maximal Lyapunov exponent $\lambda(\mathcal{T}_A(\mathcal{N}^r))$ does not depend on the canonical reduction of $(C_{\mathcal{T}})$, since all such reductions are algebraically similar (see Remark 7.17) as realizations of $y_{C_{\mathcal{T}}}$.

Proof Thanks to the obvious fact that $\gamma_2(\mathcal{T}) = \gamma_2(\mathcal{T}(\mathcal{N}^r))$, it is enough to treat the case $\mathcal{T} = \mathcal{T}(\mathcal{N}^r)$.

Assume first that $\lambda(\mathcal{T}_A) < 0$. Taking into account the definition of $\lambda(\mathcal{T}_A)$ (see (1.6)) and the boundedness of $\mathcal{N}_A$, one gets the following exponential decay estimate: there exist $K_1, \lambda > 0$ such that

$$\|\Phi_A(t,s)\| \leq K_1 e^{-\lambda(t-s)}, \qquad \text{for every } A \in \mathcal{T}_A \text{ and every } 0 \leq s \leq t.$$

As a consequence of the above and the boundedness of $\mathcal{N}$, one deduces that there exists $K_2 > 0$ such that, for every $u \in L^2$, $(A, B, C) \in \mathcal{T}$, and $t \geq 0$, one has

$$\|y_u(t)\| \leq K_2 \int_0^t e^{-\lambda(t-s)} \|u(s)\| ds. \tag{7.19}$$

If $\chi_{[0,+\infty)}$ denotes the characteristic function of $[0, +\infty)$, the integral function on the right-hand side of (7.19) can be interpreted as the convolution of

$$f_1(t) = \chi_{[0,+\infty)}(t)e^{-\lambda t} \quad \text{and} \quad f_2(t) = \chi_{[0,+\infty)}(t)\|u(t)\|.$$

That yields at once that $\|y_u\|_2 \leq \frac{K_2}{\lambda}\|u\|_2$, hence the conclusion.

Assume now that $\lambda(\mathcal{T}_A) \geq 0$. According to Proposition 1.31, up to a common linear change of coordinates, every matrix in $A \in \mathcal{N}_A$ has the upper triangular block form

$$A = \begin{pmatrix} A_{11} & A_{12} & \cdots & & \\ 0 & A_{22} & A_{23} & \cdots & \\ 0 & 0 & A_{33} & A_{34} & \cdots \\ \vdots & \ddots & \ddots & \ddots & \\ 0 & \cdots & \cdots & 0 & A_{qq} \end{pmatrix},$$

where, for $i = 1, \ldots, q$, each A_{ii} is in $M_{n_i - n_{i-1}}(\mathbb{R})$, $n_i \in \mathbb{N}$ and the set $\mathcal{A}_i := \{A_{ii} \mid A \in \mathcal{N}_A\}$ is irreducible. Consider the classes of signals $\mathcal{T}_{A,i} := \{A_{ii} = \pi_{ii}(A) \mid A \in \mathcal{T}_A\}$ and their skeletons $\mathcal{G}_{A,i} := \{A_{ii} = \pi_{ii}(A) \mid A \in \mathcal{G}_A\}$, where π_{ii} is the application introduced in (1.14), $1 \leq i \leq q$. Each $\mathcal{G}_{A,i}$ is trivially extensive and the fact that it is rich follows from Remark 6.8.

Let $\bar{\imath} \in \{1, \ldots q\}$ be the largest index such that

$$\lambda(\mathcal{T}_{A,\bar{\imath}}) = \lambda(\mathcal{T}_A) = \max_{1 \leq i \leq q} \lambda(\mathcal{T}_{A,i}),$$

whose existence is guaranteed by Proposition 1.32. Since $(C_\mathcal{T})$ is $\mathcal{G}$-controllable according to Proposition 7.13, there exists a time $\bar{t} > 0$, a switching signal $(A^0, B^0, C^0) \in \mathcal{G}_{\bar{t}}$ and control $\bar{u} \in L^2$ so that the corresponding trajectory of $(C_\mathcal{T})$ starting at 0 reaches some point $\bar{x} = (0, \ldots, 0, \bar{x}_{\bar{\imath}}, 0, \ldots, 0)^\top \neq 0$ at time $\bar{t}$.

Since $(\mathcal{T}_{A,\bar{\imath}}, \mathcal{G}_{A,\bar{\imath}})$ satisfies the hypotheses of Theorem 6.16, there exists a quasi-extremal trajectory $x_{\bar{\imath}}^{\text{q-ext}}$ starting at $\bar{x}_{\bar{\imath}}$ and corresponding to a signal $A_{\bar{\imath}\bar{\imath}}$ belonging to the weak-$\star$ closure of $(\mathcal{G}_{A,\bar{\imath}})_\infty$. Let $\bar{x}$ be a trajectory of $\dot{x} = \bar{A}x$ starting at $\bar{x}$ and corresponding to some signal $\bar{A}$ in the closure of $(\mathcal{G}_A)_\infty$ satisfying $\pi_{\bar{\imath}\bar{\imath}}(\bar{A}(\cdot)) = A_{\bar{\imath}\bar{\imath}}(\cdot)$. Notice that $\bar{x}_{\bar{\imath}} = x_{\bar{\imath}}^{\text{q-ext}}$.

Let $\overline{E}$ be the signal defined as the concatenation of A^0 and $\bar{A}$. Consider a sequence $(E^l, F^l, G^l)_{l \in \mathbb{N}}$ in $\mathcal{G}_\infty$ such that E^l weak-$\star$ converges to $\bar{A}$. The sequence $(A^0 * E^l)_{l \in \mathbb{N}}$ is in $(\mathcal{G}_A)_\infty$ and weak-$\star$ converges to $\overline{E}$. Moreover, for every positive time $t \geq \bar{t}$, the convergence of $s \mapsto \Phi_{E^l}(s-\bar{t},0)$ to $s \mapsto \Phi_{\overline{E}}(s,\bar{t})$ is uniform with respect to $s \in [\bar{t}, t]$ (Proposition 1.3).

For every $l \in \mathbb{N}$, denote by x^l the trajectory of $(C_{\mathcal{T}})$ starting at the origin and associated with the control equal to $\bar{u}$ on $[0,\bar{t}]$ and zero on $[\bar{t},+\infty)$, and with the switching signal $(A^0 * E^l, B^0 * F^l, C^0 * G^l)$. Notice that $x^l(s) = \Phi_{E^l}(s-\bar{t},0)\bar{x}$ for every $s \geq \bar{t}$. For l large enough and for every $s \in [\bar{t}, t]$, one has

$$\|x^l(s)\| \geq \|\Phi_{\pi_{\bar{t}\bar{t}}(E^l)}(s-\bar{t},0)\bar{x}_{\bar{t}}\| \geq \frac{1}{2}\|\Phi_{A_{\bar{t}\bar{t}}}(s,\bar{t})\bar{x}_{\bar{t}}\| \geq \frac{1}{2C_{\mathrm{qx}}}e^{\lambda(\mathcal{T}_A)(s-\bar{t})}\|\bar{x}_{\bar{t}}\|.$$

Hence, there exists a positive constant K_3 independent of $t \geq \bar{t}$ such that, for every l large enough and every $s \in [\bar{t}, t]$, one has

$$\|x^l(s)\| \geq K_3 e^{\lambda(\mathcal{T}_A)(s-\bar{t})}. \tag{7.20}$$

Assume now that $\lambda(\mathcal{T}_A) = 0$ and $\mathcal{T}$ is uniformly observable, and let T, γ be as in Definition 7.19. Given two positive integers l, j, let $W^o_{l,j}$ be the observability Gramian at time T associated with $(E^l(jT+\cdot), F^l(jT+\cdot), G^l(jT+\cdot))$. Let y^l be the output corresponding to x^l. Applying the uniform observability assumption to each $(E^l(jT+\cdot), F^l(jT+\cdot), G^l(jT+\cdot))$ and taking into account Eq. (7.20), we get, for $J \in \mathbb{N}$ and l large enough,

$$\begin{aligned}
\gamma_2(\mathcal{T}) \geq \frac{\|y^l\|_2^2}{\|\bar{u}\|_2^2} &\geq \frac{\int_0^{JT}\|G^l(t)x^l(\bar{t}+t)\|^2 dt}{\|\bar{u}\|_2^2} \\
&= \frac{\sum_{j=0}^{J-1}\int_{jT}^{(j+1)T}\|G^l(t)x^l(\bar{t}+t)\|^2 dt}{\|\bar{u}\|_2^2} \\
&= \frac{\sum_{j=0}^{J-1} x^l(\bar{t}+jT)^\top W^o_{l,j} x^l(\bar{t}+jT)}{\|\bar{u}\|_2^2} \geq \gamma \frac{\sum_{j=0}^{J-1}\|x^l(\bar{t}+jT)\|^2}{\|\bar{u}\|_2^2} \\
&\geq \gamma J \frac{K_3^2}{\|\bar{u}\|_2^2}.
\end{aligned}$$

This implies that the L^2-gain $\gamma_2(\mathcal{T})$ is infinite.

Let now $\lambda(\mathcal{T}_A) > 0$. Let $\bar{t}$, $\bar{u}$, and $\bar{x}$ be as above. Since the skeleton $\mathcal{G}_{A,\bar{t}}$ is Barabanov (Proposition 6.14), we can find for every $l \in \mathbb{N}$ a time $t_l \geq l$ and a switching law $(\hat{E}^l, \hat{F}^l, \hat{G}^l) \in \mathcal{G}_{t_l}$ such that

$$\|\Phi_{\hat{E}^l}(t_l,0)\bar{x}\| \geq C_{\mathrm{b}} e^{\lambda(\mathcal{T}_A)t_l}\|\bar{x}\| \geq C_{\mathrm{b}} e^{\lambda(\mathcal{T}_A)l}\|\bar{x}\| \tag{7.21}$$

for some positive constant C_{b}. According to **(T-Sk)** and because of the observability counterpart of Proposition 7.13, there exist $\tilde{s} > 0$ and a switching law $(A_*, B_*, C_*) \in \mathcal{G}_{\tilde{s}}$ such that the observability Gramian $W^o(\tilde{s})$ at time $\tilde{s}$ associated with $\dot{x}(t) = A_*(t)x(t) + B_*(t)u(t)$, $y(t) = C_*(t)x(t)$, is invertible. Up to a suitable extension on $(\tilde{s}, \infty)$, we can assume that (A_*, B_*, C_*) belongs to $\mathcal{G}_\infty$. For $l \in \mathbb{N}$, consider the switching signal

$$(A^0 * \hat{E}^l * A_*, B^0 * \hat{F}^l * B_*, C^0 * \hat{G}^l * C_*) \in \mathcal{G}_\infty.$$

Denote by $\hat{x}^l$ the corresponding trajectory of $(C_\mathcal{T})$ starting at the origin and associated with the control equal to $\bar{u}$ on $[0, \bar{t}]$ and zero for $t > \bar{t}$ and let $\hat{y}^l$ be the corresponding output. Note that $\hat{x}^l(\bar{t}+t_l) = \Phi_{\hat{E}^l}(t_l, 0)\bar{x}$. It then follows from (7.21) that there exists a positive constant K_4 independent of l so that

$$\begin{aligned}\gamma_2(\mathcal{T}) \geq \frac{\|\hat{y}^l\|_{2,t_l}^2}{\|\bar{u}\|_2^2} &\geq \frac{\int_{\bar{t}+t^l}^{\bar{t}+t^l+\tilde{s}} \|C_*(s)\hat{x}^l(s)\|^2 ds}{\|\bar{u}\|_2^2} \\ &= \frac{\hat{x}^l(\bar{t}+t_l)^\top W^o(\tilde{s})\hat{x}^l(\bar{t}+t_l)}{\|\bar{u}\|_2^2} \geq K_4 e^{2\lambda(\mathcal{T}_A)l},\end{aligned}$$

and the right-hand side clearly tends to infinity as l tends to infinity. □

Remark 7.22 Theorem 7.21 applies, in particular, to the classes introduced in Sect. 1.1.1 where $\mathcal{N}$ is star-shaped if $\mathcal{T} = \mathcal{S}_{\mathrm{con}}(\mathcal{N})$, $\mathcal{T} = \mathcal{S}_{\mathrm{lip},L}(\mathcal{N})$, or $\mathcal{T} = \mathcal{S}_{\mathrm{bv},T,\nu}(\mathcal{N})$. Indeed, the skeletons introduced in Sect. 6.1.2 are in that case rich and extensive (Lemma 6.11), and satisfy Assumption **(T-Sk)** (Lemma 7.12). Notice, moreover, that for the class of dwell-time signals with $\mathcal{N}$ compact, observability is equivalent to uniform observability (Remark 7.20) and canonical reductions are minimal realization (Theorem 7.18), hence $\gamma_2(\mathcal{S}_{\mathrm{dw},\tau}(\mathcal{N}))$ is finite if and only if $\lambda(\mathcal{S}_{\mathrm{dw},\tau}(\mathcal{N}_A^{\min})) < 0$.

7.4 Intermittent Control: Stabilization of Persistently Excited Systems

The point of view of this section is the following: we consider a linear control system

$$\dot{x} = Ax + Bu, \qquad x \in \mathbb{R}^n, \quad u \in \mathbb{R}^m, \tag{7.22}$$

which is (exponentially) stabilizable by a linear feedback, and we study the preservation of stabilizability when the control action is intermittently dimmed or switched off by exogenous time-dependent switching signals.

Mathematically, we can translate the proposed setting into the problem of finding $K \in M_{m,n}(\mathbb{R})$ which renders the switched system

$$\dot{x} = Ax + \sum_{i=1}^{m} \alpha_i b_i k_i x, \qquad x \in \mathbb{R}^n, \tag{7.23}$$

uniformly exponentially stable, where $b_1, \ldots, b_m$ and $k_1, \ldots, k_m$ represent the columns of B and the rows of K, respectively, and $\alpha = (\alpha_1, \ldots, \alpha_m)$ is a switching signal with values in $[0, 1]^m$. Denote by $\mathcal{A}$ the class of admissible switching signals α and, for every $K \in M_{m,p}(\mathbb{R})$, set

$$\mathcal{S}(K) = \Big\{A + \sum_{i=1}^{m} \alpha_i b_i k_i \mid \alpha \in \mathcal{A}\Big\} \subset \mathcal{S}_{\mathrm{arb}}(\mathcal{M}_K),$$

where

$$\mathcal{M}_K = \Big\{A + \sum_{i=1}^{m} a_i b_i k_i \mid (a_1, \ldots, a_m) \in [0, 1]^m\Big\}.$$

We assume that $\mathcal{A}$, and hence $\mathcal{S}(K)$, are shift-invariant.

Definition 7.23 We say that (7.23) is *$\mathcal{A}$-stabilizable* if there exists $K \in M_{m,n}(\mathbb{R})$ such that $(\Sigma_{\mathcal{S}(K)})$ is uniformly exponentially stable. In that case we say that K is an *$\mathcal{A}$-stabilizer.*

The stabilizability problem is clearly trivial when A is Hurwitz, independently of the class $\mathcal{A}$ (simply take $K = 0$). Another elementary observation gives rise to the following lemma.

Lemma 7.24 *Assume that A is not Hurwitz and that the signal constantly equal to zero is in $\mathcal{A}^\star$. Then (7.23) is not $\mathcal{A}$-stabilizable.*

Proof Pick any $K \in M_{m,n}(\mathbb{R})$. Since 0 is in $\mathcal{A}^\star$, then $(\Sigma_{\mathcal{S}(K)^\star})$ is not uniformly exponentially stable. By Proposition 1.14, it follows that (7.23) is not $\mathcal{A}$-stabilizable. □

Let us then consider the case where $\mathcal{A}^\star$ does not contain the signal constantly equal to zero.

Lemma 7.25 *Let $\mathcal{A}$ be such that $\mathcal{A}^\star$ does not contain the signal constantly equal to zero. Then there exist $T > 0$ and $\mu \in (0, T)$ such that $\int_t^{t+T} (\alpha_1(s) + \cdots + \alpha_m(s))ds \geq \mu$ for every $t \in [0, \infty)$ and every $\alpha \in \mathcal{A}$.*

Proof Assume by contradiction that for every $k \in \mathbb{N}$ there exist $\alpha^k \in \mathcal{A}$ and $t_k \in [0, \infty)$ such that $\int_{t_k}^{t_k+k} (\alpha_1^k(s) + \cdots + \alpha_m^k(s))ds \leq \frac{1}{k}$. By shift-invariance of $\mathcal{A}$, we can assume without loss of generality that $t_k = 0$ for every $k \in \mathbb{N}$. Then the sequence $(\alpha^k)_{k\in\mathbb{N}}$ weak-$\star$ converges to the zero signal, leading to the required contradiction. □

The previous lemma motivates the following definition.

Definition 7.26 We say that $\mathcal{A}$ is *persistently exciting* if there exist $T > 0$ and $\mu \in (0, T)$ such that $\int_t^{t+T} \alpha_i(s)ds \geq \mu$ for every $t \in [0, \infty)$, $i = 1, \ldots, m$ and $\alpha \in \mathcal{A}$.

When $\mathcal{A}$ is persistently exciting with T, μ as in Definition 7.26, we recover that $\mathcal{S}(K) \subset \mathcal{S}_{\mathrm{pe},T,\mu}(\mathcal{M}_K)$, where $\mathcal{S}_{\mathrm{pe},T,\mu}$ is defined as in Sect. 1.1.1 with $N_0 = A$ and $N_i = A + b_i k_i$ for $i = 1, \ldots, m$.

Let us now give a result of $\mathcal{A}$-stabilizability for a case where the matrix A is not Hurwitz, but *neutrally stable*, i.e., such that $t \mapsto e^{tA}$ is uniformly bounded on $[0, +\infty)$. Recall that A is neutrally stable if and only if the real part of each eigenvalue of A is nonpositive and, for eigenvalues of zero real part, the Jordan blocks are diagonal in the complex Jordan block decomposition.

Proposition 7.27 *Let A be neutrally stable and assume that* (7.22) *is stabilizable by a linear feedback. Assume that $\mathcal{A}$ is persistently exciting. Then* (7.23) *is $\mathcal{A}$-stabilizable.*

Proof Since A is neutrally stable, there exists a system of coordinates in which

$$A = \begin{pmatrix} A_1 & A_{12} \\ 0 & A_2 \end{pmatrix}, \qquad B = \begin{pmatrix} B_1 \\ B_2 \end{pmatrix},$$

where A_1 is Hurwitz and A_2 is skew-symmetric. Moreover, (A_2, B_2) is controllable, since otherwise there would exist a nontrivial subspace W of $\mathbb{R}^n$ and a projection $\Pi \colon \mathbb{R}^n \to W$ such that, for every trajectory $x(\cdot)$ of (7.22), $\Pi(x(\cdot))$ is independent of the control and has constant norm, contradicting the stabilizability assumption.

Let r be such that A_2 is a $r \times r$ matrix and write x_2 for the last r coordinates of the vector x. Set

$$K = \begin{pmatrix} 0 & -B_2^\top \end{pmatrix}$$

and denote by $\beta_1, \ldots, \beta_m$ the columns of B_2. Then x_2 satisfies

$$\dot{x}_2 = A_2 x_2 - \sum_{i=1}^{m} \alpha_i \beta_i \beta_i^\top x_2. \tag{7.24}$$

We claim that x_2 converges exponentially to zero, uniformly with respect to α in the persistently exciting class $\mathcal{A}$. The conclusion of the proposition then follows, since an autonomous linear Hurwitz system subject to a perturbation whose norm converges exponentially (with respect to time) to zero is exponentially stable at the origin.

Setting $V(x_2) = \frac{\|x_2\|^2}{2}$, we have

$$\frac{d}{dt} V(x_2(t)) \leq -\sum_{i=1}^{m} \alpha_i(t) (\beta_i^\top x_2(t))^2. \tag{7.25}$$

Let T and μ be as in Definition 7.26. By shift-invariance of $\mathcal{A}$, the proof is concluded if we show that there exists $\gamma > 0$ such that $V(x_2(T)) \leq e^{-\gamma T} V(x_2(0))$ for every solution of (7.24) corresponding to some $\alpha \in \mathcal{A}$.

Assume by contradiction that for every $k \in \mathbb{N}$ there exist $x_{2,k} \in \mathbb{R}^r$ and $\alpha^k \in \mathcal{A}$ such that the solution $x(\cdot; x_{2,k}, \alpha^k)$ of (7.24) with initial condition $x_{2,k}$ and corresponding to the signal α^k satisfies

$$V(x_2(T; x_{2,k}, \alpha^k)) > e^{-\frac{T}{k}} V(x_{2,k}).$$

Without loss of generality, by homogeneity of the previous expression, we can assume that $V(x_{2,k}) \equiv 1$. By compactness, there exist a subsequence $(\alpha^{k_j})_{j \in \mathbb{N}}$ weak-$\star$ converging to some $\alpha^\star \in \mathcal{A}^\star$ and such that $x_{2,k_j} \to y_\star$ for some $y_\star \in \{y \in \mathbb{R}^r \mid V(y) = 1\}$ and $\alpha_i^\star$ is not identically zero for each $i = 1, \ldots, m$. It follows from Proposition 1.3 that $V(x_2(T; y_\star, \alpha^\star)) = V(y_\star) = 1$. Hence, according to (7.25),

$$\sum_{i=1}^{m} \int_0^T \alpha_i^\star(t)(\beta_i^\top x_2(t; y_\star, \alpha^\star))^2 dt = 0,$$

which means that

$$\alpha_i^\star(t)\beta_i^\top x_2(t; y_\star, \alpha^\star) = 0, \qquad i = 1, \ldots, m, \tag{7.26}$$

for almost every $t \in [0, T]$. In particular, according to (7.24), $x_2(t; y_\star, \alpha^\star) = e^{tA_2} y_\star$ for $t \in [0, T]$. Let us then rewrite (7.26) as

$$\alpha_i^\star(t)\beta_i^\top e^{tA_2} y_\star = 0, \qquad i = 1, \ldots, m.$$

By analyticity of $t \mapsto \beta_i^\top e^{tA_2} y_\star$ and because $\alpha_i^\star$ is nonzero on a set of positive measure, we deduce from Lemma A.13 that $\beta_i^\top e^{tA_2} y_\star = 0$ for every $i = 1, \ldots, m$ and every $t \in \mathbb{R}$, that is,

$$B_2^\top e^{tA_2} y_\star \equiv 0,$$

contradicting the controllability of the pair (A_2, B_2). □

Proposition 7.27 deals with control systems whose uncontrolled dynamics $\dot{x} = Ax$ are stable (even though possibly not asymptotically). The next theorem shows that the result can be extended to the case where the uncontrolled dynamics is equal to a nilpotent Jordan block.

Theorem 7.28 *Let us take A equal the matrix*

$$J_n = \begin{pmatrix} 0 & 1 & 0 & \cdots & 0 \\ \vdots & \ddots & \ddots & \ddots & \vdots \\ 0 & \cdots & 0 & 1 & 0 \\ 0 & \cdots & \cdots & 0 & 1 \\ 0 & \cdots & \cdots & \cdots & 0 \end{pmatrix},$$

and assume that $b = (0, \ldots, 0, 1)^\top$ is a column of B. If $\mathcal{A}$ is persistently exciting then (7.23) *is $\mathcal{A}$-stabilizable.*

Proof The proof relies on a compactness argument and a time-contraction procedure, transforming the integral persistent excitation constraint in a pointwise one. The limit system obtained with the time-contraction procedure can be shown to be stable using a suitable Lyapunov function, and an approximation argument makes it possible to conclude the stability of a time-contracted system from the stability of the limit one.

Let T and μ be as in Definition 7.26. Without loss of generality, $m = 1$ and B is made of the single column b. System (7.23) under the feedback law $u = -Kx$ with $K = \begin{pmatrix} k_1 & \cdots & k_n \end{pmatrix}$ can be written as

$$\begin{cases} \dot{x}_j = x_{j+1}, & j = 1, \ldots, n-1, \\ \dot{x}_n = -\alpha(t)(k_1 x_1 + \cdots + k_n x_n), \end{cases} \tag{7.27}$$

with $\int_t^{T+t} \alpha(s)ds \geq \mu$ for every $t \in [0, +\infty)$.

We claim that, given $\nu > 0$, K is a $\mathcal{A}$-stabilizer if and only if $K_\nu := \begin{pmatrix} \nu^n k_1 & \cdots & \nu k_n \end{pmatrix}$ is a $\mathcal{A}_\nu$-stabilizer, where

$$\mathcal{A}_\nu = \{\alpha_\nu(\cdot) := \alpha(\nu \cdot) \mid \alpha \in \mathcal{A}\}.$$

Indeed, define, for $\nu > 0$, $D_\nu = \operatorname{diag}(\nu^{n-1}, \ldots, \nu, 1)$. Using the relations $\nu D_\nu^{-1} J_n D_\nu = J_n$, $D_\nu b = b$, and $K_\nu = \nu K D_\nu$, one easily checks that, given a solution $t \mapsto x(t)$ of (7.27), the curve $t \mapsto x_\nu(t) := D_\nu^{-1} x(\nu t)$ satisfies

$$\dot{x}_\nu = J_n x_\nu - \alpha_\nu b K_\nu x_\nu.$$

This is the same system as (7.27), but with switching signal $\alpha_\nu \in \mathcal{A}_\nu$ and linear state feedback K_ν, hence the claim is proved.

Our goal is now to identify K that is a $\mathcal{A}_\nu$-stabilizer for ν large enough. Its existence and the previous claim will conclude the proof of the theorem.

Notice that if $(\nu_k)_{k\in\mathbb{N}}$ is a sequence of positive real numbers with $\nu_k \to +\infty$ as $k \to +\infty$ and $(\alpha^{(k)})_{k\in\mathbb{N}}$ is a sequence in $\mathcal{A}$, then $\alpha_{\nu_k}^{(k)}$ weak-$\star$ converges up to subsequence to some $\alpha_\star \in L^\infty(\mathbb{R}, [\mu/T, 1])$. Indeed, the sequence $(\alpha_{\nu_k}^{(k)})_{k\in\mathbb{N}}$

is bounded in L^∞ and thus weak-$\star$ converges up to extracting a subsequence. Moreover, for every interval $[a, b]$ in $[0, +\infty)$,

$$\begin{aligned}\lim_{k\to\infty}\int_{[0,+\infty)}\alpha_{\nu_k}^{(k)}(s)1_{[a,b]}(s)ds &= \lim_{k\to\infty}\int_{\lceil\frac{a\nu_k}{T}\rceil\frac{T}{\nu_k}}^{\lfloor\frac{b\nu_k}{T}\rfloor\frac{T}{\nu_k}}\alpha_{\nu_k}^{(k)}(s)ds\\ &\geq \lim_{k\to\infty}\left(\left\lfloor\frac{b\nu_k}{T}\right\rfloor-\left\lceil\frac{a\nu_k}{T}\right\rceil\right)\frac{\mu}{\nu_k}\\ &= (b-a)\frac{\mu}{T},\end{aligned}$$

where we used that the integral of an element α_ν of $\mathcal{A}_\nu$ over an interval of length T/ν is larger than or equal to μ/ν.

We are thus led to consider the switched system $(\Sigma_{\mathcal{S}_{\mathrm{arb}}(\mathcal{N}_K)})$, where the vector K is to be chosen and

$$\mathcal{N}_K = [\![J_n - \frac{\mu}{T}bK, J_n - bK]\!],$$

which may be seen as a limit system of

$$\dot{x} = J_n x - \alpha_\nu bKx, \qquad \alpha_\mu \in \mathcal{A}_\nu, \tag{7.28}$$

as ν tends to infinity. Indeed, as a consequence of Proposition 1.3, if $(\nu_k)_{k\in\mathbb{N}}$, $(\alpha^{(k)})_{k\in\mathbb{N}}$, and $\alpha_\star$ are as above, then the flow of (7.28) with $\nu = \nu_k$ and corresponding to the switching signal $\alpha_{\mu_k}^{(k)}$ converges uniformly on compact time-intervals towards the flow of $(\Sigma_{\mathcal{S}_{\mathrm{arb}}(\mathcal{N}_K)})$ corresponding to the switching signal $\alpha_\star$.

The existence of K such that $(\Sigma_{\mathcal{S}_{\mathrm{arb}}(\mathcal{N}_K)})$ is uniformly exponentially stable can be established through Proposition 3.15. Let us fix such a stabilizing K and a common Lyapunov function V for $(\Sigma_{\mathcal{S}_{\mathrm{arb}}(\mathcal{N}_K)})$. The convergence mentioned above allows to conclude that for $\tau > 0$ there exists $\eta \in (0, 1)$ such that for every ν large enough and every trajectory $x(\cdot)$ of (7.28),

$$V(x(t+\tau)) \leq \eta V(t), \qquad t \in [0, +\infty).$$

This concludes the proof of the existence of a $\mathcal{A}_\nu$-stabilizer for ν large, and hence the proof of the theorem. □

7.5 Stabilization by Switching

In this section we deal with linear systems of the type

$$\dot{x}(t) = A(t)x(t), \qquad t \geq 0, \quad x \in \mathbb{R}^n, \tag{$\Sigma_{\mathcal{M}}$}$$

where we assume that the $\mathcal{M}$-valued switching signal $A(\cdot)$ can be used as a control parameter, and we investigate the possibility of stabilizing the system by appropriately choosing $A(\cdot)$.

We start by formalizing some basic notions of stabilizability we will deal with and some related elementary properties. In the definition below we introduce a natural notion of stabilizability related to the use of open loop switching laws.

Definition 7.29 We say that $(\Sigma_{\mathcal{M}})$ is *pointwise stabilizable* if for any $x_0 \in \mathbb{R}^n$ there exists $A_{x_0} \in \mathcal{S}_{\text{arb}}(\mathcal{M})$ such that $\lim_{t\to\infty} \Phi_{A_{x_0}}(t, 0)x_0 = 0$.

Concerning the use of closed loop switching laws, we propose the following notion of stabilizability.

Definition 7.30 For $\delta > 0$ we say that a function $\mathbf{A}\colon \mathbb{R}^n \to \mathcal{M}$ homogeneous of degree zero is a *δ-stabilizing feedback* if for every $x_0 \in \mathbb{R}^n$, the trajectory $x(\cdot)$ starting from x_0 and defined by $\dot{x}(t) = \mathbf{A}(x(\delta k))x(t)$ on each interval $[\delta k, \delta(k+1))$, $k \in \mathbb{N}$, converges to 0.

In addition we say that $(\Sigma_{\mathcal{M}})$ is *uniformly feedback stabilizable* if there exists $\bar{\delta} > 0$ such that for every $\varepsilon > 0$ there exists $T > 0$ so that for every $\delta \in (0, \bar{\delta})$ there exists a δ-stabilizing feedback satisfying the following property: for every trajectory $x(\cdot)$ it holds $\|x(t)\| \leq \varepsilon \|x(0)\|$ for $t \geq T$.

The use of sampled controls in the previous definition is justified in practice, as it is consistent with the sampling process used in computer control. Moreover, from a technical perspective, sampled controls are crucial for guaranteeing existence and uniqueness of solutions of the system, as we do not assume any regularity property on the feedback $\mathbf{A}(\cdot)$.

In order to tackle the more general problem where one is interested in the minimal achievable rate of growth of the system, we now introduce numerical values that quantify the notion of stabilizability.

Definition 7.31 We introduce the following stabilizability indices:

- The *pointwise stabilization radius* of the switched system is defined as

$$\tilde{\mu}(\mathcal{M}) = \sup_{x \in \mathbb{R}^n} \tilde{\mu}_x(\mathcal{M}),$$

  where, for every $x \in \mathbb{R}^n$,

$$\tilde{\mu}_x(\mathcal{M}) = \inf \left\{ \beta \in \mathbb{R} \;\middle|\; \begin{array}{c} \exists A \in \mathcal{S}_{\text{arb}}(\mathcal{M}),\ M > 0 \text{ s.t.} \\ \|\Phi_A(t, 0)x\| \leq M e^{\beta t} \|x\|,\ \forall t \geq 0 \end{array} \right\}.$$

- Define

$$\hat{\mu}(\mathcal{M}) = \inf \left\{ \beta \in \mathbb{R} \;\middle|\; \begin{array}{c} \exists M > 0 \text{ s.t. } \|\Phi_A(t, 0)x\| \leq M e^{\beta t} \|x\| \text{ for every } x \in \mathbb{R}^n, \\ t \geq 0, \text{ and some } A \in \mathcal{S}_{\text{arb}}(\mathcal{M}) \text{ depending on } x \end{array} \right\}.$$

In the definition of $\tilde{\mu}_x(\mathcal{M})$ the constant M may depend on the initial condition. This is no more the case for $\hat{\mu}(\mathcal{M})$.

We now provide a analogous property to (1.8) in Proposition 1.20 for the stabilizability indices introduced in Definition 7.31.

Lemma 7.32 *We have that* $\tilde{\mu}(\mathcal{M} + \gamma \mathrm{Id}_n) = \tilde{\mu}(\mathcal{M}) + \gamma$ *and* $\hat{\mu}(\mathcal{M} + \gamma \mathrm{Id}_n) = \hat{\mu}(\mathcal{M}) + \gamma$.

Proof It is easy to see that there is a one-to-one correspondence between trajectories of $(\Sigma_{\mathcal{M}})$ and those of $(\Sigma_{\mathcal{M}+\gamma \mathrm{Id}_n})$ such that the quantity β in the definition of $\tilde{\mu}$ and $\tilde{\mu}'$ becomes $\gamma + \beta$. The thesis follows immediately. □

We now establish the equality of the two quantities introduced in Definition 7.31 and we will characterize the notion of pointwise stabilizability in terms of them.

Proposition 7.33 *The following conditions are equivalent:*

(i) $\hat{\mu}(\mathcal{M}) < 0$,
(ii) $\tilde{\mu}(\mathcal{M}) < 0$,
(iii) $(\Sigma_{\mathcal{M}})$ *is pointwise stabilizable.*

Proof (i) ⇒(ii). This is an obvious consequence of the definitions of $\tilde{\mu}$ and $\hat{\mu}$.

(ii) ⇒(iii). If $\tilde{\mu}(\mathcal{M}) < 0$ then $\tilde{\mu}_x(\mathcal{M}) < 0$ for every $x \in \mathbb{R}^n$. Thus for every $\mu \in (\tilde{\mu}_x(\mathcal{M}), 0)$ we can construct $A \in \mathcal{S}_{\mathrm{arb}}(\mathcal{M})$ such that $\|\Phi_A(t,0)x\| < Me^{\mu t}\|x\|$ for every $t \geq 0$ and, in particular, $\lim_{t\to\infty} \Phi_A(t,0)x = 0$.

(iii) ⇒(i). We start by considering initial conditions in the unit sphere S^{n-1}. Since $(\Sigma_{\mathcal{M}})$ is pointwise stabilizable for any $x \in S^{n-1}$ there exists a switching signal $A_x(\cdot)$ such that

$$\|\Phi_{A_x}(t_x, 0)x\| < \frac{1}{2},$$

for a large enough time t_x. By continuity, for a small enough open neighborhood U_x of x in S^{n-1}, one has $\|\Phi_{A_x}(t_x,0)y\| < \frac{1}{2}$ for every initial $y \in U_x$. By compactness, we can select a finite covering $U_1, \ldots, U_N$ of S^{n-1} such that for every $i = 1, \ldots, N$ there exist a switching signal $A_i(\cdot)$ and a time $t_i > 0$ with $\|\Phi_{A_i}(t_i,0)y\| < \frac{1}{2}$ for $y \in U_i$. Also, there exists $R_i \geq 1$ such that $\|\Phi_{A_i}(t,0)y\| \leq R_i$, for any $t \in [0, t_i]$ and $y \in U_i$. Letting $\tilde{U}_i = \{\alpha x \mid \alpha \in \mathbb{R} \setminus \{0\},\ x \in U_i\}$, we thus get that $\|\Phi_{A_i}(t_i,0)y\| < \frac{1}{2}\|y\|$ for every $y \in \tilde{U}_i$ and $\|\Phi_{A_i}(t,0)y\| \leq R_i\|y\|$ for every $t \in [0, t_i]$ and every $y \in \tilde{U}_i$.

Starting from any point $x \in \mathbb{R}^n$, by concatenating switching signals defined on intervals of length $t_{j(1)}, t_{j(2)}, \ldots$, where $t_{j(i)} \in \{t_1, \ldots t_N\}$ for $i \geq 1$, we can thus construct a switching law $A_x(\cdot)$ such that if $t \in I_k := [\sum_{i=1}^{k-1} t_{j(i)}, \sum_{i=1}^{k} t_{j(i)}]$

then $\|\Phi_{A_x}(t,0)x\| < R_{j(k)}2^{-k+1}\|x\|$. Setting $\bar{t} = \max_{i=1,\dots,N} t_i$ and $\bar{R} = \max_{i=1,\dots,N} R_i$ we have that any time $t \in I_k$ satisfies $t \leq k\bar{t}$, so that

$$\|\Phi_{A_x}(t,0)x\| < R_{j(k)}2^{-k+1}\|x\| \leq \bar{R}2^{-\frac{t}{\bar{t}}+1}\|x\| = Me^{\mu t}\|x\|,$$

where $\mu = -(\log 2)/\bar{t}$ and $M = 2\bar{R}$. Thus $\hat{\mu}(\mathcal{M}) < 0$. □

As a direct consequence of Lemma 7.32 and Proposition 7.33 we deduce the following.

Corollary 7.34 *For every bounded set $\mathcal{M}$ of matrices in $M_n(\mathbb{R})$, $\tilde{\mu}(\mathcal{M}) = \hat{\mu}(\mathcal{M})$.*

Let us introduce the following suitable adaptation of the classical notion of control-Lyapunov function.

Definition 7.35 We say that an absolutely homogeneous function $V : \mathbb{R}^n \to [0, +\infty)$ is a *control-Lyapunov function for* $(\Sigma_{\mathcal{M}})$ if

(1) there exist two constants $0 < m \leq M$ such that $m\|x\| \leq V(x) \leq M\|x\|$ for every $x \in \mathbb{R}^n$,
(2) there exists $\mu < 0$ such that for every $x \in \mathbb{R}^n \setminus \{0\}$ there exists $A_x(\cdot)$ with $V(\Phi_{A_x}(t,0)x) < e^{\mu t}V(x)$ for $t \in [0,\infty)$.

If there exists a control-Lyapunov function for $(\Sigma_{\mathcal{M}})$ and μ is as in Item (2) above, then clearly $(\Sigma_{\mathcal{M}})$ is pointwise stabilizable and $\tilde{\mu}(\mathcal{M}) \leq \mu < 0$. If, in addition, the control-Lyapunov function is regular enough, then the system turns out to be uniformly feedback stabilizable, as stated by the result here below.

Proposition 7.36 *Assume that $\mathcal{M}$ is convex. If $(\Sigma_{\mathcal{M}})$ admits a Lipschitz continuous control-Lyapunov function V then $(\Sigma_{\mathcal{M}})$ is uniformly feedback stabilizable, and, for each $\delta > 0$ small enough, the δ-stabilizing feedback $\sigma(\cdot)$ can be taken piecewise constant in the following sense: $\mathbb{R}^n$ can be split in finitely many cones with Lipschitz continuous boundary and $\sigma(\cdot)$ is constant on each of them.*

Proof Given an initial point $x_0 \in \mathbb{R}^n$ let us consider the switching law $A_{x_0}(\cdot)$ as given by Definition 7.35. For $\delta > 0$ let us consider the matrix $\bar{A}_{x_0} = \frac{1}{\delta}\int_0^\delta A_{x_0}(s)ds$. Since $\mathcal{M}$ is convex we have that $\bar{A}_{x_0} \in \mathcal{M}$. Consider now the curve $y(\cdot) = \Phi_{A_{x_0}}(\cdot,0)x_0 - e^{t\bar{A}_{x_0}}x_0$. Then $y(\cdot)$ satisfies the equation

$$\dot{y}(t) = \bar{A}_{x_0}y(t) + (A_{x_0}(t) - \bar{A}_{x_0})\Phi_{A_{x_0}}(t,0)x_0$$

with initial condition $y(0) = 0$. The corresponding solution at time δ, given by the variation of constants formula (Lemma A.6), is

$$y(\delta) = e^{\delta\bar{A}_{x_0}}\int_0^\delta e^{-t\bar{A}_{x_0}}(A_{x_0}(t) - \bar{A}_{x_0})\Phi_{A_{x_0}}(t,0)x_0 dt.$$

Notice that

$$\begin{aligned} e^{-t\bar{A}_{x0}}(A_{x_0}(t) - \bar{A}_{x_0})\Phi_{A_{x_0}}(t,0)x_0 &= (e^{-t\bar{A}_{x0}} - \mathrm{Id}_n)(A_{x_0}(t) - \bar{A}_{x_0})\Phi_{A_{x_0}}(t,0)x_0 \\ &\quad + (A_{x_0}(t) - \bar{A}_{x_0})(\Phi_{A_{x_0}}(t,0)x_0 - x_0) \\ &\quad + (A_{x_0}(t) - \bar{A}_{x_0})x_0. \end{aligned}$$

By compactness of $\mathscr{M}$, and since the integral on $[0,\delta]$ of the last term in the above expression is zero, we have that there exists a constant $C > 0$ independent of x_0 such that, for all δ small enough,

$$\|y(\delta)\| \le C\delta^2\|x_0\|.$$

Let L be the Lipschitz constant of the Lyapunov function V and m, μ be as in Definition 7.35. Then

$$\begin{aligned} V(x_0) - V(e^{\delta\bar{A}_{x0}}x_0) &\ge V(x_0) - V(\Phi_{A_{x_0}}(\delta,0)x_0) - |V(\bar{x}(\delta)) \\ &\quad - V(\Phi_{A_{x_0}}(\delta,0)x_0)| \\ &\ge m(1 - e^{\mu\delta})\|x_0\| - CL\delta^2\|x_0\| \ge -\frac{\mu m}{2}\delta\|x_0\| \\ &\ge -\frac{\mu m}{2M}\delta V(x_0) \end{aligned}$$

for δ small enough, which implies that

$$V(e^{\delta\bar{A}_{x0}}x_0) \le \left(1 + \frac{\mu m}{2M}\delta\right)V(x_0) \le e^{\xi\delta}V(x_0)$$

for some $\xi < 0$ independent of δ (for instance, $\xi = \frac{m\mu}{2M}$). The feedback $\mathbf{A}(\cdot)$ defined as $\mathbf{A}(x) = \bar{A}_x$ is therefore δ-stabilizing. By absolute homogeneity of V, we can assume without loss of generality that $\mathbf{A}(\cdot)$ is homogeneous of degree zero.

Notice now that, by continuity of V, for every $x_0 \in S^{n-1}$, for every y in a small enough neighborhood of x_0,

$$V(e^{\delta\bar{A}_{x0}}y) \le e^{\frac{\xi}{2}\delta}V(y).$$

By absolute homogeneity of V, such inequality is true for every y in a positive cone containing x_0. By a simple compactness argument there exists a finite number of cones covering $\mathbb{R}^n$ in each of which the stabilizing feedback law may be taken constant and the boundary between such cones can be taken Lipschitz continuous. This concludes the proof of the proposition. □

We now provide a converse Lyapunov theorem for our class of systems.

Proposition 7.37 *Assume that $\mathcal{M}$ is compact and convex. For any $\mu > \tilde{\mu}(\mathcal{M})$ there exists an absolutely homogeneous, Lipschitz continuous function $V_\mu : \mathbb{R}^n \to [0, +\infty)$ which satisfies $V_\mu(\Phi_A(t, 0)x) \le e^{\mu t} V_\mu(x)$ for any $x \in \mathbb{R}^n$, $t > 0$, and some $A \in \mathcal{S}_{\text{arb}}(\mathcal{M})$ depending on x. In particular, $(\Sigma_{\mathcal{M}})$ is pointwise stabilizable if and only if it admits an absolutely homogeneous control-Lyapunov function.*

Proof We first note that, by taking $\lambda \in (\tilde{\mu}(\mathcal{M}), \mu)$, in view of the definition of $\hat{\mu}(\mathcal{M}) = \tilde{\mu}(\mathcal{M})$, there exists $C > 0$ such that

$$\inf_{B \in \mathcal{S}_{\text{arb}}(\mathcal{M})} \sup_{t \ge 0} \|\Phi_B(t, 0)x\| e^{-\lambda t} \le C\|x\|, \qquad \forall x \in \mathbb{R}^n.$$

In particular, for every $x \in \mathbb{R}^n$ there exists $A'_x \in \mathcal{S}_{\text{arb}}(\mathcal{M})$ such that

$$\|\Phi_{A'_x}(t, 0)x\| e^{-\lambda t} \le 2C\|x\|, \qquad \text{for every } t \ge 0. \tag{7.29}$$

Hence, the quantity

$$m_{x,T} := \inf_{B \in \mathcal{S}_{\text{arb}}(\mathcal{M})} \int_0^T \|\Phi_B(t, 0)x\| e^{-\mu t} dt$$

satisfies

$$m_{x,T} \le \int_0^T \|\Phi_{A'_x}(t, 0)x\| e^{-\mu t} dt \le \frac{2C}{\lambda - \mu}\|x\|, \qquad \forall x \in \mathbb{R}^n, \quad \forall T > 0. \tag{7.30}$$

Notice that $m_{x,T}$ is actually attained, as it follows from the weak-$\star$ compactness of $\mathcal{S}_{\text{arb}}(\mathcal{M})$ and Proposition 1.3. Hence, for every $T > 0$ and $x \in \mathbb{R}^n$, there exists $A^* \in \mathcal{S}_{\text{arb}}(\mathcal{M})$ such that

$$\int_0^T \|\Phi_{A^*}(s, 0)x\| e^{-\mu s} ds = m_{x,T}. \tag{7.31}$$

We claim that, for every $T > 0$ large enough and every $x \in \mathbb{R}^n$,

$$\|\Phi_{A^*}(T, 0)x\| e^{-\mu T} \le \frac{\|x\|}{2}. \tag{7.32}$$

To prove the claim, we observe that (7.31) implies the existence of $\tau \in [0, T]$ such that $\|\Phi_{A^*}(\tau, 0)x\| e^{-\mu\tau} = \frac{m_{x,T}}{T}$ so that, by (7.29), there exists a signal $A' = A'_{\Phi_{A^*}(\tau,0)x} \in \mathcal{S}_{\text{arb}}(\mathcal{M})$ such that the concatenation $\tilde{A} = A^*|_{[0,\tau]} * A'$ satisfies

$$\int_0^T \|\Phi_{\tilde{A}}(s, 0)x\| e^{-\mu s} ds \le \int_0^\tau \|\Phi_{A^*}(s, 0)x\| e^{-\mu s} ds$$

$$
\begin{aligned}
&+ \int_0^\infty \|\Phi_{A'}(s,0)\Phi_{A^*}(\tau,0)x\| e^{-\mu\tau} e^{-\mu s} ds \\
&\leq \int_0^\tau \|\Phi_{A^*}(s,0)x\| e^{-\mu s} ds + \frac{2C}{\lambda-\mu} \|\Phi_{A^*}(\tau,0)x\| e^{-\mu\tau} \\
&= \int_0^\tau \|\Phi_{A^*}(s,0)x\| e^{-\mu s} ds + \frac{2C m_{x,T}}{(\lambda-\mu)T},
\end{aligned}
$$

where we have used the second inequality in (7.30) and the definition of τ. We deduce that

$$
\begin{aligned}
\int_\tau^T \|\Phi_{A^*}(s,0)x\| e^{-\mu s} ds &= m_{x,T} - \int_0^\tau \|\Phi_{A^*}(s,0)x\| e^{-\mu s} ds \\
&\leq \int_0^T \|\Phi_{\tilde{A}}(s,0)x\| e^{-\mu s} ds - \int_0^\tau \|\Phi_{A^*}(s,0)x\| e^{-\mu s} ds \\
&\leq \frac{2C m_{x,T}}{(\lambda-\mu)T}.
\end{aligned}
$$

On the other hand, letting $\kappa := \max_{M\in\mathcal{M}} \|M\|$ and $f(s) := \|\Phi_{A^*}(s,0)x\| e^{-\mu s}$, we have $|\dot{f}(s)| \leq (\kappa + |\mu|) f(s)$ for almost every s, so that

$$
|f(T) - f(\tau)| \leq \int_\tau^T |\dot{f}(s)| ds \leq (\kappa + |\mu|) \int_\tau^T f(s) ds.
$$

Hence

$$
\begin{aligned}
\|\Phi_{A^*}(T,0)x\| e^{-\mu T} &\leq \|\Phi_{A^*}(\tau,0)x\| e^{-\mu\tau} + (\kappa + |\mu|) \int_\tau^T \|\Phi_{A^*}(s,0)x\| e^{-\mu s} ds \\
&\leq \frac{m_{x,T}}{T}\left(1 + (\kappa + |\mu|)\frac{2C}{(\lambda-\mu)}\right) \\
&\leq \frac{2C}{(\lambda-\mu)T}\left(1 + (\kappa + |\mu|)\frac{2C}{(\lambda-\mu)}\right)\|x\|,
\end{aligned}
$$

where we used (7.30) and (7.31). The claim (7.32) follows by taking T large enough, independent of x.

We now show that the function

$$
V_\mu(x) := \min_{B\in\mathcal{S}_{\mathrm{arb}}(\mathcal{M})} \int_0^T \|\Phi_B(s,0)x\| e^{-\mu s} ds, \qquad x \in \mathbb{R}^n,
$$

satisfies the conclusions of the proposition. By what precedes V_μ is well defined (that is, finite) and it is absolutely homogeneous as an immediate consequence of its definition.

Let us prove that V_μ is Lipschitz continuous. Note that the application $x \mapsto \int_0^T \|\Phi_B(s,0)x\| e^{-\mu s} ds$ is Lipschitz continuous, with Lipschitz constant bounded, uniformly with respect to $B \in \mathcal{S}_{\text{arb}}(\mathcal{M})$, by $T \max\{1, e^{-\mu T}\} e^{\kappa T}$ by (1.1). As the infimum on families of uniformly Lipschitz continuous functions is Lipschitz continuous with the same Lipschitz constant we obtain the Lipschitz continuity of V_μ.

Let us show that for every $x \in \mathbb{R}^n$ there exists a switching law $A \in \mathcal{S}_{\text{arb}}(\mathcal{M})$ such that $t \mapsto e^{-\mu t} V_\mu(\Phi_A(t,0)x)$ is nonincreasing. Notice that it is enough to prove the monotonicity for t in a right neighborhood of 0. Take A^* as in (7.31). Then, for $h \geq 0$,

$$\begin{aligned}
V_\mu(\Phi_{A^*}(h,0)x) &= \min_{B \in \mathcal{S}_{\text{arb}}(\mathcal{M})} \int_0^T \|\Phi_B(s,0)\Phi_{A^*}(h,0)x\| e^{-\mu s} ds \\
&\leq \int_0^T \|\Phi_{A^*}(s+h,0)x\| e^{-\mu s} ds \\
&= e^{\mu h} V_\mu(x) + \int_T^{T+h} \|\Phi_{A^*}(s,0)x\| e^{-\mu s} ds \\
&\quad - \int_0^h \|\Phi_{A^*}(s,0)x\| e^{-\mu s} ds \\
&\leq e^{\mu h} V_\mu(x),
\end{aligned}$$

where the last step is valid provided that h is smaller than a constant value $\bar{h} > 0$ depending on κ and μ, as it follows from (7.32).

We are left to prove the last part of the statement. We already noticed that if $(\Sigma_\mathcal{M})$ admits a control-Lyapunov function then it is pointwise stabilizable. To prove the converse, it is enough to observe that if $(\Sigma_\mathcal{M})$ is pointwise stabilizable then μ may be chosen negative (Proposition 7.33), in which case the function V_μ is a control-Lyapunov function for $(\Sigma_\mathcal{M})$. □

An immediate consequence of the above result is the equivalence between the pointwise stabilizability and the uniform feedback stabilizability as stated in the following corollary.

Corollary 7.38 *Let $\mathcal{M}$ be compact and convex. Then the three equivalent conditions in Proposition 7.33 are also equivalent to the following:*

(iv) $(\Sigma_\mathcal{M})$ is uniformly feedback stabilizable.

Proof Clearly, if $(\Sigma_\mathcal{M})$ is uniformly feedback stabilizable then it is also pointwise stabilizable. The other implication trivially follows from Propositions 7.36 and 7.37. □

7.6 Notes and References

The contents of Sect. 7.1 are partly based on [SGL02, Yan02], which contain the first results describing the structure of the reachable and observable sets from the origin associated with a switched linear control system with class of signals equal to $\mathcal{S}_{\text{pc}}(\mathcal{N})$ where $\mathcal{N}$ is made up of finitely many triples (A, B, C). The proof we provide here (see Theorem 7.7) is simpler and in the spirit of the standard proof of the Kalman criterion for linear control systems. In particular the observation made in Lemma 7.4 appears to be new in the context of switched systems. Note that the argument of [SGL02] is constructive from the motion planning point of view, unlike the proof given here. Theorem 7.10 extends Theorem 7.7 to more general classes of switching signals and appeared first in [CMS17a].

Section 7.2 is largely inspired from [Pet07], which contains more detailed results on the realization theory of linear switched systems.

Section 7.3 is borrowed from [CMS17a]. The latter paper aimed at answering an open problem posed by Hespanha in [Hes03], which was dealing more specifically with the class of dwell-time signals $\mathcal{S}_{\text{dw},\tau}(\mathcal{M})$.

The stabilization problem studied in Sect. 7.4 has been introduced in [LCBC05]. Partial solutions to this question can be found in the setting of identification and stabilization of time-dependent linear systems [ABJ+86, LPM02]. Proposition 7.27 has appeared in the form stated here in [CCLS08] (see also [ABJ+86] for en earlier equivalent result expressed as an identification property). Theorem 7.28 has been proved in the case $n = 2$ in [CCLS08] and for a general n in [CS10]. It actually holds whenever A has only eigenvalues with nonpositive real part [CS10, CMS13], even in the case of delayed state feedback [Maz14]. In [CS10] it has also been proved that, in the setting of Theorem 7.28, if the ratio μ/T is small enough, the rate of exponential decay that can be obtained by choosing the feedback K is uniformly bounded. This implies that even if the pair $(J_n + \lambda \text{Id}_n, B)$ is controllable for every $\lambda \in \mathbb{R}$, the corresponding persistently exciting system is not $\mathcal{A}$-stabilizable for $\lambda > 0$ large enough. The arbitrary rate of exponential decay can be recovered if we impose an additional Lipschitz continuity constraint on the signals, as proved in [MCS13] in the case $n = 2$ (see also [CCS14] for more remarks on stabilizability with arbitrary exponential rate in the case of intermittent damping subject to persistent excitation constraints). Infinite dimensional extensions of the stabilizability results presented in Sect. 7.4 can be found in [CMS17b, GS10, HST12].

The stabilization issues for switched linear systems dealt with in Sect. 7.5 may be considered as particular instances of stabilization problems for continuous-time nonlinear systems. In particular, let us mention that in the general framework of nonlinear control systems many noteworthy results have been obtained in the past decades, dealing with necessary and sufficient conditions for asymptotic controllability and stabilizability as well as existence and properties of control-Lyapunov functions, see for instance [Art83, Bro83, CLSS97, Son83] to cite a few. In several cases these results directly apply to the switched systems setting.

Concerning switched linear systems, we note that in dimension two and in presence of only two modes, an approach analogous to the one of Chap. 2 allows one to express stabilizability conditions in terms of finitely many quantities explicitly depending on the coefficients of the modes through elementary functions (see, e.g. [Con13]). In higher dimension, cf. [WPD98], a simple condition given next yields the design of a stabilizing state-dependent switching law for a switched linear system $(\Sigma_{\mathcal{M}})$ with $\mathcal{M} = \{M_1, \dots, M_\ell\}$: if there exists $M \in \mathrm{co}(\mathcal{M})$ which is Hurwitz, then there exists a symmetric positive-definite matrix P such that $M^\top P + PM < 0$, so that the sets $\Omega_i = \{x \in \mathbb{R}^n \mid x^\top(M_i^\top P + PM_i)x < -\varepsilon|x|^2\}$ provide an open covering of $\mathbb{R}^n$ for $\varepsilon > 0$ small enough. A stabilizing hysteresis law can then be constructed by setting $A(t) = M_i$, for some i such that $x(t) \in \Omega_i$, and assuming that switchings may only happen at the boundary of the regions Ω_i. In a similar spirit, among the approaches considered in the framework of switched linear systems, we mention the so-called min-switching strategies [MBs96, GC06], where the stabilizing state-dependent switching law is chosen so that to minimize a multiple (quadratic) Lyapunov function. Again, hysteresis switching laws may introduced also in this case in order to avoid chattering, see e.g., [SG11, Section 4.3]. Note that the approaches mentioned above are examples of logic-based methods, i.e., the corresponding controllers make use of logic-driven elements, see e.g. [Mor95, Lib03].

Section 7.5 is adapted from [JM17], although the latter reference mainly focuses on discrete-time switched linear systems. The use of *sample-and-hold* feedback controls, as in Definition 7.30, in combination with control-Lyapunov functions has been adopted for instance in [CLSS97]. A result analogous to Proposition 7.37 may be found in [SG11, Section 4.3.1], while converse Lyapunov theorems for the stabilization (or asymptotic controllability) of more general classes of nonlinear control systems are present for instance in [KT04, Rif02].

Appendix A
Complementary Definitions and Results

In this appendix we explicitly state classical results needed in the book, whose topics range from differential geometry to abstract algebra.

Let us start by recalling the definitions of Lipschitz continuous and absolutely continuous functions.

Definition A.1 (Lipschitz Continuity) A function $f: U \to \mathbb{R}^n$, with $U \subseteq \mathbb{R}^m$, is *locally Lipschitz continuous* if for every compact set $K \subset U$ there exists $L_K > 0$ such that $|f(x_1) - f(x_2)| \leq L_K|x_1 - x_2|$ if $x_1, x_2 \in K$; f is said to be *(globally) Lipschitz continuous* if there exists $L > 0$ such that one can take $L_K = L$ for every compact set $K \subset U$.

Definition A.2 (Absolute Continuity) Let I be an interval of $\mathbb{R}$. A function $\varphi: I \to \mathbb{R}^n$ is *absolutely continuous* on I if it is differentiable almost everywhere, its derivative φ' is in $L^1_{\text{loc}}(I, \mathbb{R}^n)$, and

$$\varphi(t_1) - \varphi(t_0) = \int_{t_0}^{t_1} \varphi'(s)ds$$

for every $t_0, t_1 \in I$.

An equivalent definition of absolute continuity (see e.g., [Rud87, Theorem 7.20]) goes as follows: Let I be an interval of $\mathbb{R}$. A function $\varphi: I \to \mathbb{R}^n$ is *absolutely continuous* on I if for every bounded subinterval $J \subset I$ and $\varepsilon > 0$, there exists $\delta > 0$ so that

$$\sum_{i=1}^{k} |\varphi(\beta_i) - \varphi(\alpha_i)| < \varepsilon$$

Y. Chitour et al., *Dynamics and Stability of Continuous-Time Switched Linear Systems*, PNLDE Subseries in Control 105,
https://doi.org/10.1007/978-3-031-82982-6

for any positive integer k and any disjoint collection of segments $(\alpha_1, \beta_1), \ldots,$ (α_k, β_k) in J whose lengths satisfy $\sum_{i=1}^{k}(\beta_i - \alpha_i) < \delta$. As a consequence, any locally Lipschitz continuous function $f: I \to \mathbb{R}^n$ is absolutely continuous on I and its derivative is in $L^\infty_{\mathrm{loc}}(I, \mathbb{R}^n)$. In the more general case of Lipschitz functions defined on an Euclidean space we have the following result, see, e.g., [EG15, Theorem 3.2].

Theorem A.3 (Rademacher's Theorem) *Every locally Lipschitz continuous function from an open subset of $\mathbb{R}^m$ to $\mathbb{R}^n$ is differentiable almost everywhere.*

The natural notion of solution for the controlled dynamics considered in this book is the following.

Definition A.4 (Carathéodory Solutions) Let I be an interval of $\mathbb{R}$ and $f: I \times \mathbb{R}^n \to \mathbb{R}^n$. Then, any absolutely continuous function $x: J \to \mathbb{R}^n$, with J subinterval of I, satisfying

$$\dot{x}(t) = f(t, x(t)) \tag{A.1}$$

for almost every $t \in J$ is said to be a Carathéodory solution of the differential equation (A.1) in J.

A general result concerning existence and uniqueness of solutions of non-autonomous differential equations (see, e.g., [Son98, Appendix C]) is the following.

Theorem A.5 (Local Existence and Uniqueness of Carathéodory Solutions) *Let $0 \leq t_0 < t_1 < +\infty$ and consider $f: [t_0, t_1] \times \mathbb{R}^n \to \mathbb{R}^n$ satisfying:*

- *$t \mapsto f(t, x)$ is measurable for every $x \in \mathbb{R}^n$;*
- *$x \mapsto f(t, x)$ is locally Lipschitz continuous for every $t \in [t_0, t_1]$;*
- *f is locally bounded, i.e., for every compact $K \subset \mathbb{R}^n$ there exists $C_K > 0$ such that $|f(t, x)| \leq C_K$ if $x \in K$ and $t \in [t_0, t_1]$.*

Then, for every $\bar{t} \in [t_0, t_1]$ and every $\bar{x} \in \mathbb{R}^n$ there exists an open neighborhood I of $\bar{t}$ in $[t_0, t_1]$ such that the ordinary differential equation

$$\dot{x}(t) = f(t, x(t))$$

admits a unique Carathéodory solution in I satisfying $x(\bar{t}) = \bar{x}$. Furthermore, the solution $t \mapsto x(t)$ is defined on the whole interval $[t_0, t_1]$ if C_K and L_K can be chosen independently of the compact K.

The previous result applies in particular in the case in which $f(t, x) = A(t)x + b(t)$, with $A \in L^\infty([t_0, t_1], M_n(\mathbb{R}))$, $b \in L^\infty([t_0, t_1], \mathbb{R}^n)$. Next, we state a classical result which is used extensively throughout the book.

Lemma A.6 (Variation of Constants Formula) *Let* $0 \leq t_0 < t_1 < +\infty$ *and consider* $A \in L^\infty([t_0, t_1], M_n(\mathbb{R}))$, $b \in L^\infty([t_0, t_1], \mathbb{R}^n)$, *and* $x_0 \in \mathbb{R}^n$. *Then the solution* $x\colon [t_0, t_1] \to \mathbb{R}^n$ *of* $\dot{x}(t) = A(t)x(t) + b(t)$ *such that* $x(t_0) = x_0$ *satisfies*

$$x(t) = \Phi_A(t, t_0)x_0 + \int_{t_0}^{t} \Phi_A(t, s)b(s)ds, \qquad t \in [t_0, t_1],$$

where $t \mapsto \Phi_A(t, t_0) \in M_n(\mathbb{R})$ *is such that for every* $x \in \mathbb{R}^n$ *the curve* $t \mapsto \Phi_A(t, t_0)x$ *is the solution of the Cauchy problem* $\dot{x}(t) = A(t)x(t)$, $x(t_0) = x$.

A classical result used to estimate the growth of solutions of differential equations is the following.

Theorem A.7 (Grönwall's Lemma) *Let* $f\colon [0, T] \to [0, \infty)$ *be a continuous function,* $a, b \geq 0$, *and assume that*

$$f(t) \leq a + b \int_0^t f(s)ds, \quad \textit{for } t \in [0, T].$$

Then

$$f(t) \leq ae^{bt}, \quad \textit{for } t \in [0, T].$$

The following result is usually applied to prove the convergence to the origin of solutions of differential equations.

Theorem A.8 (Barbalat's Lemma) *Let* $f\colon [0, \infty) \to \mathbb{R}$ *be Lipschitz continuous and such that the improper integral* $\int_0^\infty f(t)\,dt$ *exists and is finite. Then* $\lim_{t\to\infty} f(t) = 0$.

At several places in the book (see for instance Fenichel's lemma 1.25), we need compactness results in functional spaces (see Theorem A.11 below). For that purpose, one usually relies on specific notions of convergence such as the one defined next.

Definition A.9 (Weak-⋆ Convergence and Weak-⋆ Closure) A sequence $(\varphi_k)_{k\in\mathbb{N}}$ of elements in $L^\infty([0, +\infty), \mathbb{R}^n)$ is said to *weak-⋆ converge* to $\varphi \in L^\infty([0, +\infty), \mathbb{R}^n)$ if, for every $\psi \in L^1([0, +\infty), \mathbb{R}^n)$, the sequence $(\int_0^{+\infty} \varphi_k(t)^\top \psi(t)dt)_{k\in\mathbb{N}}$ converges to $\int_0^{+\infty} \varphi(t)^\top \psi(t)dt$. The *weak-⋆ closure* of a subset $\mathcal{F}$ of $L^\infty([0, +\infty), \mathbb{R}^n)$ is the smallest set $\mathcal{F}^\star \supset \mathcal{F}$ such that for every weak-⋆ converging sequence of elements of $\mathcal{F}^\star$ the corresponding limit is in $\mathcal{F}^\star$.

We next provide a useful criterion to check weak-⋆ convergence, which easily follows from the density of piecewise constant functions with compact support in L^1.

Lemma A.10 *Consider a bounded sequence* $(\varphi_k)_{k\in\mathbb{N}}$ *and* φ *in* $L^\infty([0,+\infty),\mathbb{R}^n)$. *Then* φ_k *weak-*$\star$ *converges to* φ *if and only if*

$$\lim_{k\to\infty}\int_{t_1}^{t_2}\varphi_k(s)\,ds=\int_{t_1}^{t_2}\varphi(s)\,ds,\qquad \text{for every } t_2>t_1\geq 0.$$

In particular, given a finite interval $I\subset[0,\infty)$, *a bounded sequence* $(\varphi_k)_{k\in\mathbb{N}}$ *and* φ *in* $L^\infty(I,\mathbb{R}^n)$, *if* φ_k *converges to* φ *in* L^1 *then it also weak-*$\star$ *converges to* φ *in* L^∞.

Here follows the main weak-$\star$ compactness result needed in the book. While we state it in the case of L^∞ functions defined on $[0,+\infty)$, notice that it holds in much greater generality (see, e.g., Corollary 3.30 and Theorems 4.13–4.14 in [Bre11]).

Theorem A.11 (Banach–Alaoglu) *Let* Ω *be a compact convex subset of* $\mathbb{R}^n$ *and consider a sequence* $(\varphi_k)_{k\in\mathbb{N}}$ *in* $L^\infty([0,+\infty),\Omega)$. *Then there exists a subsequence of* $(\varphi_k)_{k\in\mathbb{N}}$ *that weak-*$\star$ *converges to some* $\varphi\in L^\infty([0,+\infty),\Omega)$.

The next theorem provides a convergence result (up to a subsequence) for continuous functions.

Theorem A.12 (Ascoli–Arzelà) *Let* $T>0$ *and consider a sequence* $(f_k)_{k\in\mathbb{N}}$ *in* $C([0,T],\mathbb{R})$. *Assume that* $(f_k)_{k\in\mathbb{N}}$ *is uniformly bounded (i.e., bounded in* $C([0,T],\mathbb{R})$ *equipped with the norm* $\|\cdot\|_\infty$*) and that it is equicontinuous (i.e., for every* $\varepsilon>0$ *there exists* $\delta>0$ *such that for every* $k\in\mathbb{N}$, $|f_k(s)-f_k(t)|<\varepsilon$ *whenever* $|s-t|<\delta$ *for* $s,t\in[0,T]$*). Then there exists a subsequence of* $(f_k)_{k\in\mathbb{N}}$ *converging uniformly to some function in* $C([0,T],\mathbb{R})$.

The following lemma states an elementary property about analytic functions, easily following from the fact that if all the derivatives of an analytic function vanish at a common point, then the function is identically equal to zero.

Lemma A.13 *Let* Ω *be a nonempty connected open subset of* $\mathbb{R}^n$ *and consider an analytic function* $\varphi\colon\Omega\to\mathbb{R}$. *Then either* φ *is constant or every level set of* φ *(i.e., every set of the type* $\{x\in\Omega\mid\varphi(x)=c\}$ *for some* $c\in\mathbb{R}$*) is of zero measure (and, in the case* $n=1$, *made of isolated points).*

The following geometrical result is used at several places in the book and is a finite-dimensional version of Hahn–Banach theorem.

Theorem A.14 (Hyperplane Separation Theorem) *Let* Ω_1,Ω_2 *be two disjoint nonempty convex subsets of* $\mathbb{R}^n$. *Then*

1. *if* Ω_1,Ω_2 *are closed and at least one of the two is compact, then there exist* $v\in\mathbb{R}^n$, $\eta_1,\eta_2\in\mathbb{R}$ *with* $\eta_1<\eta_2$ *such that* $v^\top x_1<\eta_1$ *and* $v^\top x_2>\eta_2$ *for every* $x_1\in\Omega_1$ *and* $x_2\in\Omega_2$;
2. *if* Ω_1 *is open, then there exist* $v\in\mathbb{R}^n$ *and* $\eta\in\mathbb{R}$ *such that* $v^\top x_1<\eta$ *and* $v^\top x_2\geq\eta$ *for every* $x_1\in\Omega_1$ *and* $x_2\in\Omega_2$.

The following proposition collects several basic properties of convex analysis used throughout the book. They can be retrieved from [Roc70].

Proposition A.15 *Let $V : \mathbb{R}^n \to \mathbb{R}$ be a convex function. For every $x \in \mathbb{R}^n$, define the* subdifferential of V at x *as*

$$\partial V(x) = \{l \in \mathbb{R}^n \mid l^\top(y - x) \leq V(y) - V(x), \ \forall y \in \mathbb{R}^n\}.$$

1. *[Roc70, Theorem 25.1] V is differentiable at $x \in \mathbb{R}^n$ if and only if $\partial V(x)$ is a singleton, in which case $\partial V(x) = \{\nabla V(x)\}$.*
2. *[Roc70, Theorem 10.4] V is locally Lipschitz continuous (hence almost everywhere differentiable, by Theorem A.3 above).*
3. *[Roc70, Page 215] For every $x \in \mathbb{R}^n$, $\partial V(x)$ is compact and convex.*
4. *[Roc70, Theorem 24.5] If x is a differentiability point of V and $(x_k)_{k\in\mathbb{N}}$ is a sequence of points converging to x, then every sequence $(l_k)_{k\in\mathbb{N}}$ with $l_k \in \partial V(x_k)$ converges to $\nabla V(x)$.*
5. *[Roc70, Theorem 25.6] Given $x \in \mathbb{R}^n$, the set $\{x\} \times \partial V(x)$ is equal to the closure of the convex hull of the limit points of sequences $(y_k, \nabla V(y_k))_{k\in\mathbb{N}}$, where each y_k is a differentiability point of V.*
6. *[Roc70, Theorem 24.7] For every non-empty compact set $K \subset \mathbb{R}^n$ the set $\cup_{x\in K} \partial V(x)$ is non-empty and compact.*
7. *[Roc70, Theorem 10.8] If a sequence of convex functions $(W_k)_{k\in\mathbb{N}}$ converges to V pointwise on a dense subset of $\mathbb{R}^n$ then $(W_k)_{k\in\mathbb{N}}$ converges to V uniformly on every compact set.*

Let us now recall some definitions and results of differential geometry used in the book.

The next result guarantees pathwise connectedness by analytic paths in a connected analytic manifold. Its proof can be found in [Tho37, Sections 6–9].

Lemma A.16 *Let $\mathcal{M}$ be a finite-dimensional connected analytic manifold. Then, given any two points $p_0, p_1 \in \mathcal{M}$, there exists an analytic curve $\gamma : [0, 1] \to \mathcal{M}$ such that $\gamma(0) = p_0$ and $\gamma(1) = p_1$.*

Let us recall here below the definition of Lie algebra of vector fields, which generalizes the notion of Lie algebra of matrices.

Definition A.17 Let $\mathcal{M}$ be a smooth (finite-dimensional) manifold. Denote by $\mathrm{Vec}(\mathcal{M})$ the linear space of smooth vector fields on $\mathcal{M}$. Given $f, g \in \mathrm{Vec}(\mathcal{M})$, the *Lie bracket of f and g* is the vector field $[f, g]$ defined, in any coordinate system, by $[f, g](x) = \frac{d}{dx} f(x)g(x) - \frac{d}{dx} g(x)f(x)$. The Lie algebra generated by a subset $\mathcal{F}$ of $\mathrm{Vec}(\mathcal{M})$, denoted by $\mathrm{Lie}(\mathcal{F})$, is the minimal subspace of $\mathrm{Vec}(\mathcal{M})$ containing $\mathcal{F}$ and such that $[f, g] \in \mathrm{Lie}(\mathcal{F})$ for every $f, g \in \mathrm{Lie}(\mathcal{F})$.

Given a vector field f on $\mathcal{M}$, an initial condition $q \in \mathcal{M}$, and a time $t \in \mathbb{R}$, we denote by $e^{tf}(q)$ the integral curve of f starting at q evaluated at time t, if well-defined. We say that f is *complete* if $e^{tf} : \mathcal{M} \to \mathcal{M}$ is defined for all $t \in \mathbb{R}$. Using this definition, we can state the following two classical results in geometric control theory (see, e.g., [AS04]).

Theorem A.18 (Krener's Theorem) *Let $\mathcal{M}$ be a smooth manifold of dimension $n \in \mathbb{N}$ and $\mathcal{F}$ be a family of complete smooth vector fields on $\mathcal{M}$. Fix $x \in \mathcal{M}$ and assume that $\{f(x) \mid f \in \mathrm{Lie}(\mathcal{F})\} = T_x\mathcal{M}$. Then, for every $T > 0$ there exist $f_1, \dots, f_n \in \mathcal{F}$ and $t_1, \dots, t_n > 0$ such that $t_1 + \cdots + t_n < T$ and the differential of $(s_1, \dots, s_n) \mapsto e^{s_n f_n} \circ \cdots \circ e^{s_1 f_1}(x)$ is invertible at $(t_1, \dots, t_n)$. In particular, the set of points of the type $e^{s_n f_n} \circ \cdots \circ e^{s_1 f_1}(x)$, $s_1, \dots, s_n > 0$ has nonempty interior in $\mathcal{M}$.*

Theorem A.19 (Orbit Theorem) *Let $\mathcal{M}$ be an analytic finite-dimensional manifold and $\mathcal{F}$ be a family of complete analytic vector fields on $\mathcal{M}$. Fix $x \in \mathcal{M}$ and define*

$$O_x = \{e^{s_k f_k} \circ \cdots \circ e^{s_1 f_1}(x) \mid k \in \mathbb{N},\ f_1, \dots, f_k \in \mathcal{F},\ s_1, \dots, s_k \in \mathbb{R}\}.$$

Endow O_x with the orbit topology *having as basis the sets of the form $e^{s_k f_k} \circ \cdots \circ e^{s_1 f_1}(\Omega)$, where $k \in \mathbb{N}$, $f_1, \dots, f_k \in \mathcal{F}$, $s_1, \dots, s_k \in \mathbb{R}$, and Ω is an open subset of $\mathbb{R}^k$. Then O_x is an immersed analytic submanifold of $\mathcal{M}$ and $T_y O_x = \{f(y) \mid f \in \mathrm{Lie}(\mathcal{F})\}$ for every $y \in O_x$.*

In the special case where $f : x \mapsto Mx$ is a linear vector field, e^{tf} is the linear map associated with the matrix e^{tM}. We can associate with a set of matrices $\mathscr{M}$ the set

$$G_{\mathscr{M}} = \{e^{s_k M_k} \cdots e^{s_1 M_1} \mid k \in \mathbb{N},\ M_1, \dots, M_k \in \mathscr{M},\ s_1, \dots, s_k \in \mathbb{R}\},$$

which is the orbit through the identity of the system $\dot{X} = M(t)X$ in $M_n(\mathbb{R})$, where the control $M(\cdot)$ takes values in $\mathscr{M}$. Then $G_{\mathscr{M}}$ is a connected Lie subgroup of $\mathrm{GL}(n)$. The action of the group $G_{\mathscr{M}}$ on S^{n-1} has the following useful property.

Theorem A.20 (Theorem 2 in [BS20]) *Let G be a connected Lie subgroup of* $\mathrm{GL}(n)$. *Let $\varphi : G \times S^{n-1} \to S^{n-1}$ be defined by*

$$\varphi(M, x) = \frac{Mx}{\|Mx\|}.$$

Then there exists $x_0 \in S^{n-1}$ such that $\{\varphi(M, x_0) \mid M \in G\}$ is compact in S^{n-1}.

Measures can be easily defined on Lie groups, seen as submanifolds of linear matrix spaces. Furthermore, the group structure suggests to seek for measures which additionnally satisfy an invariance condition with respect to (left and/or right) multiplication by an element of the group. Such measures are called Haar measures, and they can be more generally defined for groups endowed with an Hausdorff topology, see e.g. [DS14]. The following result is discussed, e.g., in [HN12, Theorem 10.4.4, Definition 10.4.5 and Remark 10.4.6].

Theorem A.21 *Let G be a Lie group. Then, up to a positive scalar multiple, there exists a unique nontrivial measure m on G, endowed with the σ-algebra of its*

Borel sets, which is finite over compact sets and invariant with respect to right translations, i.e., such that $m(\mathcal{N}M) = m(\mathcal{N})$ for every $M \in G$ and every Borel set $\mathcal{N}$ of G. The measure m is said to be the (right) Haar measure *of G.*[1]

The following statement of the Pontryagin maximum principle can be found in various textbooks dealing with optimal control theory such as [Lib11, Section 4.1.2 and Section 4.3.1] and [BP07, Theorem 6.4.1].

Theorem A.22 (Pontryagin Maximum Principle for Optimal Control Problems in Mayer Form) *Let $U \subset \mathbb{R}^m$. Assume that $f\colon \mathbb{R}^n \times U \to \mathbb{R}^n$ is continuous and continuously differentiable with respect to the first variable, $K\colon \mathbb{R}^n \to \mathbb{R}$ is continuously differentiable, and $x_0 \in \mathbb{R}^n$. Let $u_*\colon [0, T_*] \to U$ be an optimal control for the maximization problem*

$$\max_{u \in L^\infty([0,T],U)} K(x_u(T)),$$

where x_u denotes the solution of $\dot{x}(t) = f(x(t), u(t))$ with initial condition $x(0) = x_0$ and the final time T is free. Then, there exists a nontrivial Lipschitz row-vector $p_*\colon [0, T_*] \to \mathbb{R}^n$ such that the following conditions are satisfied:

- $\dot{p}_*(t) = -p_*(t)\frac{d}{dx} f(x_{u_*}(t), u_*(t))$ *for a.e.* $t \in [0, T_*]$;
- $p_*(t) f(x_{u_*}(t), u_*(t)) = \max_{u \in U} \{p_*(t) f(x_{u_*}(t), u)\} = 0$ *for a.e.* $t \in [0, T_*]$;
- $p_*(T_*) = \nabla K(x_{u_*}(T_*))$.

We finally list some miscellaneous results used in the book.

Lemma A.23 (Zorn's Lemma) *A partially ordered set containing upper bounds for every totally ordered subset necessarily contains at least one maximal element.*

Theorem A.24 (Cantor Intersection Theorem) *Let $(C_i)_{i \in \mathbb{N}}$ be a sequence of closed, compact, and nonempty subsets of a given topological space such that $C_{i+1} \subset C_i$ for every $i \in \mathbb{N}$. Then $\bigcap_{i \in \mathbb{N}} C_i \neq \emptyset$.*

Theorem A.25 (Gelfand's Formula) *For every $M \in M_n(\mathbb{R})$ one has*

$$\rho(M) = \lim_{k \to +\infty} \|M^k\|^{1/k}.$$

The result is actually independent of the choice of the matrix norm in the right-hand side.

Theorem A.26 (Cayley–Hamilton Theorem) *Let $p(\lambda) = \sum_{i=0}^n \alpha_i \lambda^i$ be the characteristic polynomial of a matrix $M \in M_n(\mathbb{C})$. Then $\sum_{i=0}^n \alpha_i M^i = 0$.*

---

[1] In the case where G is compact, m is normalized so that $m(G) = 1$.

Theorem A.27 (Sylvester's Criterion) *An* $n \times n$ *Hermitian matrix* $M = (m_{ij})_{i,j=1}^{n}$ *is positive definite if and only if, for every* $l = 1, \ldots, n$*, the determinant of* $(m_{ij})_{i,j=1}^{l}$ *is positive.*

Theorem A.28 (Jordan–Hölder Theorem for R**-Modules, See, e.g., Theorem 13.7 in [CR62])** *Let* $\mathcal{M}$ *be a module over a ring* R.[2] *Consider a chain of inclusions*

$$\{0\} = \mathcal{M}_0 \subsetneq \mathcal{M}_1 \subsetneq \cdots \subsetneq \mathcal{M}_{r-1} \subsetneq \mathcal{M}_r = \mathcal{M} \tag{A.2}$$

among submodules of $\mathcal{M}$ *and assume it to be a composition series for* $\mathcal{M}$*, that is, for* $i = 1, \ldots, r$*, the modules* $\mathcal{M}_i/\mathcal{M}_{i-1}$*, called factors of the chain, do not contain nontrivial submodules. Then any other composition series for* $\mathcal{M}$ *is equivalent to* (A.2)*, in the sense that they have the same length and their factors can be paired off in such a way that they are isomorphic.*

---

[2] i.e., there exists an operation

$$R \times \mathcal{M} \to \mathcal{M}$$
$$(r, m) \mapsto rm$$

such that for every $r, r_1, r_2 \in R$ and $m, m_1, m_2 \in \mathcal{M}$

$$r(m_1 + m_2) = rm_1 + rm_2, \qquad (r_1 + r_2)m = r_1 m + r_2 m,$$
$$(r_1 r_2)m = r_1(r_2 m), \qquad 1m = m,$$

where 1 is the unity of R.

References

[ABJ+86] B.D.O. Anderson, R.R. Bitmead, C.R. Johnson, Jr., P.V. Kokotović, R.L. Kosut, I.M.Y. Mareels, L. Praly, B.D. Riedle, *Stability of Adaptive Systems: Passivity and Averaging Analysis*. MIT Press Series in Signal Processing, Optimization, and Control, vol.8 (MIT Press, Cambridge, 1986)

[ABL12] A.A. Agrachev, Y. Baryshnikov, D. Liberzon, On robust Lie-algebraic stability conditions for switched linear systems. Syst. Control Lett. **61**(2), 347–353 (2012)

[AC84] J.P. Aubin, A. Cellina, *Differential Inclusions*. Grundlehren der Mathematischen Wissenschaften [Fundamental Principles of Mathematical Sciences], vol. 264 (Springer, Berlin, 1984). Set-valued maps and viability theory

[ADLS09] D. Angeli, P. De Leenheer, E.D. Sontag, Chemical networks with inflows and outflows: a positive linear differential inclusions approach. Biotechnol. Prog. **25**(3), 632–642 (2009)

[AKO86] L. Arnold, W. Kliemann, E. Oeljeklaus, Lyapunov exponents of linear stochastic systems, in *Lyapunov Exponents (Bremen, 1984)*. Lecture Notes in Mathematics, vol. 1186 (Springer, Berlin, 1986), pp. 85–125

[AL01] A.A. Agrachev, D. Liberzon, Lie-algebraic stability criteria for switched systems. SIAM J. Control Optim. **40**(1), 253–269 (2001)

[Ang99] D. Angeli, A note on stability of arbitrarily switched homogeneous systems (preprint, 1999)

[Art83] Z. Artstein, Stabilization with relaxed controls. Nonlinear Anal. Theory Methods Appl. **7**(11), 1163–1173 (1983)

[AS04] A.A. Agrachev, Y.L. Sachkov, *Control Theory from the Geometric Viewpoint*. Encyclopaedia of Mathematical Sciences (Springer, Berlin, 2004). Control Theory and Optimization, II

[Bar88a] N.E. Barabanov, An absolute characteristic exponent of a class of linear nonstationary systems of differential equations. Sibirsk. Mat. Zh. **29**(4), 12–22, 222 (1988)

[Bar88b] N.E. Barabanov, Lyapunov indicator of discrete inclusions, I–III. Autom. Remote Control **49**, 152–157, 2823–287, 558–565 (1988)

[Bar89] N. Barabanov, Method for the computation of the Lyapunov exponent of a differential inclusion. Autom. Remote Control **50**(4), 475–479 (1989)

[Bar94] N.E. Barabanov, On the Aizerman problem for a class of third-order time-dependent systems. Dokl. Akad. Nauk **334**(2), 154–155 (1994)

Y. Chitour et al., *Dynamics and Stability of Continuous-Time Switched Linear Systems*, PNLDE Subseries in Control 105,
https://doi.org/10.1007/978-3-031-82982-6

[Bar08] N.E. Barabanov, Asymptotic behavior of extremal solutions and structure of extremal norms of linear differential inclusions of order three. Linear Algebra Appl. **428**(10), 2357–2367 (2008)

[BB08] M. Balde, U. Boscain, Stability of planar switched systems: the nondiagonalizable case. Commun. Pure Appl. Anal. **7**(1), 1–21 (2008)

[BBM09] M. Balde, U. Boscain, P. Mason, A note on stability conditions for planar switched systems. Int. J. Control **82**(10), 1882–1888 (2009)

[BBMZ14] M. Benaïm, S.L. Borgne, F. Malrieu, P.A. Zitt, On the stability of planar randomly switched systems. Ann. Appl. Probab. **24**(1), 292–311 (2014)

[BC99] A. Bacciotti, F. Ceragioli, Stability and stabilization of discontinuous systems and nonsmooth Lyapunov functions. ESAIM Control Optim. Calc. Var. **4**, 361–376 (1999)

[BCK95] S. Bhattacharyya, H. Chapellat, L. Keel, *Robust Control: The Parametric Approach*. Prentice-Hall Information and System Sciences Series (Prentice Hall, Saddle River, 1995)

[BCS06] U. Boscain, G. Charlot, M. Sigalotti, Stability of planar nonlinear switched systems. Discrete Contin. Dyn. Syst. **15**(2), 415–432 (2006)

[BCV+15] F. Blanchini, P. Colaneri, M.E. Valcher, et al., Switched positive linear systems. Found. Trends® Syst. Control **2**(2), 101–273 (2015)

[BEGFB94] S. Boyd, L. El Ghaoui, E. Feron, V. Balakrishnan, *Linear Matrix Inequalities in System and Control Theory*. SIAM Studies in Applied Mathematics (Society for Industrial and Applied Mathematics, Philadelphia, 1994)

[BJ11] M. Balde, P. Jouan, Geometry of the limit sets of linear switched systems. SIAM J. Control Optim. **49**(3), 1048–1063 (2011)

[BM99] F. Blanchini, S. Miani, A new class of universal Lyapunov functions for the control of uncertain linear systems. IEEE Trans. Autom. Control **44**(3), 641–647 (1999)

[BM02] T. Bousch, J. Mairesse, Asymptotic height optimization for topical IFS, Tetris heaps, and the finiteness conjecture. J. Am. Math. Soc. **15**(1), 77–111 (2002)

[BM05] A. Bacciotti, L. Mazzi, An invariance principle for nonlinear switched systems. Syst. Control Lett. **54**(11), 1109–1119 (2005)

[Boh13] P. Bohl, Über Differentialgleichung. J. Reine Angew. Math. **144**, 284–318 (1913)

[Bos02] U. Boscain, Stability of planar switched systems: the linear single input case. SIAM J. Control Optim. **41**(1), 89–112 (2002)

[BP07] A. Bressan, B. Piccoli, *Introduction to the Mathematical Theory of Control*. On Applied Mathematics, vol. 2 (American Institute of Mathematical Sciences, Springfield, 2007)

[BR05] A. Bacciotti, L. Rosier, *Liapunov Functions and Stability in Control Theory* (Springer, Berlin, 2005)

[Bra98] M.S. Branicky, Multiple Lyapunov functions and other analysis tools for switched and hybrid systems. IEEE Trans. Autom. Control **43**(4), 475–482 (1998)

[Bre11] H. Brezis, *Functional Analysis, Sobolev Spaces and Partial Differential Equations*. Universitext (Springer, New York, 2011)

[Bro83] R.W. Brockett, Asymptotic stability and feedback stabilization, in *Differential Geometric Control Theory* (Citeseer, 1983)

[BS20] F. Boarotto, M. Sigalotti, Dwell-time control sets and applications to the stability analysis of linear switched systems. J. Differ. Equ. **268**(4), 1345–1378 (2020)

[BTV03] V.D. Blondel, J. Theys, A.A. Vladimirov, An elementary counterexample to the finiteness conjecture. SIAM J. Matrix Anal. Appl. **24**(4), 963–970 (2003)

[BW92] M.A. Berger, Y. Wang, Bounded semigroups of matrices. Linear Algebra Appl. **166**, 21–27 (1992)

[CC17] G. Chesi, P. Colaneri, Homogeneous rational Lyapunov functions for performance analysis of switched systems with arbitrary switching and dwell time constraints. IEEE Trans. Autom. Control **62**(10), 5124–5137 (2017)

[CCG+12] G. Chesi, P. Colaneri, J.C. Geromel, R. Middleton, R. Shorten, A nonconservative LMI condition for stability of switched systems with guaranteed dwell time. IEEE Trans. Autom. Control **57**(5), 1297–1302 (2012)

[CCLS08] A. Chaillet, Y. Chitour, A. Loría, M. Sigalotti, Uniform stabilization for linear systems with persistency of excitation: the neutrally stable and the double integrator cases. Math. Control Signals Syst. **20**(2), 135–156 (2008)

[CCS14] Y. Chitour, F. Colonius, M. Sigalotti, Growth rates for persistently excited linear systems. Math. Control Signals Syst. **26**(4), 589–616 (2014)

[CGM15] Y. Chitour, M. Gaye, P. Mason, Geometric and asymptotic properties associated with linear switched systems. J. Differ. Equ. **259**(11), 5582–5616 (2015)

[CGPS21] Y. Chitour, N. Guglielmi, V.Y. Protasov, M. Sigalotti, Switching systems with dwell time: computing the maximal Lyapunov exponent. Nonlinear Anal. Hybrid Syst. **40**, 101021 (2021)

[CGTV09] G. Chesi, A. Garulli, A. Tesi, A. Vicino, *Homogeneous Polynomial Forms for Robustness Analysis of Uncertain Systems*, vol. 390 (Springer, Berlin, 2009)

[CK93] F. Colonius, W. Kliemann, Minimal and maximal Lyapunov exponents of bilinear control systems. J. Differ. Equ. **101**(2), 232–275 (1993)

[CK00] F. Colonius, W. Kliemann, *The Dynamics of Control*. Systems & Control: Foundations & Applications (Birkhäuser, Boston, 2000). With an appendix by Lars Grüne

[CK14] F. Colonius, W. Kliemann, *Dynamical Systems and Linear Algebra*. Graduate Studies in Mathematics, vol. 158 (American Mathematical Society, Providence, 2014)

[CL93] N. Cohen, I. Lewkowicz, A necessary and sufficient criterion for the stability of a convex set of matrices. IEEE Trans. Autom. Control **38**(4), 611–615 (1993)

[CLR97] N. Cohen, I. Lewkowicz, L. Rodman, Exponential stability of triangular differential inclusion systems. Syst. Control Lett. **30**(4), 159–164 (1997)

[CLSS97] F.H. Clarke, Y.S. Ledyaev, E.D. Sontag, A.I. Subbotin, Asymptotic controllability implies feedback stabilization. IEEE Trans. Autom. Control **42**(10), 1394–1407 (1997)

[CMS12] Y. Chitour, P. Mason, M. Sigalotti, On the marginal instability of linear switched systems. Syst. Control Lett. **61**(6), 747–757 (2012)

[CMS13] Y. Chitour, G. Mazanti, M. Sigalotti, Stabilization of persistently excited linear systems, in *Hybrid Systems with Constraints*, ed. by J. Daafouz, S. Tarbouriech, M. Sigalotti. Automation — Control and Industrial Engineering Series, chap. 4 (Wiley-ISTE, London, 2013), pp. 85–120

[CMS17a] Y. Chitour, P. Mason, M. Sigalotti, A characterization of switched linear control systems with finite L_2-gain. IEEE Trans. Autom. Control **62**(4), 1825–1837 (2017)

[CMS17b] Y. Chitour, G. Mazanti, M. Sigalotti, Persistently damped transport on a network of circles. Trans. Am. Math. Soc. **369**(6), 3841–3881 (2017)

[CMS22] Y. Chitour, P. Mason, M. Sigalotti, Characterization of linear switched systems admitting a Barabanov norm. Math. Rep. (Bucur.) **24**(74), 125–137 (2022)

[Con13] S. Cong, Characterising the stabilisability for second-order linear switched systems. Int. J. Control **86**(3), 519–528 (2013)

[CP14] M.J. Cáceres, B. Perthame, Beyond blow-up in excitatory integrate and fire neuronal networks: refractory period and spontaneous activity. J. Theor. Biol. **350**, 81–89 (2014)

[CR62] C. Curtis, I. Reiner, *Representation Theory of Finite Groups and Associative Algebras*. Pure and Applied Mathematics, vol. XI (Interscience Publishers, New York, 1962)

[CS10] Y. Chitour, M. Sigalotti, On the stabilization of persistently excited linear systems. SIAM J. Control Optim. **48**(6), 4032–4055 (2010)

[CUB20] Y. Chitour, R. Ushirobira, H. Bouhemou, Stabilization for a perturbed chain of integrators in prescribed time. SIAM J. Control Optim. **58**(2), 1022–1048 (2020)

[DL92] I. Daubechies, J.C. Lagarias, Sets of matrices all infinite products of which converge. Linear Algebra Appl. **161**, 227–263 (1992)

[DM99] W.P. Dayawansa, C.F. Martin, A converse Lyapunov theorem for a class of dynamical systems which undergo switching. IEEE Trans. Autom. Control **44**(4), 751–760 (1999)

[DS14] J. Diestel, A. Spalsbury, *The Joys of Haar Measure*. Graduate Studies in Mathematics, vol. 150 (American Mathematical Society, Providence, 2014)

[EG15] L.C. Evans, R.F. Gariepy, *Measure Theory and Fine Properties of Functions*. Textbooks in Mathematics, revised edition (CRC Press, Boca Raton, 2015)

[Fen74] N. Fenichel, Asymptotic stability with rate conditions. Ind. Univ. Math. J. **23**, 1109–1137 (1973–1974)

[Fil80] A. Filippov, Stability conditions in homogeneous systems with arbitrary regime switching. Autom. Remote Control **41**(8), 1078–1085 (1980)

[Fil88] A.F. Filippov, *Differential Equations with Discontinuous Righthand Sides*. Mathematics and its Applications (Soviet Series), vol. 18 (Kluwer Academic, Dordrecht, 1988). Translated from the Russian

[GC06] J.C. Geromel, P. Colaneri, Stability and stabilization of continuous-time switched linear systems. SIAM J. Control Optim. **45**(5), 1915–1930 (2006)

[GC10] J.C. Geromel, P. Colaneri, H_∞ and dwell time specifications of continuous-time switched linear systems. IEEE Trans. Autom. Control **55**(1), 207–212 (2010)

[GK94] J.P. Gauthier, I.A.K. Kupka, Observability and observers for nonlinear systems. SIAM J. Control Optim. **32**(4), 975–994 (1994)

[GP13] N. Guglielmi, V. Protasov, Exact computation of joint spectral characteristics of linear operators. Found. Comput. Math. **13**(1), 37–97 (2013)

[GS10] M. Gugat, M. Sigalotti, Stars of vibrating strings: switching boundary feedback stabilization. Netw. Heterog. Media **5**(2), 299–314 (2010)

[GSM07] L. Gurvits, R. Shorten, O. Mason, On the stability of switched positive linear systems. IEEE Trans. Autom. Control **52**(6), 1099–1103 (2007)

[GST08] R. Goebel, R.G. Sanfelice, A.R. Teel, Invariance principles for switching systems via hybrid systems techniques. Syst. Control Lett. **57**(12), 980–986 (2008)

[GST12] R. Goebel, R.G. Sanfelice, A.R. Teel, *Hybrid Dynamical Systems* (Princeton University Press, Princeton, 2012). Modeling, stability, and robustness

[Gur03] L. Gurvits, What is the finiteness conjecture for linear continuous time inclusions? in *42nd IEEE International Conference on Decision and Control*, vol. 2 (2003), pp. 1165–1169

[GWZ05] N. Guglielmi, F. Wirth, M. Zennaro, Complex polytope extremality results for families of matrices. SIAM J. Matrix Anal. Appl. **27**(3), 721–743 (2005)

[GZ01] N. Guglielmi, M. Zennaro, On the asymptotic properties of a family of matrices. Linear Algebra Appl. **322**(1–3), 169–192 (2001)

[Hes03] J.P. Hespanha, $\mathcal{L}_2$-induced gains of switched linear systems, in *Unsolved Problems in Mathematical Systems & Control Theory*, ed. by V.D. Blondel, A. Megretski (Princeton University Press, Princeton, 2003), ppp. 131–133

[Hes04] J.A.P. Hespanha, Uniform stability of switched linear systems: extensions of LaSalle's invariance principle. IEEE Trans. Autom. Control **49**(4), 470–482 (2004)

[HMST11] K.G. Hare, I.D. Morris, N. Sidorov, J. Theys, An explicit counterexample to the Lagarias-Wang finiteness conjecture. Adv. Math. **226**(6), 4667–4701 (2011)

[HN12] J. Hilgert, K.H. Neeb, *Structure and Geometry of Lie Groups*. Springer Monographs in Mathematics (Springer, New York, 2012)

[HS95] C. Heil, G. Strang, Continuity of the joint spectral radius: application to wavelets, in *Linear Algebra for Signal Processing (Minneapolis, MN, 1992)*. The IMA Volumes in Mathematics and its Applications, vol. 69 (Springer, New York, 1995), pp. 51–61

[HS11] F.M. Hante, M. Sigalotti, Converse Lyapunov theorems for switched systems in Banach and Hilbert spaces. SIAM J. Control Optim. **49**(2), 752–770 (2011)

[HST12] F.M. Hante, M. Sigalotti, M. Tucsnak, On conditions for asymptotic stability of dissipative infinite-dimensional systems with intermittent damping. J. Differ. Equ. **252**(10), 5569–5593 (2012)

[ISW03] B. Ingalls, E. Sontag, Y. Wang, An infinite-time relaxation theorem for differential inclusions. Proc. Am. Math. Soc. **131**(2), 487–499 (2003)

[JM17] R.M. Jungers, P. Mason, On feedback stabilization of linear switched systems via switching signal control. SIAM J. Control Optim. **55**(2), 1179–1198 (2017)

[JN13] P. Jouan, S. Naciri, Asymptotic stability of uniformly bounded nonlinear switched systems. Math. Control Relat. Fields **3**(3), 323–345 (2013)

[Jun09] R. Jungers, *The Joint Spectral Radius*. Lecture Notes in Control and Information Sciences, vol. 385 (Springer, Berlin, 2009). Theory and applications

[Kli87] W. Kliemann, Recurrence and invariant measures for degenerate diffusions. Ann. Probab. **15**(2), 690–707 (1987)

[KN06] C. King, M. Nathanson, On the existence of a common quadratic Lyapunov function for a rank one difference. Linear Algebra Appl. **419**(2–3), 400–416 (2006)

[Koz05] V. Kozyakin, A dynamical systems construction of a counterexample to the finiteness conjecture, in *Proceedings of the 44th IEEE Conference on Decision and Control* (2005), pp. 2338–2343

[KP87] V.A. Kamenetskiy, Y.S. Pyatnitskiy, An iterative method of Lyapunov function construction for differential inclusions. Syst. Control Lett. **8**(5), 445–451 (1987)

[Kra63] N.N. Krasovskii, *Stability of Motion: Applications of Lyapunov's Second Method to Differential Systems and Equations with Delay* (Stanford University Press, Stanford, 1963)

[KT04] C.M. Kellett, A.R. Teel, Weak converse Lyapunov theorems and control-Lyapunov functions. SIAM J. Control Optim. **42**(6), 1934–1959 (2004)

[Kur56] J. Kurzweil, On the inversion of Lyapunov's second theorem on stability of motion. Am. Math. Soc. Transl. **24**(2), 19–77 (1956)

[LA22] H. Lin, P.J. Antsaklis, *Hybrid Dynamical Systems—Fundamentals and Methods*. Advanced Textbooks in Control and Signal Processing (Springer, Cham, 2022). ©2022

[LCBC05] A. Loría, A. Chaillet, G. Besançon, Y. Chitour, On the PE stabilization of time-varying systems: open questions and preliminary answers, in *44th IEEE Conference on Decision and Control, 2005 and 2005 European Control Conference* (IEEE, Piscataway, 2005), pp. 6847–6852

[LHM99] D. Liberzon, J.P. Hespanha, A.S. Morse, Stability of switched systems: a Lie-algebraic condition. Syst. Control Lett. **37**(3), 117–122 (1999)

[Lib01] M.R. Liberzon, Lur'e problem of absolute stability - a historical essay. IFAC Proc. Vol. **34**(6), 25–28 (2001). 5th IFAC Symposium on Nonlinear Control Systems 2001, St Petersburg, Russia, 4–6 July 2001

[Lib03] D. Liberzon, *Switching in Systems and Control*. Systems & Control: Foundations & Applications (Birkhäuser, Boston, 2003)

[Lib11] D. Liberzon, *Calculus of Variations and Optimal Control Theory: A Concise Introduction* (Princeton University Press, Princeton, 2011)

[Lib23] Liberzon, D. (2023). Commutation relations and stability of switched systems: a personal history (2023). arXiv:2304.11155

[LP44] A.I. Lourie, V.N. Postnikov, Concerning the theory of stability of regulating systems. Appl. Math. Mech. [Akad. Nauk SSSR. Prikl. Mat. Mech.] **8**, 246–248 (1944)

[LPM02] A. Loría, E. Panteley, K. Melhem, UGAS of skew-symmetric time-varying systems: application to stabilization of chained form systems. Eur. J. Control **8**(1), 33–43 (2002)

[LSW96] Y. Lin, E.D. Sontag, Y. Wang, A smooth converse Lyapunov theorem for robust stability. SIAM J. Control Optim. **34**(1), 124–160 (1996)

[LW95] J.C. Lagarias, Y. Wang, The finiteness conjecture for the generalized spectral radius of a set of matrices. Linear Algebra Appl. **214**:17–42 (1995)

[Lya92] A.M. Lyapunov, *The General Problem of the Stability of Motion*. Commun. Soc. Math. Kharkow (CRC Press, Boca Raton, 1892)

[MAG06] J.L. Mancilla-Aguilar, R.A. García, An extension of LaSalle's invariance principle for switched systems. Syst. Control Lett. **55**(5), 376–384 (2006)

[Mar06] M. Margaliot, Stability analysis of switched systems using variational principles: an introduction. Autom. J. IFAC **42**(12), 2059–2077 (2006)

[Mas56] J.L. Massera, Contributions to stability theory. Ann. Math. **64**, 182–206 (1956)

[Maz14] G. Mazanti, Stabilization of persistently excited linear systems by delayed feedback laws. Syst. Control Lett. **68**, 57–67 (2014)

[MBC06] P. Mason, U. Boscain, Y. Chitour, Common polynomial Lyapunov functions for linear switched systems. SIAM J. Control Optim. **45**(1), 226–245 (2006)

[MBs96] J. Malmborg, B. Bernhardsson, K.J.A. Ström, A stabilizing switching scheme for multi controller systems. IFAC Proc. Vol. **29**(1), 2627–2632 (1996). 13th World Congress of IFAC, 1996, San Francisco USA, 30 June–5 July

[MCS13] G. Mazanti, Y. Chitour, M. Sigalotti, Stabilization of two-dimensional persistently excited linear control systems with arbitrary rate of convergence. SIAM J. Control Optim. **51**(2), 801–823 (2013)

[MCS23] P. Mason, Y. Chitour, M. Sigalotti, On universal classes of Lyapunov functions for linear switched systems. Autom. J. IFAC **155**, 111155 (2023)

[Mei79] A. Meilakhs, Design of stable control systems subject to parametric perturbation. Autom. Remote Control **39**(10), 1409–1418 (1979)

[Mej20] T. Mejstrik, Algorithm 1011: improved invariant polytope algorithm and applications. ACM Trans. Math. Softw. **46**(3), 1–26 (2020)

[MMK97] Y. Mori, T. Mori, Y. Kuroe, A solution to the common Lyapunov function problem for continuous-time systems, in *Proceedings of the 36th IEEE Conference on Decision and Control*, vol. 4 (1997), pp. 3530–3531

[Mor93] A. Morse, Dwell-time switching, in *Proceedings of the 2nd European Control Conference* (1993), pp. 176–181

[Mor95] A.S. Morse, Control using logic-based switching, in *Trends in Control*, ed. by A. Isidori (Springer, London, 1995), pp. 69–113

[Mor12] I.D. Morris, A new sufficient condition for the uniqueness of Barabanov norms. SIAM J. Matrix Anal. Appl. **33**(2), 317–324 (2012)

[Mor22] I.D. Morris, Marginally unstable discrete-time linear switched systems with highly irregular trajectory growth. Syst. Control Lett. **163**, 105216 (2022)

[MP86a] A.P. Molchanov, E.S. Pyatnitskiĭ, Lyapunov functions that define necessary and sufficient conditions for absolute stability of nonlinear nonstationary control systems. I. Autom. Remote Control **47**(3), 343–354 (1986)

[MP86b] A.P. Molchanov, E.S. Pyatnitskiĭ, Lyapunov functions that define necessary and sufficient conditions for absolute stability of nonlinear nonstationary control systems. II. Autom. Remote Control **47**(4), 443–451 (1986)

[PD91] P. Peleties, R. DeCarlo, Asymptotic stability of m-switched systems using Lyapunov-like functions, in *1991 American Control Conference* (1991), pp. 1679–1684

[Pet07] M. Petreczky, Realization theory for linear switched systems: formal power series approach. Syst. Control Lett. **56**(9–10), 588–595 (2007)

[PJ15] V.Y. Protasov, R.M. Jungers, Resonance and marginal instability of switching systems. Nonlinear Anal. Hybrid Syst. **17**, 81–93 (2015)

[PP03] S. Prajna, A. Papachristodoulou, Analysis of switched and hybrid systems—beyond piecewise quadratic methods, in *Proceedings of the American Control Conference, Denver* (2003)

[PR91a] E. Pyatnitskii, L. Rapoport, Existence of periodic motions and test for absolute stability of nonlinear nonstationary systems in the three-dimensional case. Autom. Remote Control **52**(5), 648–658 (1991)

[PR91b] E. Pyatnitskii, L. Rapoport, Periodic motions and criteria for absolute stability of nonlinear time-dependent systems. Autom. Remote Control **52**(10), 1379–1387 (1991)

[PR96] E. Pyatnitskiy, L. Rapoport, Criteria of asymptotic stability of differential inclusions and periodic motions of time-varying nonlinear control systems. IEEE Trans. Circuits Syst. I Fund. Theory Appl. **43**(3), 219–229 (1996)

[Pro16] V.Y. Protasov, Linear switching systems with slow growth of trajectories. Syst. Control Lett. **90**, 54–60 (2016)

[Pro22] V.Y. Protasov, The Barabanov norm is generically unique, simple, and easily computed. SIAM J. Control Optim. **60**(4), 2246–2267 (2022)

[PTS77] R.V. Patel, M. Toda, B. Sridhar, Robustness of linear quadratic state feedback designs in the presence of system uncertainty. IEEE Trans. Autom. Control **22**(6), 945–949 (1977)

[Pya70] E. Pyatnitsky, Absolute stability of nonstationary nonlinear systems. Autom. Remote Control **1**, 5–15 (1970)

[Pya71] E. Pyatnitsky, Criterion for absolute stability of nonlinear feedback systems of order two with one nonlinear element. Autom. Remote Control **1**, 5–16 (1971)

[Rap93] L. Rapoport, Antiperiodic motions and an algebraic criterion for the absolute stability of nonlinear time-dependent systems in the three-dimensional case. Autom. Remote Control **54**(7), 1063–1076 (1993)

[Rif02] L. Rifford, Semiconcave control-Lyapunov functions and stabilizing feedbacks. SIAM J. Control Optim. **41**(3), 659–681 (2002)

[Roc70] R. Rockafellar, *Convex Analysis*. Princeton Landmarks in Mathematics and Physics (Princeton University Press, Princeton, 1970)

[RS60] G.C. Rota, G. Strang, A note on the joint spectral radius. Indag. Math. **22**(4), 379–381 (1960)

[RSD10] P. Riedinger, M. Sigalotti, J. Daafouz, On the algebraic characterization of invariant sets of switched linear systems. Autom. J. IFAC **46**(6), 1047–1052 (2010)

[Rud87] W. Rudin, *Real and complex analysis*, 3rd edn. (McGraw-Hill, New York, 1987)

[SG05] Z. Sun, S.S. Ge, *Switched Linear Systems: Control and Design*. Communications and Control Engineering (Springer, London, 2005)

[SG11] Z. Sun, S.S. Ge, *Stability Theory of Switched Dynamical Systems*. Communications and Control Engineering Series (Springer, London, 2011)

[SGL02] Z. Sun, S.S. Ge, T.H. Lee, Controllability and reachability criteria for switched linear systems. Autom. J. IFAC **38**(5), 775–786 (2002)

[SGT07] R.G. Sanfelice, R. Goebel, A.R. Teel, Invariance principles for hybrid systems with connections to detectability and asymptotic stability. IEEE Trans. Autom. Control **52**(12), 2282–2297 (2007)

[SMOC04] R.N. Shorten, O. Mason, F. O'Cairbre, P. Curran, A unifying framework for the SISO circle criterion and other quadratic stability criteria. Int. J. Control **77**(1), 1–8 (2004)

[SMT95] L.A.B. San Martin, P.A. Tonelli, Semigroup actions on homogeneous spaces. Semigroup Forum **50**(1), 59–88 (1995)

[SN98] R. Shorten, K. Narendra, On the stability and existence of common Lyapunov functions for stable linear switching systems, in *Proceedings of the 37th IEEE Conference on Decision and Control*, vol. 4 (1998), pp. 3723–3724

[SN99] R. Shorten, K. Narendra, Necessary and sufficient conditions for the existence of a common quadratic Lyapunov function for two stable second order linear time–invariant systems, in *Proceedings 1999, American Control Conference* (1999), pp. 1410–1414

[SN02] R.N. Shorten, K.S. Narendra, Necessary and sufficient conditions for the existence of a common quadratic Lyapunov function for a finite number of stable second order linear time-invariant systems. Int. J. Adap. Control Signal Process. **16**(10), 709–728 (2002)

[SNM03] R. Shorten, K.S. Narendra, O. Mason, A result on common quadratic Lyapunov functions. IEEE Trans. Autom. Control **48**(1), 110–113 (2003)

[Son83] E.D. Sontag, A Lyapunov-like characterization of asymptotic controllability. SIAM J. Control Optim. **21**(3), 462–471 (1983)

[Son98] E.D. Sontag, *Mathematical Control Theory: Deterministic Finite-Dimensional Systems*. Texts in Applied Mathematics, vol. 6, 2nd edn. (Springer, New York, 1998)

[Spi99] M. Spivak, *A Comprehensive Introduction to Differential Geometry*, vol. 2, 3rd edn. (Publish or Perish, Houston, 1999)

[SVR11] U. Serres, J.C. Vivalda, P. Riedinger, On the convergence of linear switched systems. IEEE Trans. Autom. Control **56**(2), 320–332 (2011)

[SWM+07] R. Shorten, F. Wirth, O. Mason, K. Wulff, C. King, Stability criteria for switched and hybrid systems. SIAM Rev. **49**(4), 545–592 (2007)

[Tes12] G. Teschl, *Ordinary Differential Equations and Dynamical Systems*. Graduate Studies in Mathematics, vol. 140 (American Mathematical Society, Providence, 2012)

[Tho37] T.Y. Thomas, Arcs in affinely connected spaces. Ann. Math. **38**(1), 120–130 (1937)

[vdSS00] A. van der Schaft, H. Schumacher, *An Introduction to Hybrid Dynamical Systems*. Lecture Notes in Control and Information Sciences, vol. 251 (Springer, London, 2000)

[Wil69] F.W. Wilson, Smoothing derivatives of functions and applications. Trans. Am. Math. Soc. **139**, 413–428 (1969)

[Wir02] F. Wirth, The generalized spectral radius and extremal norms. Linear Algebra Appl. **342**, 17–40 (2002)

[Wit66] H. Witsenhausen, A class of hybrid-state continuous-time dynamic systems. IEEE Trans. Autom. Control **11**(2), 161–167 (1966)

[WPD98] M. Wicks, P. Peleties, R. DeCarlo, Switched controller synthesis for the quadratic stabilisation of a pair of unstable linear systems. Eur. J. Control **4**(2), 140–147 (1998)

[Yan02] Z. Yang, An algebraic approach towards the controllability of controlled switching linear hybrid systems. Automatica **38**(7), 1221–1228 (2002)

Index

Y. Chitour et al., *Dynamics and Stability of Continuous-Time Switched Linear Systems*, PNLDE Subseries in Control 105,
https://doi.org/10.1007/978-3-031-82982-6

The manufacturer's authorised representative in the EU is Springer Nature Customer Service Centre GmbH, Europaplatz 3, 69115 Heidelberg, Germany. If you have any concerns regarding our products, please contact ProductSafety@springernature.com

Printed and bound by CPI Group (UK) Ltd, Croydon, CR0 4YY

01/10/2026

02225960-0005